E. Kreyszig / E. J. Norminton

MAPLE COMPUTER GUIDE

A Self-Contained Introduction

For

Erwin Kreyszig

ADVANCED ENGINEERING MATHEMATICS

Eighth Edition

JOHN WILEY & SONS, INC.

New York • Chichester • Brisbane • Toronto • Singapore

PUBLISHER Peter Janzow
EDITOR Barbara Holland
ASSOCIATE EDITOR Mary Johenk
MARKETING MANAGER Julie Lindstrom
ASSOCIATE PRODUCTION DIRECTOR Lucille Buonocore
SENIOR PRODUCTION EDITOR Monique Calello
COVER DESIGNER Madelyn Lesure
COVER PHOTO Chris Rogers/The Stock Market
ELECTRONIC PREPRESS SuperScript

This book was set in 10/12 Computer Modern
and printed and bound by Von Hoffmann Graphics.
The cover was printed by Brady Palmer Printing Co.

This book is printed on acid-free paper.

The paper in this book was manufactured by a mill whose forest management
programs include sustained yield harvesting of its timberlands. Sustained yield
harvesting principles ensure that the numbers of trees cut each year does not exceed
the amount of new growth.

To order books or for customer service call 1-800-CALL-WILEY (225-5945).

ISBN 0-471-38668-5 (paperback)

Printed in the United States of America

10 9 8 7 6 5 4 3 2 1

PREFACE

This **MAPLE COMPUTER GUIDE** supplements the new edition (the Eighth Edition of 1999) of **ADVANCED ENGINEERING MATHEMATICS** by Erwin Kreyszig (J. Wiley & Sons, Inc., 605 Third Avenue, New York NY 10016; to order, call 1-800-225-5945 or 212-850-6000; see also

http://www.wiley.com/college/mat/kreyszig154962/).

In this Guide, ADVANCED ENGINEERING MATHEMATICS will be quoted as **AEM**.

CAS's (Computer Algebraic Systems) have meanwhile become very popular, and Maple is one of the most successful ones, used by many engineers, mathematicians, physicists, and computer scientists. In connection with AEM the algebraical, numerical, and graphical power of Maple can help the student in working-out class notes, in doing homework and exams, as well as in pursuing self-study. This will generally enhance the student's understanding, skill, and motivation.

This Guide *supplements* **AEM** but, of course, *does not replace* it. Indeed, its use is not needed in studying from AEM itself. Thus, this Guide just invites the student to gain experience with a prominent CAS.

This Guide has two new main features, not shared by its predecessor, the Maple Computer Manual of 1994.

1. **It is much simpler than the latter**. It is an elementary introduction that presents Maple as simply as possible, in contrast to most other books on Maple.

2. **It is self-contained**, so that it can be used without referring to AEM, once the student has learned how to master the subject matter as such. Thus, the reference to AEM given after each example and problem is only for the case that a student needs help with the subject matter or wants to know where to find the same problem or similar problems in AEM.

This Guide contains over 130 carefully selected worked-out examples and about 400 problems for solution, covering all fields discussed in AEM.

This Guide is written in Student's Maple 6, which for the present purpose is practically as useful as general Maple 6 would be. For most problems you can also use Maple 5 if 6 is not yet available to you.

Maple (and Student's Maple) can do **numerical calculations** (calculation with numbers), **symbolic calculations** (calculations with formulas consisting of letters and other symbols, such as $+$, $-$, $=$, etc.), and **graphing** (plotting of curves and surfaces and other figures). This includes differentiation and integration, solution of differential equations, matrix and vector algebra and calculus, complex analysis, numerical methods, and statistical problems. All this is shown in this Guide.

Commands that you have to type are printed in BLUE, responses by the computer in BLACK.

For efficient help we wish to thank Prof. John Todd (Pasadena), Herbert E. Kreyszig (New York), and the personnel of J. Wiley, as listed on page *iv*, in particular Ms. Barbara Holland for a pleasant continuation of the cooperation on the Eighth Edition of AEM itself.

ERWIN KREYSZIG
EDWARD J. NORMINTON

Contents

INTRODUCTION, GENERAL COMMANDS

Familiarity with Maple or other CAS's (Computer Algebraic Systems) will not be assumed.

Colors. Commands and data that you have to type are given in BLUE , computer responses in BLACK.

Chapters in this Guide correspond to those in AEM.

Information on the screen. This is often the best way of learning or recalling Maple. For instance, type **?matrix** , press ENTER . The screen will give you information and show examples. Press the down arrow key because there may be more. When you are done, press Ctrl and F4 simultaneously (or click the mouse on the x in the upper right corner of the help screen. This will bring you back where you were before you typed **?matrix** .

Information from books. See Appendix 1.

Collecting your own experience. Make up your personal manual, recording commands used, typical examples calculated, and positive and negative experience gained by trial and error.

Answers to odd-numbered problems. See Appendix 2.

Instructor's Maple Manual is available from the publisher upon request.

Beginning. Suppose that Student's Maple* has been loaded into your computer and that you are ready to start your first Maple session. The computer will display a prompt >, showing that Maple is ready to take your commands. A **command** is what you tell the computer by typing it on the keyboard. This can be data or the request to add a few numbers or to solve a differential equation that you typed before, etc. Commands are printed in BLUE . The computer shows something on the screen; this is called the **response** to your command, and is printed in BLACK in this Guide.

For instance, to calculate $2 + 4 + 3.5 - 0.6 + 3 \times 2 + (1/8)^5$, type this with * instead of the multiplication cross or dot, and with ^ for exponentiation, and then press Enter . Thus,

```
> 2 + 4 + 3.5 - 0.6 + 3*2 + (1/8)^5
>
```

```
Warning, premature end of input
```

---

* This Guide is written in Student's Maple 6, which for the present purpose is practically as useful as general Maple 6 would be. For most problems you can also use Maple 5 if Maple 6 is not yet available to you.

The response tells you that each command must be followed by **a semicolon** ; .
This shows the computer that the command is complete. Thus type a semicolon ;
after your command and then press again Enter .

> 2 + 4 + 3.5 - 0.6 + 3*2 + (1/8)^5;

$$14.90003052$$

(If you don't want the response to appear on the screen, type **colon** : instead of ; .
See later for an illustration.)

> 1/4 - 3/8;

$$-1/8$$

If you want a decimal fraction as the response, type evalf (suggesting evaluate
floating),

> evalf(1/4 - 3/8);   # Resp. $-.1250000000$

If you want fewer digits, e.g., five, type

> evalf(1/4 - 3/8, 5);   # Resp. $-.12500$

means **response** by the computer and is used to save space. The symbol # will also
be used to precede a **comment** (for instance an explanation) as the following line
shows.

> evalf(sqrt(2), 70);   # sqrt means **square root.**

$$1.4142135623730950488016887242096980785696718753769480731766679737990732$$

> evalf(2*Pi, 30);   # Note: **Pi** (not pi; try it).

$$6.28318530717958647692528676656$$

Sequences, Sums. The sequence 1, x, $\frac{1}{2}x^2$, $\frac{1}{6}x^3$, $\frac{1}{24}x^4$, $\frac{1}{120}x^5$ is obtained by typing

> seq(x^n/n!, n = 0..5);

$$1, \quad x, \; \frac{1}{2}\,x^2, \; \frac{1}{6}\,x^3, \; \frac{1}{24}\,x^4, \; \frac{1}{120}\,x^5$$

The sum of these terms is obtained by typing

> sum(x^n/n!, n = 0..5);

$$1 + x + \frac{1}{2}\,x^2 + \frac{1}{6}\,x^3 + \frac{1}{24}\,x^4 + \frac{1}{120}\,x^5$$

Similarly, $1/2^2 + 1/2^4 + 1/2^6 + \ldots + 1/2^{20}$ is obtained by

> sum(1/2^(2*n), n = 1..10);   # Resp. $\dfrac{349525}{1048576}$

> evalf(%);   # % avoids retyping.

$$.3333330154$$

Give names to expressions that you will need later. For instance,

```
> f:= sin(3Pi*x);
Error, missing operator or ';'
> f := sin (3*Pi*x);
```

$$f := \sin(3\pi x)$$

```
> g := (2/Pi)*f^2;                      # Here f is used and substituted.
```

$$g := 2\frac{\sin(3\pi x)^2}{\pi}$$

```
> g1 := subs(x = 0.5, g);               # Value of g at x = 0.5, unevaluated
```

$$g1 := 2\frac{\sin(1.5\pi)^2}{\pi}$$

```
> g2:= evalf(%, 6);          # Same expression evaluated as decimal fraction
```

$$g2 := .636620$$

Removal of assigned values. This is important. For instance, type

```
> y := 4;                               # This is a value assigned to y.
```
$$y := 4$$
```
> exp(y);                    # Resp. e^4
```

But if you want e^y with arbitrary y, first remove the assignment $y = 4$ ("**unassign** y") by typing

```
> y := 'y':              # No response shown because we used : instead of ;
> exp(y);                # Resp. e^y
```

You now have what you want. If the computer shows you unexpected responses, be they questionable or outright nonsense, chances are that it uses **previous assignments**, so that you first have to unassign variables x, y, f, g, ..., constants a, c, k, n,... or whatever.

Parentheses (...), braces {...}, brackets [...]. Carefully distinguish among these three. They are for different purposes. Parentheses (...) are the most usual ones – look again at what you have just read. Braces {...} are for sets. Brackets [...] are for vectors, matrices, subscripts, etc. See the examples, where all three kinds occur in numerous places.

Further commands for symbolic calculations

```
> p := x^2 + 2*x - 3:            # Response not shown because of the colon
> factor(%);
```

$$(x + 3)(x - 1)$$

```
> expand(%);
```

$$x^2 + 2x - 3$$

```
> q := p/(x - 1);                          # Division by x − 1
```

$$q := \frac{x^2 + 2\,x - 3}{x - 1}$$

```
> q := simplify(%);              # This is a very useful command.
```

$$q := x + 3$$

Equations can be solved by the command **solve**. For instance,

```
> solve(cos(x) = 1/sqrt(2));               # sqrt means square root; see before
```

$$\frac{1}{4}\,\pi$$

```
> evalf(%, 6);
```

$$.785398$$

Note that **solve** may give you only some of the solutions, as in the present case. Type **?solve** and **?fsolve** for further information on these commands.

Special functions are known to Maple. This includes all the functions of calculus (e^x, sin x, tan x, arccos x, etc.) as well as higher functions. For the latter, type, for instance **?GAMMA**, **?Bessel**, **?erf**, **?P**.

Useful commands are

```
> expand (cos(a + b));              # Resp. cos(a) cos(b) − sin(a) sin(b)
> combine(%, trig);                          # Resp. cos(a + b)
> evalf(combine(exp(2)*exp(6)));             # Resp. 2980.957987
> expand((a + b)^4);
```

$$a^4 + 4\,a^3\,b + 6\,a^2\,b^2 + 4\,a\,b^3 + b^4$$

Differentiation and integration. The commands are **diff** and **int**, For instance,

```
> diff(cos(2*x), x);
```

$$-2\,\sin(2\,x)$$

```
> f := x^2 + sin(3*x);
```

$$f := x^2 + \sin(3\,x)$$

```
> diff(f, x);                              # Resp. 2 x + 3 cos(3 x)
> diff(f, x, x);                           # Resp. 2 − 9 sin(3 x)
> diff(exp(x)*cos(y), y);          # Partial derivative with respect to y
```

$$-e^x\,\sin(y)$$

```
> int(tan(x), x);
```

$$-\ln(\cos(x))$$

```
> int(f, x);
```

$$\frac{1}{3}\,x^3 - \frac{1}{3}\,\cos(3\,x)$$

```
> int(f, x = 0..Pi);
```

$$\frac{2}{3} + \frac{1}{3}\,\pi^3$$

Further details on differentiation and integration are given at the beginning of Part A, and applications follow in various examples in this Guide.

Packages, such as the linalg package (for linear algebra), the DEtool package (for ordinary differential equations), the stats package (for statistics), the plot package, etc., contain many commands for special areas, as their names indicate. Load a package by with(linalg); , with(DEtools); , etc., or followed by a colon instead of ; if you don't want to see the content.

Incorrect and incomplete commands have been included in the examples from time to time with explanations, to help you in finding out what may go wrong and for what reason.

Responses differing from those shown in the examples may occur occasionally in connnection with arbitrary constants (for instance, $\_C1\,\cos(x) + \_C2\,\sin(x)$ may have the constants interchanged) or with arbitrary vectors (notably, eigenvectors, which are determined only up to a nonzero constant). Be prepared that this may happen – it is not your fault. And in repeating a calculation, some quantities may again appear in a different order.

Graphing. A simple plot is

```
> plot(1 - exp(-2*x), x = 0..3, scaling = constrained);
```

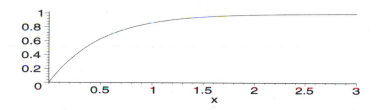

Here, scaling = constrained gives equal scales on both axes. This is **optional**, that is, you can drop it. Try it.

Type ?plot for information on various kinds of plots. (Use the down arrow key to see all the information and examples!)

Various commands related to plotting are included in the examples that contain figures. To facilitate such a search through this book for specific tasks and commands related to plotting, we include on the next page a list as a guide that shows where to find an answer.

PLOTTING GUIDE

Task	Example	Page
Animation	11.1	119, 129
Bar graphs	22.3	228
Box plots	22.2	227
Complex numbers	12.1	132
Conformal mappings	12.4, 12.5	136, 139
Curves given parametrically	3.3	35
Curves plotted jointly	1.5, 4.2	13, 45
Curves in space	9.1	97
Direction fields	1.2, 3.5	10, 37
Fluid flows	16.3	168
Functions given "piecewise"	5.4	60
Graphs	21.1	219
Grids in the plane	19.8	210
Histograms	22.2	227
Labeling coordinate axes	1.3	11
Numbering coordinate axes	1.1	9
Points	12.1, 12.6	132, 141
Polygons	17.2, 19.9	173, 211
Sequence, plotting terms	12.6, 14.1	140, 148
Series, plot from it	4.1	43
Surfaces	11.6, 12.5, 20.1	127, 138, 216
Title of a figure	1.3	11

PART A. ORDINARY DIFFERENTIAL EQUATIONS (ODE's)

Content. First-order ODE's (Chap. 1)
Second and higher order ODE's (Chap. 2)
Systems of ODE's (Chap. 3)
Series solution of ODE's (Chap. 4)
Solution of ODE's by Laplace transforms (Chap. 5)

DEtools package. Load it by typing `with(DEtools):`. Typing `?DEtools` shows that the package contains commands for plotting, for solving special ODE's, etc. Click on any of its keywords listed, for instance, `linearsol`, to see what it means and how you can use it. You will ***not*** need the package all the time, but it can be helpful, for instance, for more complicated ODE's. For further help type `?odeadvisor` and then click on `separable` or `exact`, etc., to see in detail what it can do for you.

Derivatives $y', y'', y''',\dots$ may be typed as `diff(y(x), x)`, `diff(y(x), x, x)`, `diff(y(x), x, x, x)`,... . For instance,

```
> y := exp(-x)*sin(2*x);                                    # 2*x, not 2x
```
$$y := e^{(-x)}\sin(2x)$$

```
> diff(y, x);                     # Resp. −e^(−x) sin(2x) + 2 e^(−x) cos(2x)
```

```
> diff(y, x, x);                  # Resp. −3 e^(−x) sin(2x) − 4 e^(−x) cos(2x)
```

D-notation for derivatives. Those derivatives may also be typed as `D(y)(x)`, `(D@@2)(y)(x)`, `(D@@3)(y)(x)`, etc. For instance (type `?D` for information)

```
> D(sin@(2*x));                                  # @ is essential. Try without.
```
$$2\,(\cos@(2\,x))\ \ D(x)$$

```
> D(exp@(-x)*(sin@(2*x)));
```
$$-(\exp@(-x))\ D(x)\ (\sin@(2\,x))\ +\ 2\ (\exp@(-x))\ (\cos@(2\,x))\ D(x)$$

```
> factor(%);
```
$$-(\exp@(-x))\ \ D(x)((\sin@(2\,x))\ -\ 2\ (\cos@(2\,x)))$$

Thus the derivative is $e^{-x}(-\sin 2x + 2\cos 2x)$, as before. We shall mainly use the first of these two notations.

Integration. `int(f, x)` gives the indefinite integral (the antiderivative) and `int(f, x = a..b)` the definite integral. For instance,

```
> f := 4*x*exp(2*x);                                      # Resp. f := 4 x e^(2 x)
> int(f, x);                              # Don't forget the last x. Try without.
```
$$2\,x\,e^{(2\,x)}\ -\ e^{(2\,x)}$$

```
> int(f, x = -infinity..3);                                       # Resp. 5 e^6
> evalf(%, 7);                                              # Resp. 2017.144
```

Command for solving ODE's and systems. `dsolve` gives general solutions as well as particular solutions of initial value problems. See the various examples.

Chapter 1

First-Order ODE's

Content. General solutions (Ex. 1.1)
          Direction fields (Ex. 1.2, Prs. 1.1, 1.2)
          Separable ODE's (Prs. 1.3-1.8)
          Exact ODE's, integrating factors (Ex. 1.4, Prs. 1.9, 1.10)
          Linear ODE's, mixing problems, electric circuits (Exs. 1.3, 1.6,
               Prs. 1.11, 1.13, 1.17, 1.18)
          Bernoulli ODE, Verhulst population model (Ex. 1.5, Prs. 1.14, 1.15)
          Picard iteration, do-loop (Prs. 1.19, 1.20)

Examples for Chapter 1

EXAMPLE 1.1   **GENERAL SOLUTIONS**

General solutions are obtained by `dsolve`. For instance,

```
> y := 'y':                              # Unassign y that has just been used.
```

Type the ODE. (Note that `diff(y(x), x)` denotes $\dfrac{d}{dx}$ by $\dfrac{\partial}{\partial x}$.)

```
> ode := diff(y(x), x) = 3*y(x);
```

$$ode := \frac{\partial}{\partial x}\, \mathrm{y}(x) = 3\,\mathrm{y}(x)$$

```
> dsolve(ode);
```

$$\mathrm{y}(x) = \_C1\, \mathrm{e}^{(3\,x)}$$

Give a name to the solution by which you can use it further, say, `sol`.

```
> sol := dsolve(ode);
```

$$sol := \mathrm{y}(x) = \_C1\, \mathrm{e}^{(3\,x)}$$

_C1 is the Maple notation for the **arbitrary constant** throughout.

   Initial value problems are also solved by `dsolve`. For instance,

```
> ypartic := dsolve({ode, y(0) = 2));
Error, ')' unexpected
```

Carefully distinguish between braces {} and parentheses (). Here,

```
> ypartic := dsolve({ode, y(0) = 2});          # Particular solution
```

$$ypartic := \mathrm{y}(x) = 2\,\mathrm{e}^{(3\,x)}$$

Checking solutions obtained on the computer is at least as important as it is in working with paper and pencil – the computer will sometimes fool you. Type

```
> y1 := c*exp(3*x):          # Your general solution just obtained
```

```
> subs(y(x) = y1, ode);          # ode is the ODE from which you started.
```

$$\frac{\partial}{\partial x}\,c\,\mathrm{e}^{(3\,x)} = 3\,c\,\mathrm{e}^{(3\,x)}$$

```
> eval(%);          # This checks the solution.
```

$$3\,c\,\mathrm{e}^{(3\,x)} = 3\,c\,\mathrm{e}^{(3\,x)}$$

```
> subs(x = 0, ypartic);
```

$$\mathrm{y}(0) = 2\,\mathrm{e}^{0}$$

```
> simplify(%);          # This verifies the initial condition.
```

$$\mathrm{y}(0) = 2$$

```
> plot(ypartic, x = 0..1, ytickmarks = [2, 10, 20, 30, 40]);
```
Plotting error, empty plot
```
> plot(rhs(ypartic), x = 0..1, ytickmarks = [2, 10, 20, 30, 40]);
```

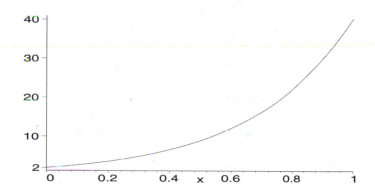

Example 1.1. Particular solution $y(x) = 2\,e^{3\,x}$

rhs means **right-hand side** and gives the function $2\,e^{3\,x}$, which you want to plot, whereas **ypartic** alone gives the whole equation $y = 2\,e^{3\,x}$.

    Similar Material in AEM: pp. 6, 9 (#25), 15

EXAMPLE 1.2   DIRECTION FIELDS

Direction fields and approximate solution curves of ODE's can be plotted on the computer by first loading the **DEtools package**, typing

```
> with(DEtools):
```

and then typing the ODE and points (x, y) through which you want to have approximate solution curves. Show this for the ODE $y' = xy$ and the two points $(0, 1)$ and $(0, 2)$.

Solution. Type the ODE

```
> ode := diff(y(x), x) = x*y(x);
```

$$ode := \frac{\partial}{\partial x}\, \mathrm{y}(x) = x\,\mathrm{y}(x)$$

and the given points (initial conditions for particular solutions represented by those curves)

```
> inits := {[0, 1], [0, 2]};          # Resp. inits := {[0, 1], [0, 2]}
```

The plot command for the direction field and solution curves is DEplot. It must contain y(x) as shown and x and y ranges. scaling = constrained (equal scales on both axes) is optional. Try without.

```
> DEplot(ode, y(x), x = -3..3, y = 0..3, inits, scaling = constrained);
```

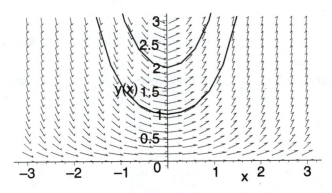

Example 1.2. Direction field for $y' = xy$

The equation can be solved by separating variables, $y'/y = x$, and integration, $y = c \exp(x^2/2)$.

    Similar Material in AEM: pp. 11, 13

| EXAMPLE 1.3 | **MIXING PROBLEMS** |

Mixing problems involve a tank into which brine flows, the content of the tank is stirred (this is the 'mixing'), and the mixture flows out. The model is the ODE

$$y' = \text{Salt inflow rate minus Salt outflow rate},$$

where $y(t)$ is the amount of salt in the tank, and $y' = dy/dt$ is the time rate of change of $y(t)$. Assume the following. At $t = 0$ the tank contains 200 gal of water in which 40 lb of salt are dissolved. The inflow is 10 lb/min (5 gal of brine, each containing 2 lb of salt). 5 gal/min of mixture flows out. Hence the model is

$$y' = 10 - (5/200)y, \qquad y(0) = 40.$$

Solve this initial value problem by typing

```
> ode := diff(y(t), t) = 10 - (5/200)*y(t);
```

$$ode := \frac{\partial}{\partial t}\, \text{y}(t) = 10 - \frac{1}{40}\, \text{y}(t)$$

```
> ypartic := dsolve({ode, y(0) = 40});
```

$$ypartic := \text{y}(t) = 400 - 360\, \text{e}^{\left(-\frac{1}{40}\, t\right)}$$

Plot this particular solution as shown. In the first command, rhs (right-hand side) is missing. In the second command a parenthesis after ypartic is missing.

```
> plot(ypartic, t = 0..250, labels = [t, y], title = 'Salt content');
```

Plotting error, empty plot

```
> plot(rhs(ypartic, t = 0..250, labels = [t, y], title = 'Salt content');
```

Error, ';' unexpected

```
> plot(rhs(ypartic), t = 0..250, y = 0..500, labels = [t, y],
   xtickmarks = [0, 50, 100, 150, 200, 250], ytickmarks =
   [0, 40, 100, 200, 300, 400], title = 'Salt content');
```

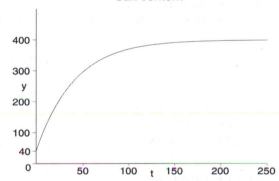

Salt content

Example 1.3. Salt content $y(t)$ in the tank

The plot illustrates that $y(t)$ approaches the limit 400 lb.
         Similar Material in AEM: pp. 20, 23

EXAMPLE 1.4   **INTEGRATING FACTORS**

Integrating factors convert nonexact ODE's into exact ODE's. Let the given ODE be

$$(1) \qquad P\, dx + Q\, dy = 2\, \sin\left(y^2\right) dx + x\, y\, \cos\left(y^2\right) dy = 0.$$

Thus

```
> P := 2*sin(y^2);                    # Resp. P := 2 (sin(y²)
> Q := x*y*cos(y^2);                  # Resp. Q := x y cos (y²)
```

The exactness test fails. Indeed,

```
> diff(P, y) - diff(Q, x);
```
                                                        # Resp. $3\cos\left(y^2\right)y$

In the case of exactness the response would be zero.

Try for an integrating factor $F(x)$ depending only on x. (Ordinarily, an integrating factor (if it exists) would depend on both x and y.) The exactness condition for (1) multiplied by $F(x)$ is

```
> eq1 := diff(F(x)*P, y) - diff(F(x)*Q, x) = 0;
```

$$eq1 := 3\,\mathrm{F}(x)\cos\left(y^2\right)y - \left(\frac{\partial}{\partial x}\,\mathrm{F}(x)\right)x\,y\cos\left(y^2\right) = 0$$

Division by $y\cos\left(y^2\right)$ gives an equation no longer containing y. Hence this becomes a first-order ODE for $F(x)$, which you can solve by dsolve , so that you will get an integrating factor.

```
> simplify(eq1/(y*cos(y^2)));
```
                                 # Resp. $3\,\mathrm{F}(x) - \left(\dfrac{\partial}{\partial x}\mathrm{F}(x)\right)x = 0$

```
> sol := dsolve(%);
```
                                        # Resp. $sol := \mathrm{F}(x) = \_C1\,x^3$

(If you write F on the left, you would get a warning. Try it.) Hence x^3 is an integrating factor. (You can choose $\_C1 = 1$.) You can now obtain an implicit solution $u(x, y) = const$ by integrating $x^3\,P$ with respect to x and $x^3\,Q$ with respect to y, typing

```
> int(x^3*P, x);
```
                                        # Resp. $\dfrac{1}{2}x^4\sin\left(y^2\right)$

```
> int(x^3*Q, y);
```
                                        # Resp. $\dfrac{1}{2}x^4\sin\left(y^2\right)$

Hence a solution is $x^4\sin\left(y^2\right) = const$.

Similar Material in AEM: pp. 30, 32

EXAMPLE 1.5   BERNOULLI'S EQUATION

Bernoulli's equation includes as a special case an important population model, the **Verhulst equation** $y' - Ay = -By^2$, where A and B are positive constants. Type this equation as

```
> ode := diff(y(x), x) - A*y(x) = -B*y(x)^2;
```

$$ode := \left(\frac{\partial}{\partial x}\,\mathrm{y}(x)\right) - A\,\mathrm{y}(x) = -B\,\mathrm{y}(x)^2$$

where x is time. Solve it by dsolve(...) ,

```
> sol := dsolve(ode);
```
                        # Resp. $sol := \mathrm{y}(x) = \dfrac{A}{B + \mathrm{e}^{(-A\,x)}\_C1\,A}$

The DEtools package has a special command for Bernoulli equations, which confirms your result. Type

```
> with(DEtools):
> bernoullisol(ode);
```
                        # Resp. $\left\{\mathrm{y}(x) = \dfrac{A}{B + \mathrm{e}^{(-A\,x)}\_C1\,A}\right\}$

For plotting you must choose specific values of A and B. Set $A = B = 1$, for simplicity.

```
> sol2 := subs(A = 1, B = 1, sol);
```
    # Resp. $sol2 := y(x) = \dfrac{1}{1 + e^{(-x)} \_C1}$

From this obtain and plot three typical particular solutions

```
> y1 := subs(_C1 = 10, sol2);
```
    # Resp. $y1 := y(x) = \dfrac{1}{1 + 10\, e^{(-x)}}$

```
> y2 := subs(_C1 = 0, sol2):
```

```
> y3 := subs(_C1 = -0.5, sol2):
```

Use ; instead of : to see all three responses. Now plot the three solutions on common axes by typing

```
> plot({rhs(y1), rhs(y2), rhs(y3)}, x = 0..5, y = 0..2.3, labels = [x, y],
   ytickmarks = [0, 0.5, 1., 1.5, 2], title = 'Verhulst population model');
```

Verhulst population model

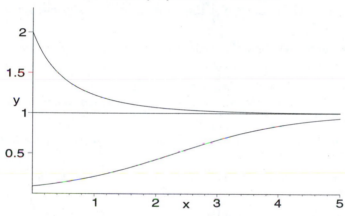

Example 1.5. Typical solution curves of the Verhulst ODE

Similar Material in AEM: p. 43

EXAMPLE 1.6   *RL-*CIRCUIT

The current $i(t)$ in an RL-circuit with $R = 50$ ohms, $L = 1$ henry, and electromotive force $110 \cos 3t$ volts is obtained by solving the ODE

```
> ode := diff(i(t), t) + 50*i(t) = 110*cos(3*t);
```

$$ode := \left(\frac{\partial}{\partial t}\, i(t)\right) + 50\, i(t) = 110\, \cos(3\, t)$$

Assume that $i(0) = 0$. The solution obtained by dsolve is

```
> sol := dsolve({ode, i(0) = 0});
```

$$sol := i(t) = \frac{5500}{2509}\, \cos(3\, t) + \frac{330}{2509}\, \sin(3\, t) - \frac{5500}{2509}\, e^{(-50\, t)}$$

```
> sol2 := evalf(sol, 5);          # Coefficients written as decimal fractions
```

$$sol2 := \mathrm{i}(t) = 2.1921\,\cos(3.\,t) + .13153\,\sin(3.\,t) - 2.1921\,e^{(-50.\,t)}$$

The last result shows a reasonable number of digits. The steady-state solution is a harmonic motion with the frequency of the electromotive force. The exponential term dies out very fast because $R/L = 50$ is large.

Similar Material in AEM: pp. 43, 47

Problem Set for Chapter 1

Pr.1.1 (Direction field) Plot the direction field of the ODE $y' = y^2$ and approximate solution curves through the points $(0, 1)$, $(0, 2)$, and $(0, 3)$.
(*AEM Ref.* pp. 11, 13 (#9))

Pr.1.2 (Direction field) Plot the direction field of $y' = -4x/9y$ and approximate solution curves through $(0, 2)$ and $(0, 4)$. (*AEM Ref.* p. 11)

Pr.1.3 (Exponential growth) Find and plot the solution of $y' = 0.2y$ satisfying $y(0) = 2$.
(*AEM Ref.* p. 9 (#25))

Pr.1.4 (Exponential approach) Solve the initial value problem $y' + y = 2$, $y(0) = 0$. Plot the solution for $t = 0 \ldots 5$. (*AEM Ref.* pp. 23, 24)

Pr.1.5 (Exponential decay) Find the particular solution of $y' = ky$ satisfying $y(0) = 4$. Determine k such that at $t = 1$ the solution $y(t)$ has decreased to half its initial value.
(*AEM Ref.* pp. 6, 23, 24)

Pr.1.6 (Initial value problem) Solve $y' = 1 + y^2$, $y(0) = 0$ and plot the solution curve.
(*AEM Ref.* p. 15)

Pr.1.7 (Checking solutions) Check whether $y = \tan x$ satisfies $y' = 1 + y^2$.
(*AEM Ref.* p. 15)

Pr.1.8 (Separable equation) Solve the initial value problem $y^3 y' + x^3 = 0$, $y(0) = 1$.
(*AEM Ref.* p. 18 (#14))

Pr.1.9 (Test for exactness) Is the following equation exact?

$$(x^3 + 3\,x\,y^2)\,dx + (3\,x^2\,y + y^3)\,dy = 0.$$

Solve this equation. (*AEM Ref.* p. 27)

Pr.1.10 (Integrating factor) Show that $\cos(x + y)$ is an integrating factor of
$y\,dx + (y + \tan(x + y))\,dy = 0$ and solve the exact equation. (*AEM Ref.* p. 32 (#26))

Pr.1.11 (Linear differential equation) Find the general solution of $y' = 2(y - 1)\tan 2x$ and from it the particular solution y_p satisfying the initial condition $y(0) = 4$. Plot y_p. (*AEM Ref.* p. 39 (#17))

Pr.1.12 (Beats) Find and plot the solution of the initial value problem

$$\csc x\,dy - (y\cot x\,\csc x + 20\cos 20x)\,dx = 0, \qquad y(\pi/2) = 0.$$

Pr.1.13 (Linear differential equation) The general solution of $y' + p(x)y = r(x)$ is

$$y(x) = e^{-h}\left[\int e^h\,r\,dx + c\right], \qquad h = \int p(x)\,dx.$$

Solve $y' + ky = e^{-kx}$ by this integral formula. (*AEM Ref.* p. 34)

Pr.1.14 (Bernoulli equation) Solve $y' + (1/3)y = (1/3)(1 - 2x)y^4$ (a) directly by `dsolve`, and (b) by setting $u = 1/y^3$, simplifying the ODE in u, and then applying `dsolve`. (*AEM Ref.* p. 39 (#33))

Pr.1.15 (Verhulst equation) Solve the Verhulst equation $y' - 3y = -5y^2$. Find three initial conditions such that the corresponding solutions are (1) increasing, (2) constant, (3) decreasing. Plot these solutions. (*AEM Ref.* pp. 13, 37, 41)

Pr.1.16 (Orthogonal trajectories) Plot some of the hyperbolas $xy = c = const$. Find and plot some of their orthogonal trajectories, all curves and trajectories on common axes. (*AEM Ref.* p. 51 (#16))

Pr.1.17 (RC-circuit) The current $i(t)$ in an RC-circuit is governed by the ODE

$$R\,di/dt + i/C = dE/dt.$$

Solve this ODE for a general resistance R, capacitance C, and electromotive force $E(t) = E_0 \sin(\omega t)$. Plot $i(t)$, assuming that $R = 1$ ohm, $C = 1$ farad, $\omega = 1\,\mathrm{sec}^{-1}$, $E_0 = 220$ volts, and $i(0) = 0$ ampere. (*AEM Ref.* p. 46)

Pr.1.18 (RL-circuit) Model the current in an RL-circuit with $L = 0.1$ henry, $R = 5$ ohms, and a 12-volt battery. Determine and plot (on common axes) the current $i(t)$ when $i(0)$ equals 5, 2.5, 1, 0 amps. (*AEM Ref.* p. 43)

Pr.1.19 (Picard iteration) Integrating $y' = f(x, y)$, $y(x_0) = y_0$ gives

$$y(x) = y_0 + \int_{x_0}^{x} f(t,\, y(t))\, dt.$$

This suggests the Picard iteration

$$y_n(x) = y_0 + \int_{x_0}^{x} f(t,\, y_{n-1}(t))\, dt, \qquad n = 1, 2, \dots .$$

Solve $y' = 1 + y^2$, $y(0) = 0$ by Picard iteration. (*AEM Ref.* p. 57)

Pr.1.20 (Experiment on Picard iteration by a do-loop) Obtain and plot the solution of Pr.1.19 (and a few initial value problems of your choice) by the do-loop with $N = 5\text{-}10$ steps

```
> N := 5:                                           # for instance
> y0 := 0:                                           # for instance
> pic(0) := 0:
> for n from 1 to N do
>    pic(n) := sort(y0 + int(1 + (subs(x = t, pic(n - 1)))^2, t = 0..x));
> od:
> S := seq(pic(n), n = 1..N):
> with(plots):
> plot({S}, x = 0..1);
```

Explanation. `sort` has the effect that the powers come out in their natural order. The do-loop begins with the line that contains `do` and terminates with the line `od` (the `do` written in reverse order). Call this the opening line and the closing line of the do-loop. The closing line ends with a colon : , not with a ; . The present do-loop consists of a single command (except for the opening line and the closing line) because Picard's iteration can be written as a single formula. (*AEM Ref.* p. 57)

Chapter 2

Linear ODE's of Second and Higher Order

Content. General solutions, initial value problems (Ex. 2.1, Prs. 2.1, 2.2)
Vibrating mass on a spring (Exs. 2.2, 2.3, Prs. 2.3, 2.7, 2.8, 2.10)
Euler-Cauchy equations (Ex. 2.4, Prs. 2.12, 2.13)
Wronskian (Ex. 2.5, Prs. 2.4, 2.14)
Nonhomogeneous linear ODE's (Exs. 2.6-2.8, Prs. 2.6, 2.16, 2.18, 2.19)
Resonance, beats, electric circuits (Exs. 2.9, 2.10, Prs. 2.15, 2.17, 2.20)

DEtools package, commands for **derivatives**, **integrals**, **solution of ODE's** see the opening of Part A.

Examples for Chapter 2

EXAMPLE 2.1 **GENERAL SOLUTION. INITIAL VALUE PROBLEM**

Find a general solution y of the given ODE. Find and plot the particular solution y_p satisfying the given initial conditions.

$$y'' + y' - 2y = 0, \qquad y(0) = 4, \qquad y'(0) = -5.$$

Solution. Type the ODE as

```
> ode := diff(y(x), x, x) + diff(y(x), x) - 2*y(x) = 0;
```

$$ode := \left(\frac{\partial^2}{\partial x^2}\, \mathrm{y}(x) \right) + \left(\frac{\partial}{\partial x}\, \mathrm{y}(x) \right) - 2\,\mathrm{y}(x) = 0$$

or, equivalently, as

```
> ode2 := (D@@2)(y)(x) + D(y)(x) - 2*y(x) = 0;
```

$$ode2 := \left(\mathrm{D}^{(2)} \right)(y)(x) + \mathrm{D}(y)(x) - 2\,\mathrm{y}(x) = 0$$

Obtain a general solution by `dsolve` from either one of these two equations, the results being the same:

```
> sol := dsolve(ode);          # Resp. sol := y(x) = _C1 e^x + _C2 e^(-2x)
> sol2 := dsolve(ode2);        # Resp. sol2 := y(x) = _C1 e^x + _C2 e^(-2x)
```

Obtain the particular solution of the initial value problem directly by the command

```
> ypartic := dsolve({ode, y(0) = 4, D(y)(0) = -5});
```

$$ypartic := \mathrm{y}(x) = \mathrm{e}^x + 3\,\mathrm{e}^{(-2x)}$$

You can check this answer by determining the arbitrary constants in the general solution as follows. For this you need the derivative of `sol`

```
> yprime := diff(sol, x);
```

$$yprime := \frac{\partial}{\partial x}\,\mathrm{y}(x) = \_C1\,\mathrm{e}^x - 2\,\_C2\,\mathrm{e}^{(-2\,x)}$$

and from this the values of y and y' at $x = 0$,

```
> y0 := subs(x = 0, rhs(sol));          # Resp. y0 := _C1 e^0 + _C2 e^0
> y0 := eval(%);                        # Resp. y0 := _C1 + _C2
> yprime0 := eval(subs(x = 0, rhs(yprime)));
```

$$yprime0 := \_C1 - 2\,\_C2$$

(You needed `eval` because `subs` usually does not evaluate by itself.) Equating `y0` to 4 and `yprime0` to -5 (the given initial values), you obtain the corresponding values of $\_C1$ and $\_C2$ by the command

```
> S := solve({y0 = 4, yprime0 = -5},  {_C1, _C2});
```

$$S := \{\_C2 = 3,\ \_C1 = 1\}$$

```
> subs(S, sol);
```

$$\mathrm{y}(x) = \mathrm{e}^x + 3\,\mathrm{e}^{(-2\,x)}$$

(If you are familiar with matrices and want to determine the two constants by using matrices, type `?linsolve`.)

You get the figure of y_p by the commands

```
> plot(ypartic, x = 0..2);
```

Plotting error, empty plot

```
> plot(rhs(ypartic), x = 0..2, y = 0..9,
   xtickmarks = [0, 0.5, 1, 1.5, 2], ytickmarks = [0, 2, 4, 6, 8]);
```

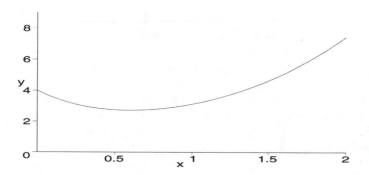

Example 2.1. Particular solution $y_p = e^x + 3\,e^{-2\,x}$

Similar Material in AEM: p. 73

EXAMPLE 2.2 **MASS-SPRING SYSTEM.**
COMPLEX CHARACTERISTIC ROOTS.
DAMPED OSCILLATIONS

Solve

$$y'' + 0.2\,y' + 4.01\,y = 0, \qquad y(0) = 0, \qquad y'(0) = 2.$$

Solution. This is the model of the vertical vibrations of a weight (of mass 1) attached to the lower end of an elastic spring (of spring constant 4.01), whose upper end is fixed. This mechanical mass-spring system has damping (damping constant 0.2). It begins its motion at the displacement 0 (the position of static equilibrium) with an initial velocity 2.

Type the equation

```
> ode := diff(y(t), t, t) + 0.2*diff(y(t), t) + 4.01*y(t) = 0;
```

$$ode := \left(\frac{\partial^2}{\partial t^2}\,\mathrm{y}(t) \right) + .2 \left(\frac{\partial}{\partial t}\,\mathrm{y}(t) \right) + 4.01\,\mathrm{y}(t) = 0$$

where t is time. Solve the initial value problem by dsolve ,

```
> yp := dsolve({ode, y(0) = 0, D(y)(0) = 2});
```

$$yp := \mathrm{y}(t) = \mathrm{e}^{\left(-\frac{1}{10}\,t\right)}\,\sin(2\,t)$$

```
> plot(rhs(yp), t = 0..20);
```

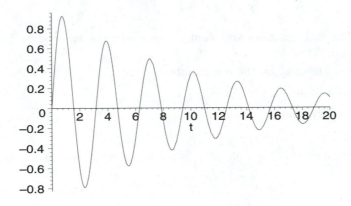

Example 2.2. Damped oscillations

This is the typical form of damped vibrations governed by a linear ODE. The oscillations lie between the exponential curves $\exp(-0.1t)$ and $-\exp(-0.1t)$. Damping takes energy from the system, so that the maximum amplitudes of the motion decrease with time and eventually go to zero. Note that the motion begins at 0 and with a positive slope, in agreement with the initial conditions.

Similar Material in AEM: p. 78

EXAMPLE 2.3 **THE THREE CASES OF DAMPING**

Consider the ODE $y'' + cy' + y = 0$ with $c = 1/2$, 2, 3. The term cy' is the damping term. No damping, $c = 0$, gives harmonic oscillations. For reasons of continuity you should expect (decreasing) oscillations when $c > 0$ is small ("**underdamping**"), but non-oscillatory behavior when c is large ("**critical damping**" and "**overdamping**"). Show this by solving the equation with $c = 1/2$, 2, 3 for the case that the motion starts from rest at $y = 1$.

Solution. Type the equation in the three cases as

```
> y1 :- 'y1': 	y2 := 'y2': 	y3 := 'y3':
> ode1 := diff(y1(t), t, t) + (1/2)*diff(y1(t), t) + y1(t) = 0;
```

$$ode1 := \left(\frac{\partial^2}{\partial t^2}\, y1(t) \right) + \frac{1}{2} \left(\frac{\partial}{\partial t}\, y1(t) \right) + y1(t) = 0$$

```
> ode2 := diff(y2(t), t, t) + 2*diff(y2(t), t) + y2(t) = 0;
```

$$ode2 := \left(\frac{\partial^2}{\partial t^2}\, y2(t) \right) + 2 \left(\frac{\partial}{\partial t}\, y2(t) \right) + y2(t) = 0$$

```
> ode3 := diff(y3(t), t, t) + 3*diff(y3(t), t) + y3(t) = 0;
```

$$ode3 := \left(\frac{\partial^2}{\partial t^2}\, y3(t) \right) + 3 \left(\frac{\partial}{\partial t}\, y3(t) \right) + y3(t) = 0$$

The initial conditions are $y(0) = 1$, $y'(0) = 0$. This gives the solutions

```
> yp1 := dsolve({ode1, y1(0) = 1, D(y1)(0) = 0});
```

$$yp1 := y1(t) = e^{\left(-\frac{1}{4}t\right)} \cos\left(\frac{1}{4}\sqrt{15}\,t \right) + \frac{1}{15}\sqrt{15}\,e^{\left(-\frac{1}{4}t\right)} \sin\left(\frac{1}{4}\sqrt{15}\,t \right)$$

```
> yp2 := dsolve({ode2, y2(0) = 1, D(y2)(0) = 0});
```

$$yp2 := y2(t) = e^{(-t)} + e^{(-t)}\,t$$

```
> yp3 := dsolve({ode3, y3(0) = 1, D(y3)(0) = 0});
```

$$yp3 := y3(t) = \left(\frac{1}{2} + \frac{3}{10}\sqrt{5} \right) e^{\left(\frac{1}{2}\left(-3+\sqrt{5}\right)t\right)} + \left(\frac{1}{2} - \frac{3}{10}\sqrt{5} \right) e^{\left(-\frac{1}{2}\left(3+\sqrt{5}\right)t\right)}$$

Plot these three particular solutions on common axes by the command

```
> plot({rhs(yp1), rhs(yp2), rhs(yp3)}, t = 0..20);
```

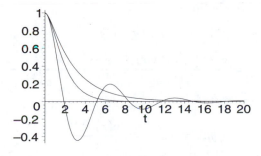

Example 2.3. Typical solutions for the three cases of damping

$c = 1/2$ (**underdamping**) gives the oscillatory solution. $c = 2$ (**critical damping**) corresponds to the **double root** of the characteristic equation $\lambda^2 + 2\lambda + 1 = 0$, and

$c = 3$ (**overdamping**) gives a monotone decreasing solution (as does $c = 2$ in the present case).

If you change the initial velocity from 0 to -2, you obtain a critical solution that has a zero at $t = 1$ and is no longer monotone. Indeed, then the solution is

```
> yp2b := dsolve({ode2, y2(0) = 1, D(y2)(0) = -2});
```

$$yp2b := \mathrm{y2}(t) = e^{(-t)} - e^{(-t)}\, t$$

```
> fsolve(rhs(yp2b) = 0);
```
                                                    # This gives the zero at $t = 1$.

Similar Material in AEM: pp. 72, 78

EXAMPLE 2.4   **THE THREE CASES FOR AN EULER-CAUCHY EQUATION**

Solve

$$x^2 y'' + axy' + y = 0.$$

Solution. The three cases correspond to, say, $a - 1 = 1/2$, 2, 3, thus, $a = 3/2$, 3, 4. Indeed, type the Euler-Cauchy equations as

```
> ode1 := x^2*diff(y1(x), x, x) + (3/2)*x*diff(y1(x), x) + y1(x) = 0;
```

$$ode1 := x^2 \left(\frac{\partial^2}{\partial x^2}\, \mathrm{y1}(x) \right) + \frac{3}{2}\, x \left(\frac{\partial}{\partial x}\, \mathrm{y1}(x) \right) + \mathrm{y1}(x) = 0$$

```
> ode2 := x^2*diff(y2(x), x, x) + 3*x*diff(y2(x), x) + y2(x) = 0;
```

$$ode2 := x^2 \left(\frac{\partial^2}{\partial x^2}\, \mathrm{y2}(x) \right) + 3\, x \left(\frac{\partial}{\partial x}\, \mathrm{y2}(x) \right) + \mathrm{y2}(x) = 0$$

```
> ode3 := x^2*diff(y3(x), x, x) + 4*x*diff(y3(x), x) + y3(x) = 0;
```

$$ode3 := x^2 \left(\frac{\partial^2}{\partial x^2}\, \mathrm{y3}(x) \right) + 4\, x \left(\frac{\partial}{\partial x}\, \mathrm{y3}(x) \right) + \mathrm{y3}(x) = 0$$

Choose initial conditions $y(1) = 1$, $y'(1) = 0$. You cannot choose conditions at $x = 0$, where the coefficients of the equation divided by x^2, that is, a/x and $1/x^2$, are singular. Apply dsolve to obtain

```
> yp1 := dsolve({ode1, y1(1) = 1, D(y1)(1) = 0});
```

$$yp1 := \mathrm{y1}(x) = \frac{\cos\left(\frac{1}{4}\, \sqrt{15}\, \ln(x) \right)}{x^{\left(\frac{1}{4} \right)}} + \frac{\frac{1}{15}\sqrt{15}\, \sin\left(\frac{1}{4}\, \sqrt{15}\, \ln(x) \right)}{x^{\left(\frac{1}{4} \right)}}$$

```
> yp2 := dsolve({ode2, y2(1) = 1, D(y2)(1) = 0});
```

$$yp2 := \mathrm{y2}(x) = \frac{\ln(x) + 1}{x}$$

```
> yp3 := dsolve({ode3, y3(1) = 1, D(y3)(1) = 0});
```

$$yp3 := \mathrm{y3}(x) = \left(\frac{1}{2} + \frac{3}{10}\, \sqrt{5} \right) x^{\left(\frac{1}{2}\sqrt{5} - \frac{3}{2} \right)} + \left(\frac{1}{2} - \frac{3}{10}\, \sqrt{5} \right) x^{\left(-\frac{1}{2}\sqrt{5} - \frac{3}{2} \right)}$$

```
> plot({rhs(yp1), rhs(yp2), rhs(yp3)}, x = 0..6, y = -1..1,
  ytickmarks = [-1, 0, 1]);
```

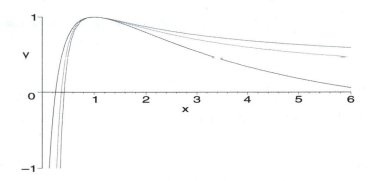

Example 2.4. The three cases of solutions for an Euler-Cauchy equation

Similar Material in AEM: pp. 94, 95

| EXAMPLE 2.5 | **WRONSKIAN**

For $y'' + p_1(x)y' + p_0(x)y = 0$ the **Wronskian** is the determinant

$$W(y_1,\, y_2) = \begin{vmatrix} y_1 & y_2 \\ y_1' & y_2' \end{vmatrix} = y_1\, y_2' - y_2\, y_1'.$$

$\{y_1,\, y_2\}$ is a basis of solutions of the equation (with continuous p_1 and p_0) if and only if $W(y_1,\, y_2)$ is different from zero. Show this for the basis $y_1 = \cos k\, x$, $y_2 = \sin k\, x$ of the ODE $y'' + k^2 y = 0$ with $k \neq 0$.

Solution. The basis is

```
> y1 := cos(k*x):    y2 := sin(k*x):
```

From this you obtain the Wronskian by typing

```
> W := y1*diff(y2, x) - y2*diff(y1, x);
```

$$\mathrm{W} := \cos(k\, x)^2\, k + \sin(k\, x)^2\, k$$

```
> simplify(%);                                              # Resp. k
```

If $k = 0$, then $y_1 = 1$, $y_2 = 0$, which is not a basis. If k is not 0, then y_1, y_2 form a basis.

  If you know about determinants and matrices, you can use them in connection with Maple. They are part of the `linalg` package. Type `?linalg[det]` for information. Load the package by typing

```
> with(linalg):
```

Ignore the warning that you will get as a response; it is not relevant for your purpose. The Wronskian is the determinant of the 2×2 matrix **A** which you can type as

```
> A := matrix([[y1, y2], [diff(y1, x), diff(y2, x)]]);
```

$$A := \begin{bmatrix} \cos(k\,x) & \sin(k\,x) \\ -\sin(k\,x)\,k & \cos(k\,x)\,k \end{bmatrix}$$

and obtain $W = \det \mathbf{A}$ (suggesting "determinant" of $\mathbf{A}$) simply by typing

```
> W := det(A);                          # Resp. W := cos(k x)² k + sin(k x)² k
> simplify(%);                                              # Resp. k
```

Even more simply, type

```
> y := [y1, y2];                        # Resp. y := [cos(kx), sin(kx)]
```

Then obtain the Wronskian matrix by typing

```
> A := wronskian(y, x);
```

$$A := \begin{bmatrix} \cos(k\,x) & \sin(k\,x) \\ -\sin(k\,x)\,k & \cos(k\,x)\,k \end{bmatrix}$$

and finally the Wronskian itself by typing

```
> W := det(A);                          # Resp. W := cos(k x)² k + sin(k x)² k
> simplify(W);                                             # Resp. k
```

For an ODE of order 2 you can get away without knowing about determinants. For an ODE of greater order you will need determinants. For instance, a basis of

$$y''' - 2y'' - y' + 2y = 0$$

is $y_1 = e^{-x}$, $y_2 = e^x$, $y_3 = e^{2x}$ (verify that these are solutions), as can be seen by typing the corresponding 3×3 Wronskian matrix (whose third row contains the second derivatives of the three solutions). Hence type the solutions

```
> y1 := exp(-x):    y2 := exp(x):    y3 := exp(2*x):
```

then

```
> y := [y1, y2, y3];                    # Resp. y := [e^(-x), e^x, e^(2 x)]
> A := wronskian(y, x);
```

$$A := \begin{bmatrix} e^{(-x)} & e^x & e^{(2x)} \\ -e^{(-x)} & e^x & 2\,e^{(2x)} \\ e^{(-x)} & e^x & 4\,e^{(2x)} \end{bmatrix}$$

```
> W := det(A);                          # Resp. W := 6 e^(-x) e^x e^(2 x)
> simplify(%);                                         # Resp. 6 e^(2 x)
```

Since this is not zero, the three solutions form a basis of solutions for the given ODE on any interval.

   Similar Material in AEM: p. 97

| EXAMPLE 2.6 | **NONHOMOGENEOUS LINEAR ODE's**

Nonhomogeneous linear ODE's can be solved by dsolve practically in the same way as homogeneous ODE's. For instance, to solve

$$y'' + 2y' + 101y = 10.4\,e^x, \qquad y(0) = 1.1, \qquad y'(0) = -0.9,$$

type the equation as before, say,

```
> y := 'y':                          # Unassign y (used in the previous example).
> ode := diff(y(x), x, x) + 2*diff(y(x), x) + 101*y(x) = 10.4*exp(x);
```

$$ode := \left(\frac{\partial^2}{\partial x^2} \, \mathrm{y}(x) \right) + 2 \left(\frac{\partial}{\partial x} \, \mathrm{y}(x) \right) + 101 \, \mathrm{y}(x) = 10.4 \, \mathrm{e}^x$$

Then dsolve, applied to the initial value problem, gives

```
> yp := dsolve({ode, y(0) = 1.1, D(y)(0) = -0.9});
```

$$yp := \mathrm{y}(x) = \frac{1}{10} \, \mathrm{e}^x + \mathrm{e}^{(-x)} \cos(10 \, x)$$

```
> yp := evalf(%,2);                          # Gives 2 digits only
```

$$yp := \mathrm{y}(x) = .10 \, \mathrm{e}^x + \mathrm{e}^{(-1.\,x)} \cos(10.\,x)$$

```
> plot(rhs(yp), x = 0..3.5, xtickmarks = [1, 2, 3], labels = [x, y]);
```

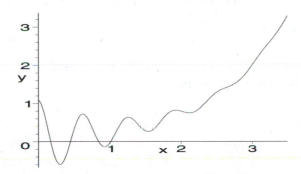

Example 2.6. Particular solution $y = 0.1 e^x + e^{-x} \cos 10\,x$

You see that the oscillatory effect of the second term decreases and the solution curve approaches the curve of the exponential function $0.1 \, e^x$.

    Similar Material in AEM: p. 103

| EXAMPLE 2.7 | **SOLUTION BY UNDETERMINED COEFFICIENTS** |

Write a given nonhomogeneous ODE, say,

$$y'' - 3y' + 2y = e^x,$$

in the form $Ly = r$, where Ly (L suggesting 'linear operator') is the left-hand side of the given ODE,

```
> y := 'y':
> Ly := diff(y(x), x, x) - 3*diff(y(x), x) + 2*y(x);
```

$$Ly := \left(\frac{\partial^2}{\partial x^2} \, \mathrm{y}(x) \right) - 3 \left(\frac{\partial}{\partial x} \, \mathrm{y}(x) \right) + 2 \, \mathrm{y}(x)$$

and r is the right-hand side,

```
> r := exp(x);
```
       # Resp. $r := e^x$

To see whether the modification rule applies, you must first solve the homogeneous equation $Ly = 0$; thus,

```
> yh := dsolve(Ly = 0);
```
     # Resp. $yh := y(x) = \_C1\, e^x + \_C2\, e^{(2\,x)}$

You see that r is a solution of the homogeneous ODE, so that the modification rule (for a simple root) does apply. That is, instead of $C\,e^x$ you have to use

```
> yp := C*x*exp(x);
```
     # Resp. $yp := C\,x\,e^x$

(In the case of a double root you would have to multiply e^x by x^2 instead of x.) The constant C is unknown. You find its value from the given nonhomogeneous equation with $y = yp$; that is,

```
> sol := eval(subs(y(x) = yp, Ly = r));
```

$$sol := -C\,e^x = e^x$$

You see that $C = -1$. The command for obtaining this (in a more involved case) would be

```
> C0 := solve(sol, C);
```
     # Resp. $C0 := -1$

Substituting this into yp and adding yh, you obtain a general solution of the given ODE. Perhaps you do this step by step.

```
> rhs(yh);
```
     # Resp. $\_C1\, e^x + \_C2\, e^{(2\,x)}$
```
> yp;
```
     # Resp. $C\,x\,e^x$
```
> yp2 := subs(C = C0, yp);
```
     # Resp. $yp2 := -x\,e^x$
```
> ygen := rhs(yh) + yp2;
```
     # Resp. $ygen := \_C1\, e^x + \_C2\, e^{(2\,x)} - x\,e^x$

This example served to explain the method. Clearly, you can obtain the answer directly by typing

```
> dsolve(Ly = r);
```

$$y(x) = (-x + \_C1\, e^x + \_C2)\, e^x$$

This agrees with your previous result, except for the notation.

     Similar Material in AEM: p. 105

| EXAMPLE 2.8 | **SOLUTION BY VARIATION OF PARAMETERS** |

To solve

$$y'' + y = \sec x$$

by variation of parameters, write it as $Ly = r$, typing

```
> y := 'y':
> Ly := diff(y(x), x, x) + y(x);
```
     # Resp. $Ly := \left(\dfrac{\partial^2}{\partial x^2}\, y(x) \right) + y(x)$

```
> r := sec(x);
```
     # Resp. $r := \sec(x)$

Obtain a general solution of the homogeneous ODE by dsolve,

```
> yh := dsolve(Ly = 0);                          # General solution of Ly = 0
```

$$yh := y(x) = \_C1 \, \cos(x) + \_C2 \, \sin(x)$$

Hence a basis of solutions of the homogeneous ODE, as needed in the present method, is

```
> y1 := cos(x):    y2 := sin(x):
```

For this basis the Wronskian is (see Example 2.5 in this Guide, if necessary)

```
> W := y1*diff(y2, x) - y2*diff(y1, x);
```

$$W := \cos(x)^2 + \sin(x)^2$$

```
> simplify(%);                                   # Resp. 1
```

The formula for a particular solution is

$$y_p = -y_1 \int \frac{y_2 r}{W} \, dx + y_2 \int \frac{y_1 r}{W} \, dx.$$

For the present ODE you obtain from it

```
> yp := -y1*int(y2*r/W, x) + y2*int(y1*r/W, x);
```

$$yp := \cos(x) \, \ln\left(\cos(x)\right) + \sin(x) \, x$$

Hence a general solution of the given nonhomogeneous equation is

```
> ygen := yh + (yp);
Error, (in simpl/relopsum) invalid terms in sum
> ygen := rhs(yh) + yp;
```

$$ygen := \_C1 \, \cos(x) + \_C2 \, \sin(x) + \cos(x) \ln(\cos(x)) + \sin(x) \, x$$

This example served to explain the method. Clearly, you can obtain the solution directly by typing

```
> dsolve(Ly = r);
```

$$y(x) = \_C1 \, \cos(x) + \_C2 \, \sin(x) + \cos(x) \ln(\cos(x)) + \sin(x) \, x$$

Similar Material in AEM: p. 109

| EXAMPLE 2.9 | **FORCED VIBRATIONS. RESONANCE. BEATS**

Resonance occurs in an **undamped vibrating system** if the frequency of the driving force equals the natural frequency of the free vibrations of the system. For instance, let

$$y'' + y = \cos t.$$

Type this ODE and solve it.

Solution.

```
> y := 'y':
> ode := diff(y(t), t, t) + y(t) = cos(t);
```

$$ode := \left(\frac{\partial^2}{\partial t^2} \, \mathrm{y}(t) \right) + \mathrm{y}(t) = \cos(t)$$

```
> sol := dsolve(ode);
```

$$sol := \mathrm{y}(t) = \frac{1}{2} \, \sin(t) \, t + \_C1 \, \sin(t) + \_C2 \, \cos(t)$$

The first term on the right will be growing without bound as time t increases to infinity. You can single it out by choosing $\_C1 = 0$ and $\_C2 = 0$,

```
> y1 := subs(_C1 = 0, _C2 = 0, rhs(sol));
```

$$y1 := \frac{1}{2} \, \sin(t) \, t$$

```
> plot(y1, t = 0..50, title = 'Resonance');
```

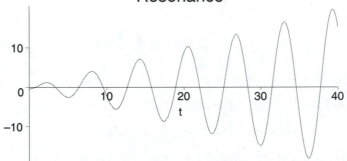

Resonance

Example 2.9. Resonance

Beats occur if the frequency of the driving force of the undamped system is close to the natural frequency of the free vibrations. For example, consider the equation $y'' + y = 0.19 \cos 0.9t$. (The factor 0.19 has been included to have a simpler final answer; it is of no other importance.) Type and solve this ODE.

Solution.

```
> ode2 := diff(y2(t), t, t) + y2(t) = 0.19*cos(0.9*t);
```

$$ode2 := \left(\frac{\partial^2}{\partial t^2} \, \mathrm{y2}(t) \right) + \mathrm{y2}(t) = .19 \, \cos(.9 \, t)$$

```
> sol2 := dsolve(ode2);
```

$$sol2 := \mathrm{y2}(t) = \cos\left(\frac{9}{10} \, t \right) + \_C1 \, \sin(t) + \_C2 \, \cos(t)$$

```
> sol3 := subs(_C1 = 0, _C2 = -1, sol2);        # Retain the two cosine terms.
```

$$sol3 := \mathrm{y2}(t) = \cos\left(\frac{9}{10} \, t \right) - \cos(t)$$

To comprehend this form of the solution ("**beats**"), note that the difference of two cosine functions can be written as the product of two sine functions, one with a high frequency (this gives the rapid oscillations shown in the figure) and one with a low frequency $(0.05/(2\pi)$, giving a period of about 126). The formula is obtained by

```
> 2*combine(sin(0.95*t)*sin(0.05*t));        # Resp. cos(.90 t) - cos(1.00 t)
```

```
> plot(rhs(sol3), t = 0..200, title = 'Beats');
```

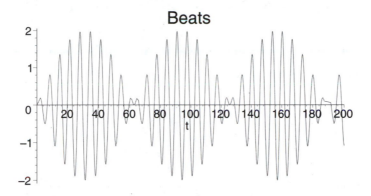

Example 2.9. Beats given by $y = 2 \sin 0.95t \, \sin 0.05t$

Similar Material in AEM: pp. 114, 115

| **EXAMPLE 2.10** | ***RLC*-CIRCUIT** |

The current $i(t)$ in an *RLC*-circuit with sinusoidal electromotive force is obtained by solving

$$Li' + Ri + \frac{1}{C} \int i \, dt = E_0 \sin \omega t$$

where $Q = \int i \, dt$ is the charge on the capacitor. Differentiation gives the ODE

$$Li'' + Ri' + \frac{1}{C}i = \omega \, E_0 \cos \omega t.$$

(Write i since Maple uses I for $\sqrt{-1}$.) For instance, let $L = 0.1$ henry, $R = 100$ ohms, $C = 0.001$ farad, $E_0 = 155$, and $\omega = 377$ (thus 60 hertz). Solve the ODE for these data.

Solution. Write the ODE as $Mi = r$, where Mi denotes the left-hand side. (Note that L here is used for the inductance.)

```
> Mi := 0.1*diff(i(t), t, t) + 100*diff(i(t), t) + 1000*i(t);
```

$$Mi := .1 \left(\frac{\partial^2}{\partial t^2} i(t) \right) + 100 \left(\frac{\partial}{\partial t} i(t) \right) + 1000 \, i(t)$$

```
> r := 377*155*cos(377*t);                    # Resp. r := 58435 cos(377 t)
```

A general solution of the homogeneous ODE is

```
> igen := evalf(dsolve(Mi = 0), 5);
```

$$igen := i(t) = \_C1 \, e^{(-10.100\,t)} + \_C2 \, e^{(-989.90\,t)}$$

A general solution of the nonhomogeneous ODE is

```
> igen := evalf(dsolve(Mi = r), 5);
```

$$igen := i(t) = -.48382 \, \cos(377.\,t) + 1.3804 \, \sin(377.\,t) + \_C1 \, e^{(-10.100\,t)} + \_C2 \, e^{(-989.90\,t)}$$

The particular solution satisfying $i(0) = 0$, $i\,'(0) = 0$ is (see the figure)

```
> ipart := evalf(dsolve({Mi = r, i(0) = 0, D(i)(0) = 0}), 5);
```

$ipart := i(t) =$
$$-.48382\,\cos(377.\,t) + 1.3804\,\sin(377.\,t) - .04236\,e^{(-10.100\,t)} + .52616\,e^{(-989.90\,t)}$$

```
> plot(rhs(ipart), t = 0..0.13);
```

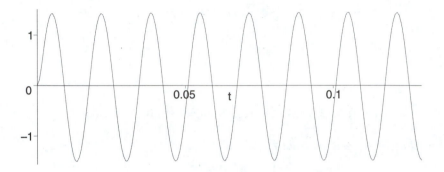

Example 2.10. Current in the RLC-circuit

The oscillation is very rapid. The **transient solution** becomes practically zero after a very short time, so that a transient period is practically not visible in the figure. $i(0) = 0$ means no current at $t = 0$. Assume that the capacitor is uncharged at $t = 0$; thus $Q(0) = 0$ in the first formula in this example. Solve that formula algebraically for i', obtaining

$$i\,' = \frac{1}{L}(E_0\,\sin\,\omega t - R\,i - \frac{Q}{C}).$$

For $t = 0$ all three terms on the right are 0. This shows that the initial condition $i'(0) = 0$ physically means that the capacitor is initially uncharged.

    Similar Material in AEM: pp. 118-121

Problem Set for Chapter 2

Pr.2.1 (General solution, initial value problem, harmonic oscillations) Find a general solution of $y'' + 3y' + 2y = 0$. Find and plot the particular solution satisfying the initial conditions $y(0) = 4$, $y'(0) = -2$. (*AEM Ref.* pp. 72, 73)

Pr.2.2 (Maximum of solution) Find and plot the solution y_p of the initial value problem $y'' - 2y' + y = 0$, $y(0) = 4$, $y'(0) = 3$. Find the location and the value of the maximum of y_p, first from the graph and then by calculation. (*AEM Ref.* p. 74)

Pr.2.3 (Experiment on damped oscillations, dependence on initial conditions) Experiment with Example 2.2 in this Guide by systematically changing the initial conditions (e.g., by keeping $y'(0)$ constant and varying $y(0)$, etc.) in order to find out whether and how solutions depend on initial conditions. (*AEM Ref.* pp. 87-90)

Pr.2.4 (Linear independence, Wronskian) Show that the solution in Pr.2.1 is a general solution. (*AEM Ref.* pp. 73, 97)

Pr.2.5 (**Verification of solution**) Show that a linear combination of e^{-x}, $x\,e^{-x}$, $x^2\,e^{-x}$ is a solution of $y''' + 3y'' + 3y' + y = 0$. (*AEM Ref.* p. 135)

Pr.2.6 (**Nonhomogeneous equation, complex roots**) Find a general solution of

$$y'' + 2y' + 226y = e^{-0.001\,t}.$$

Find and plot the particular solution which starts at $y = 0$ with initial velocity 0. (*AEM Ref.* pp. 79, 101)

Pr.2.7 (**Experiment on overdamping**) Find a general solution of $y'' + 10y' + 16y = 0$. Find and plot (on common axes) the three particular solutions starting from $y = 1$ with different initial velocities, namely, -1, 0, 1. Can you find an initial velocity such that the particular solution becomes 0 for some positive t? For two different positive t? What is the effect of changing the initial displacement? (*AEM Ref.* p. 87)

Pr.2.8 (**Critical damping**) Find the value of the damping constant c such that the mass-spring system
$$y'' + cy' + 37.21y = 0$$
is critically damped. Characterize the other two cases (underdamping, overdamping) in terms of the values of c. (*AEM Ref.* p. 88)

Pr.2.9 (**Pendulum**) Find the motion of a pendulum of mass m and length 1 meter for small angle $|T|$ (so that $\sin T$ can be approximated by T rather accurately). Assume that the pendulum starts from the equilibrium position ($T = 0$) with velocity 1. (*AEM Ref.* p. 91 (#10))

Pr.2.10 (**Logarithmic decrement**) Prove on the computer that the ratio of two consecutive maximum amplitudes of a damped oscillation $y = C \exp(-at) \cos(wt - d)$ is constant, the natural logarithm of this ratio being called the *logarithmic decrement* and being given by $LD = 2\pi a/w$. (C, a, w, d are constant.) (*AEM Ref.* p. 92 (#18))

Pr.2.11 (**Boundary value problem**) Solve $y'' + y = 0$, $y(0) = 3$, $y(\pi) = -3$. (*AEM Ref.* p. 80)

Pr.2.12 (**Euler-Cauchy equation**) Determine a and b in the Euler-Cauchy equation $x^2y'' + axy' + by = 0$ so that the auxiliary equation has a double root and determine a general solution for this "critical case". (*AEM Ref.* p. 93)

Pr.2.13 (**Euler-Cauchy equation**) Find the Euler-Cauchy equation with $x \cos(\ln x)$ and $x \sin(\ln x)$ as a basis of solutions. (*AEM Ref.* p. 95)

Pr.2.14 (**Wronskian**) Show that $\{e^{-2x}, e^{2x}, e^{4x}\}$ is a basis of solutions of the ODE $y''' - 4y'' - 4y' + 16y = 0$. (*AEM Ref.* p. 127)

Pr.2.15 (**Resonance**) Find and plot the solution of $y'' + 4y = -12 \sin 2t$, $y(0) = 1$, $y'(0) = 3$. (*AEM Ref.* p. 114)

Pr.2.16 (**Nonhomogeneous ODE**) Show that $-e^{-2x} \ln x$ is a particular solution of

$$y'' + 4y' + 4y = e^{-2x}/x^2.$$

Find a general solution of this ODE and determine its arbitrary constants so that you obtain a particular solution satisfying the initial conditions $y(1) = 1/e^2$, $y'(1) = -2/e^2$. (*AEM Ref.* p. 104 (#15))

Pr.2.17 **(Beats)** Find and plot the solution of the initial value problem $y'' + 100y = 36\cos 8t$, $y(0) = 0$, $y'(0) = 0$. Can you see by looking at the ODE that you will obtain beats, because the frequency of the driving force is close to that of the free harmonic oscillations? (*AEM Ref.* p. 115)

Pr.2.18 **(Undetermined coefficients)** Find a particular solution of $y'' + 3y' - 18y = 9\sinh 3x$ by the method of undetermined coefficients. Check your result by dsolve.
(*AEM Ref.* pp. 104, 107 (#12))

Pr.2.19 **(Variation of parameters)** Find a particular solution of the ODE

$$y'' - 2y' + y = 3x^{3/2}\, e^x$$

by the method of variation of parameters. (*AEM Ref.* pp. 108, 111 (#7))

Pr.2.20 **(*RLC*-circuit)** Find the current in the *RLC*-circuit with $R = 16$ ohms, $L = 8$ henrys, $C = 1/8$ farad, and $E = 100\cos 2t$ volts. Also find and plot the particular solution satisfying $i(0) = 0$, $i'(0) = 0$. (*AEM Ref.* pp. 118-122)

Chapter 3

Systems of Differential Equations.

Phase Plane, Qualitative Methods

Content. Solution of systems, use of matrices (Exs. 3.1, 3.2)
Critical points, pendulum ODE (Exs. 3.3-3.5, Prs. 3.1-3.4, 3.6, 3.11)
Nonhomogeneous systems (Exs. 3.6, 3.7)
Van der Pol and related ODE's (Ex. 3.8, Prs. 3.12-3.15)
Mixing problems, networks (Prs. 3.8-3.10)

Writing and solving systems see Ex. 3.1, etc. **Use of matrices** needs the linalg package; see Ex. 3.2. **Plotting trajectories** see Exs. 3.3, 3.5, 3.8.

Examples for Chapter 3

EXAMPLE 3.1	**HOW TO WRITE A SYSTEM OF ODE's?**
	INITIAL VALUE PROBLEM

A system of ODE's

$$y_1' = -3y_1 + y_2$$
$$y_2' = y_1 - 3y_2$$

can be written (ignore the warning that you get after writing the first line)

```
> sys := D(y1)(t) = -3*y1(t) + y2(t),
>        D(y2)(t) = y1(t) - 3*y2(t);
```

$$sys := D(y1)(t) = -3\,y1(t) + y2(t),\ D(y2)(t) = y1(t) - 3\,y2(t)$$

If for some reason you need the two equations separately, you can get them by sys[1] and sys[2] ; thus,

```
> sys[1];                    # Resp. D(y1)(t) = -3 y1(t) + y2(t)
> sys[2];                    # Resp. D(y2)(t) = y1(t) - 3 y2(t)
```

You can solve this system by dsolve ,

```
> sol := dsolve({sys});
```

$$sol := \{y2(t) = \_C1\,e^{(-2\,t)} - \_C2\,e^{(-4\,t)},\ y1(t) = \_C1\,e^{(-2\,t)} + \_C2\,e^{(-4\,t)}\}$$

Note that the solutions may appear in a different order and with the notations for the arbitrary constants interchanged. Observe the use of braces in {sys} (this is the set of the two equations).

31

Initial value problems can be solved as for a single equation. Thus, let $y_1(0) = 1$ and $y_2(0) = 5$. You obtain the solution from `dsolve` and can plot the two curves as usual.

```
> yp := dsolve({sys, y1(0) = 1, y2(0) = 5});
```

$$yp := \{y2(t) = 2\,e^{(-4\,t)} + 3\,e^{(-2\,t)}, \ \ y1(t) = 3\,e^{(-2\,t)} - 2\,e^{(-4\,t)}\}$$

```
> plot({rhs(yp[1]), rhs(yp[2])}, t = 0...5);
```

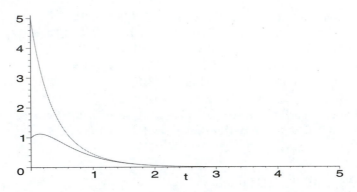

Example 3.1. Solutions y_1 (lower curve) and y_2 of the initial value problem

Similar Material in AEM: p. 163

| EXAMPLE 3.2 | USE OF MATRICES IN SOLVING SYSTEMS OF ODE's |

A general solution of a homogeneous linear system of ODE's turns out to involve the eigenvalues and eigenvectors of the coefficient matrix of the system and can readily be obtained from these eigenvalues and vectors. We explain this for the system in Example 3.1 in this Guide, $y_1' = -3y_1 + y_2$, $y_2' = y_1 - 3y_2$. To treat matrices and vectors with Maple, load the `linalg` package by

```
> with(linalg):
```

Ignore the warning. Type the coefficient matrix **A**,

```
> A := matrix([[-3, 1], [1, -3]]);
```

$$A := \begin{bmatrix} -3 & 1 \\ 1 & -3 \end{bmatrix}$$

An **eigenvalue** of **A** is a number λ such that the vector equation $\mathbf{Ax} = \lambda\mathbf{x}$ has a vector solution **x** not the zero vector. (A zero vector is a solution of this equation for any value of λ, and is called the '***trivial solution***'.) This nonzero **x** is called an **eigenvector** of **A** corresponding to that eigenvalue λ. The eigenvalues are the solutions of the **characteristic equation** $\det(\mathbf{A} - \lambda\mathbf{I}) = 0$. Here, **I** is the **unit matrix**. If you want the eigenvalues, type

```
> eigenvalues(A);                                    # Resp. -2, -4
```

If you want the eigenvalues as well as eigenvectors (as it will usually be the case), type

```
> eig := eigenvectors(A);
```

$$eig := [-2, 1, \{[1, 1]\}], [-4, 1, \{[-1, 1]\}]$$

This reads: -2 is an eigenvalue, of algebraic multiplicity 1, and $\mathbf{x_1} = [1, 1]^T$ is a corresponding eigenvector. -4 is an eigenvalue, of algebraic multiplicity 1, and $\mathbf{x_2} = [-1, 1]^T$ is a corresponding eigenvector. (Here, T indicates 'transpose', that is, these are column vectors, written here as row vectors, to save space.) You should be aware that in `eig` the eigenvalues (and corresponding eigenvectors) may appear in a different order, and the eigenvectors may differ from the response shown here by a nonzero multiplicative constant. This gives the general solution of the system

$$\mathbf{y} = c_1 \mathbf{x_1} e^{-2t} + c_2 \mathbf{x_2} e^{-4t}.$$

You can pick from `eig` the items you need for composing this solution,

```
> eig[1];                    # Resp. [-2, 1, {[1, 1]}]
> L1 := eig[1][1];           # Resp. L1 := -2    (First eigenvalue)
> eig[1][3];                 # Resp. {[1, 1]}
> x1 := eig[1][3][1];        # Resp. x1 := [1, 1]  (Braces gone)
> L2 := eig[2][1];           # Resp. L2 := -4    (Second eigenvalue)
> x2 := eig[2][3][1];        # Resp. x2 := [-1, 1]
```

With these notations you can now write the corresponding general solution as

```
> y := c1*x1*exp(L1*t) + c2*x2*exp(L2*t);
```

$$y := c1 \; x1 \; \mathrm{e}^{(-2t)} + c2 \; x2 \; \mathrm{e}^{(-4t)}$$

In components by the command `evalm` (suggesting 'evaluate matrix')

```
> y := evalm(%);
```

$$y := \left[c1 \; \mathrm{e}^{(-2t)} - c2 \; \mathrm{e}^{(-4t)}, \; c1 \; \mathrm{e}^{(-2t)} + c2 \; \mathrm{e}^{(-4t)} \right]$$

This agrees with the answer in Example 3.1 of this Guide, except for the notations for the arbitrary constants.

      Similar Material in AEM: pp. 151, 163

| EXAMPLE 3.3 | **CRITICAL POINTS. NODE** |

From a given linear system of ODE's

$$\mathbf{y'} = \mathbf{Ay}, \qquad \text{in components} \qquad \begin{array}{l} y_1{'} = a_{11} \, y_1 + a_{12} \, y_2 \\ y_2{'} = a_{21} \, y_1 + a_{22} \, y_2 \end{array} \qquad (' = d/dt)$$

we obtain

$$\frac{d \, y_2}{d \, y_1} = \frac{d \, y_2}{d \, t} \Big/ \frac{d \, y_1}{d \, t} = \frac{a_{21} \, y_1 + a_{22} \, y_2}{a_{11} \, y_1 + a_{12} \, y_2}.$$

At each point (y_1, y_2) in the $y_1 y_2$-plane (the **phase plane**) this determines a tangent direction of the solution graphed as a curve in the $y_1 y_2$-plane. Such a curve is called a **trajectory** (or *path* or *orbit*). An exception is the point $(y_1, y_2) = (0, 0)$, at which the numerator and the denominator on the right are both zero. This is called a **critical point** of the system. It can be classified in terms of three quantities related to the coefficient matrix $\mathbf{A} = [a_{jk}]$ of the system. These quantities are

$$p = a_{11} + a_{22} \qquad \text{the \textbf{trace} of the matrix}$$

$$q = \det \mathbf{A} = a_{11} a_{22} - a_{12} a_{21} \quad \text{the \textbf{determinant} of the matrix}$$

$$\delta = p^2 - 4 q.$$

The critical point is a:

Node	if $q > 0$ and $\delta \geq 0$,
Saddle point	if $q < 0$,
Center	if $p = 0$ and $q > 0$,
Spiral point	if $p \neq 0$ and $\delta < 0$.

For example, let the matrix of the system be (see the previous example)

```
> with(linalg):                                    # Ignore the warning.

> A := matrix([[-3, 1], [1, -3]]);
```

$$A := \begin{bmatrix} -3 & 1 \\ 1 & -3 \end{bmatrix}$$

Then the quantities needed for classifying the critical point are

```
> p := trace(A);                                   # Resp. p := -6
> q := det(A);                                      # Resp. q := 8
> delta := p^2 - 4*q;                               # Resp. δ := 4
```

Since q and δ are positive, the critical point at the origin is a node. More precisely, it is an **improper node,** characterized by the property that at that point all trajectories, except for two of them, have the same tangent direction, as the figure shows. (A **proper node** is one at which for each direction there is a trajectory having it as the tangent direction at the node.) Each trajectory shown is obtained by choosing a pair of constants c_1, c_2 in the general solution in the preceding example. For $c_1 = 0$ you get one of the two straight lines and for $c_2 = 0$ the other. For any other choice of c_1, c_2 you get a quadratic parabola because in that general solution, $e^{-4t} = \left(e^{-2t}\right)^2$. In the plot commands, set $e^{-t} = \tau$ and write again t for τ, for simplicity.

```
> with(plots):
> p1 := plot([t, t], t = -10..10):
> p2 := plot([t, -t], t = -10..10):
> p3 := plot([t + t^2/10, t - t^2/10], t = -10..10):
> p4 := plot([t - t^2/10, t + t^2/10], t = -10..10):
> p5 := plot([t + t^2/20, t - t^2/20], t = -10..10):
> p6 := plot([t - t^2/20, t + t^2/20], t = -10..10):
> display(p1, p2, p3, p4, p5, p6, title = 'Trajectories near a node',
    scaling = constrained);
```

Trajectories near a node

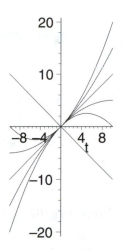

Example 3.3. Trajectories near an improper node

Similar Material in AEM: pp. 163, 172

| EXAMPLE 3.4 | **PROPER NODE, SADDLE POINT, CENTER, SPIRAL POINT** |

Improper nodes were discussed and shown in the previous example. Phase plane plots (**"phase portraits"**) of the other types of critical points are given on pp. 165 and 166 of AEM. We present here the corresponding calculations for typical examples.

A ***proper node*** with trajectories $y_2 = k\,y_1$ and $y_1 = 0$ (straight lines) is obtained for the system $y_1{}' = y_1$, $y_2{}' = y_2$, whose matrix is the unit matrix.

A ***saddle point*** with trajectories $y_1\,y_2 = const$ (hyperbolas) is obtained for $y_1{}' = y_1$, $y_2{}' = -y_2$ with matrix

```
> with(linalg):                              # Ignore the warning.

> A := matrix([[1, 0], [0, -1]]);
```

$$A := \begin{bmatrix} 1 & 0 \\ 0 & -1 \end{bmatrix}$$

for which $q = \det \mathbf{A} = -1 < 0$, so that we have a saddle (see the criteria in the previous example). The eigenvalues of $\mathbf{A}$ have opposite signs, as is typical, their values being 1 and -1.

A ***center*** with trajectories $y_1{}^2 + y_2{}^2/4 = const$ (ellipses) is obtained for $y_1{}' = y_2$, $y_2{}' = -4y_1$ with matrix

```
> A := matrix([[0, 1], [-4, 0]]);
```

$$A := \begin{bmatrix} 0 & 1 \\ -4 & 0 \end{bmatrix}$$

for which $p = 0$ and $q = 4 > 0$, which gives a center. The eigenvalues of **A** are pure imaginary, which is typical, the values being $-2i$ and $2i$.

```
> eigenvalues(A);                                    # Resp. 2I, -2I
```

A **spiral point** with trajectories $r = c \exp(-\theta)$ (spirals, in polar coordinates) is obtained for the system with matrix

```
> A := matrix([[-1, 1], [-1, -1]]);
```

$$A := \begin{bmatrix} -1 & 1 \\ -1 & -1 \end{bmatrix}$$

for which $p = -2 \neq 0$ and $\delta = p^2 - 4q = 4 - 8 < 0$, which gives a spiral point. In this case, **A** has complex eigenvalues,

```
> eigenvalues(A);                                    # Resp. -1 + I, -1 - I
```

         Similar Material in AEM: pp. 164-167

| EXAMPLE 3.5 | **PENDULUM EQUATION** |

The ODE governing the motion of a pendulum (e.g., in a pendulum clock) is $y'' + k \sin y = 0$ (here we have disregarded damping) and is a classical case of an ODE not exactly solvable in terms of finitely many elementary functions. Physically, $y(t)$ is the angle of displacement from rest (the position when the pendulum is vertical), and t is time. k is a positive constant, and we choose $k = 1$ for simplicity. We set $y = y_1$ and $y' = y_1' = y_2$. Then $y'' = y_2' = -k \sin y = -\sin y = -\sin y_1$. Hence you can write the given equation (with $k = 1$) as a system $y_1' = y_2$, $y_2' = -\sin y_1$. This illustrates the standard method of **transforming an ODE of second order into a system of two first-order ODE's.** On the computer, type the system as

```
> with(DEtools):                                     # Ignore the warning.
> sys := D(y1)(t) = y2(t),
>        D(y2)(t) = -sin(y1(t));
```

$$sys := \mathrm{D}(y1)(t) = y2(t), \mathrm{D}(y2)(t) = -\sin(y1(t))$$

Now plot a direction field of the system and in it three approximate trajectories (phase plane representations of three solutions) that will turn out to be typical; specify each of them by one of the three initial conditions

```
> inits := [0, 0, 1], [0, 0, 2], [0, 0, 3];
```

$$inits := [0, 0, 1], [0, 0, 2], [0, 0, 3]$$

where $[0, 0, 1]$ means $t = 0$, $y_1 = 0$, $y_2 = 1$, etc.; thus, each condition specifies a t-value (0) and a point in the phase plane through which the trajectory should pass. Then plot. (For information, type ?DEplot .)

```
> DEplot([sys[1], sys[2]], [y1(t), y2(t)], t = -10..10, y1 = -6..6,
  y2 = -3..3, [inits], scaling = constrained, title = 'Phase portrait
  of the undamped pendulum');
```

Phase portrait of the undamped pendulum

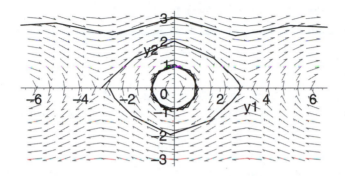

Example 3.5. Direction field with typical trajectories for the undamped pendulum

Critical points are at 0 and at $\pm 2\pi$, $\pm 4\pi$,... by periodicity. These are centers. One of the three solutions is typical of the circular-shaped closed curves surrounding 0. Further critical points are at π and at $-\pi$, $\pm 3\pi$, $\pm 5\pi$, ... by periodicity. These are saddle points. Your second curve passes (approximately) through two of them. The third kind of solution corresponds to swings of the pendulum through 360 degrees (full rotations), which, of course, an ordinary pendulum could not do.

    Similar Material in AEM: p. 176

| EXAMPLE 3.6 | **NONHOMOGENEOUS SYSTEM**

Find a general solution of the nonhomogeneous linear system

$$\mathbf{y}' = \mathbf{A}\mathbf{y} + \mathbf{g} = \begin{bmatrix} 2 & -4 \\ 1 & -3 \end{bmatrix} \mathbf{y} + \begin{bmatrix} 2t^2 + 10t \\ t^2 + 9t + 3 \end{bmatrix}.$$

Solution. Type the system as shown and obtain a general solution by the command dsolve.

```
> y1 := 'y1':      y2 := 'y2':
> sys := D(y1)(t) = 2*y1(t) - 4*y2(t) + 2*t^2 + 10*t,
>        D(y2)(t) = y1(t) - 3*y2(t) + t^2 + 9*t + 3;
```

$sys := \mathrm{D}(y1)(t) = 2\,y1(t) - 4\,y2(t) + 2\,t^2 + 10\,t,\ \mathrm{D}(y2)(t) = y1(t) - 3\,y2(t) + t^2 + 9\,t + 3$

```
> sol := dsolve({sys},  {y1(t), y2(t)});
```

$sol := \{y2(t) = 3\,t + \dfrac{1}{4}\,\_C1\,e^t + \_C2\,e^{(-2\,t)},\ y1(t) = -t^2 + \_C1\,e^t + \_C2\,e^{(-2\,t)}\}$

From sol you see that a particular solution of the *nonhomogeneous* system is $y_{1p} = -t^2$, $y_{2p} = 3\,t$.

    Similar Material in AEM: p. 184

EXAMPLE 3.7   METHOD OF UNDETERMINED COEFFICIENTS

Find a particular solution of the nonhomogeneous system in the previous example by the method of undetermined coefficients.

Solution. The system is $\mathbf{y'} = \mathbf{Ay} + \mathbf{g}$, where

$$\mathbf{g} = \begin{bmatrix} 2\,t^2 + 10\,t \\ t^2 + 9\,t + 3 \end{bmatrix}.$$

The nonhomogeneous part $\mathbf{g}$ suggests choosing $\mathbf{y}_p = \mathbf{u} + \mathbf{v}t + \mathbf{w}t^2$, with vectors $\mathbf{u} = [u_1,\, u_2]$, $\mathbf{v} = [v_1,\, v_2]$, $\mathbf{w} = [w_1,\, w_2]$ to be determined by substitution. Thus, type

```
> y1 := 'y1':     y2 := 'y2':
> yp := [y1(t) = u1 + v1*t + w1*t^2,
>        y2(t) = u2 + v2*t + w2*t^2];
```

$$yp := [y1(t) = u1 + v1\,t + w1\,t^2,\ y2(t) = u2 + v2\,t + w2\,t^2]$$

Here, `y1(t)` and `y2(t)` are the components of the vector `yp` to be determined, and you get them individually by typing

```
> r1 := rhs(yp[1]);                    # Resp. r1 := u1 + v1t + w1t²
> r2 := rhs(yp[2]);                    # Resp. r2 := u2 + v2t + w2t²
```

Accordingly, type the system as

```
> sys := diff(r1, t) = 2*r1 - 4*r2 + 2*t^2 + 10*t,
>        diff(r2, t) = r1 - 3*r2 + t^2 + 9*t + 3;
```

$$sys := v1 + 2\,w1\,t = 2\,u1 + 2\,v1\,t + 2\,w1\,t^2 - 4\,u2 - 4\,v2\,t - 4\,w2\,t^2 + 2\,t^2 + 10\,t,$$
$$v2 + 2\,w2\,t = u1 + v1\,t + w1\,t^2 - 3\,u2 - 3\,v2\,t - 3\,w2\,t^2 + t^2 + 9\,t + 3$$

Determine the six unknown components of the three unknown vectors $\mathbf{u}$, $\mathbf{v}$, $\mathbf{w}$. From `sys` get three vector equations `eq1`, `eq2`, `eq3` (involving the components of those unknown vectors) by setting $t = 0,\ 1,\ -1$,

```
> eq1 := eval(subs(t = 0,  {sys}));
```

$$eq1 := \{v1 = 2\,u1 - 4\,u2,\ v2 = u1 + 3 - 3\,u2\}$$

```
> eq2 := eval(subs(t = 1,  {sys}));
```

$$eq2 := \{v1 + 2\,w1 = 2\,u1 + 2\,v1 + 2\,w1 - 4\,u2 - 4\,v2 - 4\,w2 + 12,$$
$$v2 + 2\,w2 = u1 + v1 + w1 - 3\,u2 - 3\,v2 - 3\,w2 + 13\}$$

```
> eq3 := eval(subs(t = -1,  {sys}));
```

$$eq3 := \{v1 - 2\,w1 = 2\,u1 - 2\,v1 + 2\,w1 - 4\,u2 + 4\,v2 - 4\,w2 - 8,$$
$$v2 - 2\,w2 = u1 - v1 + w1 - 3\,u2 + 3\,v2 - 3\,w2 - 5\}$$

Solve this linear system of six equations for the six components `eq1[1]`, `eq1[2]`, `eq2[1]`, `eq2[2]`, `eq3[1]`, `eq3[2]` of these three vector equations by the command

```
> s := solve({eq1[1], eq1[2], eq2[1], eq2[2], eq3[1], eq3[2]},
  {u1, u2, v1, v2, w1, w2});
```

$$s := \{v1 = 0,\ u1 = 0,\ w2 = 0,\ v2 = 3,\ w1 = -1,\ u2 = 0\}$$

Substitution of these six values into `yp` gives the answer, in agreement with y_p in the previous example,

```
> subs(seq(s[n], n = 1..6), yp);              # Resp. [y1(t) = -t², y2(t) = 3 t]
```

Similar Material in AEM: p. 184

| EXAMPLE 3.8 | **VAN DER POL EQUATION. LIMIT CYCLE**

The famous Van der Pol equation

$$y'' - \mu(1 - y^2)y' + y = 0 \qquad (\mu > 0)$$

models, for instance, thermionic valve circuits. It governs "**self-sustained oscillations**", that is, for small $y^2 < 1$ the second term is negative ("negative damping"), so that energy is fed into the physical system, whereas for $y^2 > 1$, we have damping that takes energy from the system and makes the amplitudes y to become smaller. This gives a **limit cycle** (corresponding to a periodic solution), which the other trajectories are approaching. For simplicity take $\mu = 1$ and consider

$$y'' - (1 - y^2)y' + y = 0.$$

Convert the equation to a system of equations by the standard procedure (explained and also used in Example 3.5 in this Guide), namely, $y_1' = y_2$, $y_2' = (1 - y_1^2)y_2 - y_1$. Type this as

```
> sys := D(y1)(t) = y2(t),
>          D(y2)(t) = (1 - y1(t)^2)*y2(t) - y1(t);
```

$$sys := \mathrm{D}(y1)(t) = y2(t), \mathrm{D}(y2)(t) = \left(1 - y1(t)^2\right) y2(t) - y1(t)$$

Pick three points (triples t, y_1, y_2) through which your trajectories should pass, say,

```
> inits := [0, 0, 4], [0, 0, 2], [0, 0, 0.5];
```

$$inits := [0, 0, 4], [0, 0, 2], [0, 0, .5]$$

These triples mean $t = 0$, $y_1 = 0$, $y_2 = 4$, etc. Then plot a phase portrait by the commands (type `?DEplot` for information)

```
> with(DEtools):
> DEplot([sys[1], sys[2]], [y1(t), y2(t)], t = 0..10, y1 = -5..5,
   y2 = -5..5, [inits], scaling = constrained, stepsize = 0.05);
```

The last trajectory starts at $(0, 0.5)$ and increases. The second seems to be periodic, giving a **limit cycle**. This is different from a center because the limit cycle is approached by trajectories from the interior (your last curve) and from the exterior (the first trajectory, which is rather inaccurate in the plot). "Small solutions" increase, "large solutions" decrease, hence the existence of a common limit curve seems plausible.

For Maple help type `?odeadvisor` and then click on `Van der Pol`.

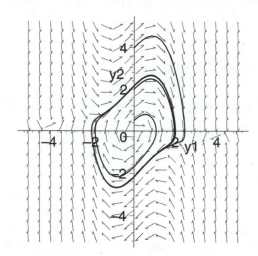

Example 3.8. Limit cycle and trajectories for the Van der Pol equation with $\mu = 1$

Similar Material in AEM: pp. 181, 182

Problem Set for Chapter 3

Pr.3.1 (Node) Using the criteria, show that $y_1' = y_1$, $y_2' = 2y_2$ has a node. Find a general solution. Plot a phase portrait. (*AEM Ref.* pp. 164, 165)

Pr.3.2 (Saddle point) Find and plot (by `DEplot`) the solution of the initial value problem $y_1' = 2y_1 - 4y_2$, $y_2' = y_1 - 3y_2$, $y_1(0) = 3$, $y_2(0) = 0$ as a trajectory.
(*AEM Ref.* p. 165)

Pr.3.3 (Center) Solve $y'' + 9y = 0$ by first converting it to a system of equations. Plot a phase portrait. (*AEM Ref.* pp. 165, 166)

Pr.3.4 (Spiral point) Find a general solution of the system $y_1' = -y_1 + 4y_2$, $y_2' = -y_1 - y_2$. From it find the particular solution satisfying the initial conditions $y_1(0) = 1$, $y_2(0) = 1$. Plot it as a trajectory in the phase plane. (*AEM Ref.* p. 166)

Pr.3.5 (No basis of eigenvectors available) Determine the eigenvalues and eigenvectors of the coefficient matrix of the system

$$y_1' = 4y_1 + y_2, \qquad y_2' = -y_1 + 2y_2.$$

Show that this gives no basis of eigenvectors, so that you get at first only one solution. It is shown in AEM, pp. 167, 168, that a second independent solution is of the form

$$\mathbf{y} = (\mathbf{x}t + \mathbf{u})e^{\lambda t}.$$

Show that `dsolve` does indeed give a general solution of the corresponding form. (*AEM Ref.* pp. 167, 169)

Pr.3.6 **(Damped pendulum)** Add a damping term cy' to the pendulum equation in Example 3.5 in this Guide, choosing $c = 1/4$. Plot a phase portrait. Compare with Example 3.5. (*AEM Ref.* pp. 177, 178)

Pr.3.7 **(Harmonic oscillations)** Indicate the kind of trajectories of $y'' + (1/9)y = 0$ in the phase plane. (*AEM Ref.* p. 174 (#12))

Pr.3.8 **(Mixing problem involving two tanks)** Tank T_1 contains initially 100 gal of pure water. Tank T_2 contains initially 100 gal of water in which 150 lb of fertilizer are solved. Liquid circulates through the tanks (by means of two tubes connecting the tanks) at a constant rate of 2 gal/min and the mixture is kept uniform by stirring. Find the amounts of fertilizer $y_1(t)$ and $y_2(t)$ in T_1 and T_2, respectively, where t is time. (*AEM Ref.* pp. 152-154)

Pr.3.9 **(Electrical network)** Find and plot the currents $i_1(t)$, $i_2(t)$ in a network governed by the equations

$$i_1' + 4(i_1 - i_2) = 12$$

$$6i_2 + 4(i_2 - i_1) + 4 \int i_2\, dt = 0,$$

assuming $i_1(0) = 0$, $i_2(0) = 0$. The equations result from the two loops of the network. Loop 1 contains an inductor of $L = 1$ henry (giving i_1'), a resistor of $R = 4$ ohms (common to both loops), and a battery of 12 volts. Loop 2 contains another resistor of 6 ohms and a capacitor of $1/4$ farad. (Differentiate the second equation to get rid of the integral.) (*AEM Ref.* pp. 154-156)

Pr.3.10 **(Electrical network)** Find a general solution in Pr.3.9 by using matrices. (*AEM Ref.* p. 155)

Pr.3.11 **(Damped oscillations)** Indicate the kind of trajectories of $y'' + 2y' + 2y = 0$. (*AEM Ref.* p. 174 (#14))

Pr.3.12 **(Duffing equation, soft spring)** Plot a phase portrait of the Duffing equation with a soft spring, say, $y'' + y - y^3 = 0$. Include some trajectories. What kind of trajectories would you expect if $|y|$ is small? If it is large? Think it over before you plot. (*AEM Ref.* p. 183)

Pr.3.13 **(Duffing equation, hard spring)** Plot a phase portrait of the Duffing equation with a hard spring, say, $y'' + y + y^3 = 0$. Include some trajectories. Do they look physically reasonable? (*AEM Ref.* p. 183)

Pr.3.14 **(Van der Pol equation)** Plot a direction field for the equation with $\mu = 1/2$. Describe how it differs from that for $\mu = 1$ in Example 3.8 in this Guide. (*AEM Ref.* p. 182)

Pr.3.15 **(Experiment on Van der Pol equation)** Experiment with the Van der Pol equation in the previous problem, to find out how the shape of the limit cycle changes when you change μ.

Series Solutions of Differential Equations

Content. Solving in series, plotting (Ex. 4.1, Prs. 4.1-4.3)
         Legendre's ODE, orthogonality, Fourier-Legendre series
            (Exs. 4.2-4.4, Prs. 4.4-4.8, 4.20)
         Frobenius method, hypergeometric ODE (Ex. 4.5, Prs. 4.9-4.13)
         Bessel's ODE (Ex. 4.6, Prs. 4.14-4.19)

Series solutions of ODE's
are obtained by `dsolve({ode, y(0) = K0, D(y)(0) = K1}, y(x), series)`, as
explained in Example 4.1, along with **plotting** and **calculating** values from series.
Obtaining series independently of ODE's. Use `series(f, x = a, O)`, `O` the order
of the error term, `a` the center. Type `?series`, `?powseries`. For instance, the
geometric series is obtained by typing

```
> series(1/(1-x), x = 0, 14);
```

$$1 + x + x^2 + x^3 + x^4 + x^5 + x^6 + x^7 + x^8 + x^9 + x^{10} + x^{11} + x^{12} + x^{13} + O(x^{14})$$

Or use `seq` for a sequence and `sum` for a sum (a polynomial). For example,

```
> sum(x^n, n = 0..14);
```

$$1 + x + x^2 + x^3 + x^4 + x^5 + x^6 + x^7 + x^8 + x^9 + x^{10} + x^{11} + x^{12} + x^{13} + x^{14}$$

Special functions, elementary and higher, are known to Maple. For instance,
`series(BesselJ(0,x), x, 20)` gives the (Maclaurin) series of the Bessel function
$J_0(x)$, etc. Try it. See Examples 4.2-4.6.

Orthopoly package. This package handles the Legendre polynomials (called by
`P(n, x)`) and other orthogonal polynomials. Type `?orthopoly`. See Examples 4.2
and 4.4 in this Guide.

Examples for Chapter 4

EXAMPLE 4.1   **POWER SERIES SOLUTIONS. PLOTS FROM THEM.
                              NUMERICAL VALUES**

Find a **power series solution** of the initial value problem $(1-x^2)\,y'' - 2x\,y' + 72y = 0$,
$y(0) = 0$, $y'(0) = 1$. Plot the solution and find its value at $x = 0.8$.

Solution. . Type the ODE.

```
> ode := (1 - x^2)*diff(y(x), x, x) - 2*x*diff(y(x), x) + 72*y(x) = 0;
```

$$ode := (1 - x^2)\left(\frac{\partial^2}{\partial x^2}\,y(x)\right) - 2x\left(\frac{\partial}{\partial x}\,y(x)\right) + 72\,y(x) = 0$$

Type the number of terms of the series you want, say, by taking a remainder $O(x^{12})$. Then type `dsolve(...,series)`, including the initial conditions. This is a set of data, hence include it in braces $\{\,\dots\,\}$.

```
> Order := 12:
> sol := dsolve({ode, y(0) = 0, D(y)(0) = 1}, y(x), series);
```

$$sol := \mathrm{y}(x) = x - \frac{35}{3}\,x^3 + 35x^5 - 35x^7 + \frac{70}{9}x^9 + \frac{14}{11}x^{11} + O(x^{12})$$

Plotting. Convert this to a polynomial by dropping the remainder (the error term).

```
> p := convert(sol, polynom);
```

$$p := \mathrm{y}(x) = x - \frac{35}{3}\,x^3 + 35x^5 - 35x^7 + \frac{70}{9}x^9 + \frac{14}{11}x^{11}$$

```
> plot(p, x = -1..1);
Plotting error, empty plot
```

Type `rhs(p)`, suggesting 'right-hand side of `p`'. You will get the polynomial without the left-hand side `y(x) =`

```
> rhs(p);                    # Resp.
```
$$x - \frac{35}{3}\,x^3 + 35x^5 - 35x^7 + \frac{70}{9}x^9 + \frac{14}{11}x^{11}$$

```
> plot(rhs(p), x = -1..1);
```

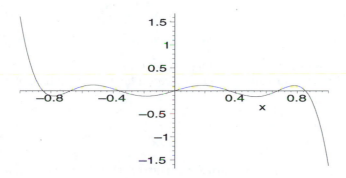

Example 4.1. Plot of the series solution

Numerical values can be obtained by the command `subs`. For instance,

```
> subs(x = 0.8, p);                    # Resp. y(.8) = .1086767683
```

The zeros of the polynomial `p` are

```
> evalf(fsolve(rhs(p) = 0, x), 6);
```

$$-1.39522, -.852542, -.675059, -.360622, 0., .360622, .675059, .852542, 1.39522$$

We mention that the ODE is Legendre's equation with parameter $n = 8$, but the series solution obtained is not a Legendre polynomial (to be discussed in the next example).

    Similar Material in AEM: p. 205

EXAMPLE 4.2   **LEGENDRE POLYNOMIALS.**
                 THE ORTHOPOLY PACKAGE.
                 PROCEDURES

The Legendre polynomials $P_0(x)$, $P_1(x)$, $P_2(x)$, ... are, after the Bessel functions, probably the most important special functions in applications. In your work you have to load the package `orthopoly`. First type `?orthopoly[legendre]` for information. Then load by typing

> `with(orthopoly):`   # Ignore the warning.

You can now obtain the Legendre polynomials by typing, for instance,

> `P3 := P(3, x);`   # Resp. $P3 := \dfrac{5}{2} x^3 - \dfrac{3}{2} x$

A formula for defining any of these polynomials is **_Rodrigues's formula_**

$$P_n(x) = \frac{1}{2^n \, n!} \frac{d^n}{d \, x^n} [(x^2 - 1)^n].$$

This can be used for obtaining individual Legendre polynomials by what is called a **_procedural definition_** or simply a **procedure.** For information, type `?proc` and `?procedure[paramtype]`. In the simplest case, type,

> `f := proc(x)`

>   `x^2`

> `end:`

> `f(2.5);`   # Resp. 6.25

Do a few examples of your own, before you go on.

     Using Rodrigues's formula, type a procedure for the Legendre polynomials.

> `Leg := proc(n)`

>   `1/(2^n*n!)*diff((x^2 - 1)^n, x$n)`

> `end:`

`x$n` means `x, x, ..., x,` that is, **differentiate _n_ times**. You can now simply type `Leg(1)`, `Leg(2)`,

> `Leg(3);`   # Resp. $x^3 + \dfrac{3}{2} (x^2 - 1) x$

> `simplify(%);`   # Resp. $\dfrac{5}{2} x^3 - \dfrac{3}{2} x$

and so on. Now if you type Leg(0), which is $P_0 = 1$, you get

> `Leg(0);`

`Error, (in Leg) wrong number (or type) of parameters in function diff`

that is, Maple does not interpret the zeroth derivative as the function itself. But `proc` is flexible; to include $n = 0$, extend the procedure to

> `Legendre := proc(n)`

> `if n = 0`

> `then`

```
>   1
> else
>   1/(2^n*n!)*diff((x^2 - 1)^n, x$n)
> fi;
> end:
```

`fi` is `if` reversed and indicates the end of the `if-then-else` statement. Note further that a procedure `proc` needs `end:` (More complicated procedures will be considered in Chapter 19 of this Guide.) You now obtain the polynomials as before, including

```
> Legendre(0);                                      # Resp. 1
```

You may now put the procedure into the command `seq` and obtain the first few polynomials at once, and plot them on common axes. That is,

```
> S := seq(Legendre(n), n = 0..4);
```

$$S := 1,\, x,\, \frac{3}{2}\,x^2 - \frac{1}{2},\, x^3 + \frac{3}{2}\left(x^2 - 1\right)x,\, x^4 + 3\left(x^2 - 1\right)x^2 + \frac{3}{8}\left(x^2 - 1\right)^2$$

To obtain these polynomials in the usual form, use the commands `simplify` or `expand`. Type

```
> seq(expand(S[m]), m = 1..5);
```

$$1,\, x,\, \frac{3}{2}\,x^2 - \frac{1}{2},\, \frac{5}{2}\,x^3 - \frac{3}{2},\, \frac{35}{8}\,x^4 - \frac{15}{4}\,x^2 + \frac{3}{8}$$

```
> plot({S}, x = -1.2..1.2, y = -1.1..1.1, xtickmarks = [-1, 0, 1],
  ytickmarks = [-1, -0.5, 0, 0.5, 1], title = 'Legendre polynomials');
```

Legendre polynomials

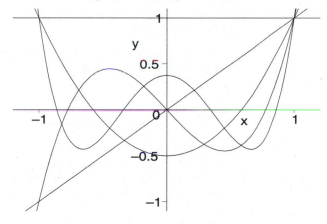

Example 4.2. Legendre polynomials $P_0, \ldots, P_4$

Similar Material in AEM: pp. 207, 208

| **EXAMPLE 4.3** | **LEGENDRE'S DIFFERENTIAL EQUATION** |

The Legendre polynomials $P_n(x), n = 0, 1, 2, \ldots$, just considered in Example 4.2 are solutions of **Legendre's differential equation**

$$(1 - x^2)y'' - 2xy' + n(n+1)y = 0$$

satisfying suitable initial conditions (depending on n). Type

```
> LEG := (1 - x^2)*diff(y(x), x, x) - 2*x*diff(y(x), x)
    + n*(n + 1)*y(x)= 0;
```

$$LEG := \left(1 - x^2\right) \left(\frac{\partial^2}{\partial x^2} y(x)\right) - 2x \left(\frac{\partial}{\partial x} y(x)\right) + n(n+1) y(x) = 0$$

For instance, consider $n = 4$,

```
> LEG4 := subs(n = 4, LEG);
```

$$LEG4 := \left(1 - x^2\right) \left(\frac{\partial^2}{\partial x^2} y(x)\right) - 2x \left(\frac{\partial}{\partial x} y(x)\right) + 20 y(x) = 0$$

To obtain the particular solution $P_4(x)$ of this ODE, note that $P_4(0) = 3/8$ and its derivative at $x = 0$ is 0. Thus, type

```
> dsolve({LEG4, y(0) = 3/8, D(y)(0) = 0});
```

$$y(x) = \frac{35}{8} x^4 - \frac{15}{4} x^2 + \frac{3}{8}$$

Legendre polynomials are very convenient in connection with **boundary value problems** because at -1 and 1 they equal 1 when n is even, or -1 and 1 when n is odd. Thus you also obtain P_4 from

```
> dsolve({LEG4, y(-1) = 1, y(1) = 1});
```

$$y(x) = \frac{35}{8} x^4 - \frac{15}{4} x^2 + \frac{3}{8}$$

Coefficient recursion. For a power series solution of the Legendre equation with arbitrary positive integer n the coefficient recursion will involve three subsequent powers x^s, x^{s+1}, x^{s+2}. Accordingly, type

```
> ser := sum(a[m]*x^m, m = s..s + 2);
```

$$ser := a_s x^s + a_{1+s} x^{(1+s)} + a_{s+2} x^{(s+2)}$$

Substitute this into the Legendre equation, where `eval` evaluates the derivatives and `simplify` multiplies out $1 - x^2$ (the first coefficient of the Legendre equation) times the very long corresponding expression and $-2x$ (the other coefficient of the equation) times the corresponding long expression.

```
> eval(subs(y(x) = ser, LEG));                    # Very long response
```

```
> simplify(%);                                    # An even longer response
```

The coefficient of x^s (a sum of expressions) is now obtained by the command

```
> coeff(lhs(%), x^s);              # Terms may appear in different order.
```

$$2 a_{s+2} - a_s s^2 - a_s s + 3 a_{s+2} s + a_{s+2} s^2 + n^2 a_s + n a_s$$

This gives the coefficient recursion

```
> a[s+2] := solve(%, a[s+2]);          # Terms may appear in different order.
```

$$a_{s+2} := \frac{a_s \left(s^2 + s - n - n^2 \right)}{2 + 3s + s^2}$$

```
> a[s+2] := factor(%);
```

$$a_{s+2} := \frac{a_s \left(s + 1 + n \right) \left(s - n \right)}{\left(s + 2 \right) \left(1 + s \right)}$$

This is the basic recursion relation for the coefficients of the power series solution of the Legendre equation with positive integer parameter n that leads to the Legendre polynomials.

    Similar Material in AEM: pp. 206, 207

EXAMPLE 4.4　ORTHOGONALITY. FOURIER-LEGENDRE SERIES

The importance of the Legendre polynomials results in part from their orthogonality. They are orthogonal on the interval from -1 to 1. By definition this means that the integral of the product of two Legendre polynomials of different orders, integrated from -1 to 1, is zero,

$$\int_{-1}^{1} P(m, x) \, P(n, x) \, dx = 0 \qquad (m \neq n).$$

It can be shown that for $n = m$ the integral equals $2/(2m + 1)$, as is stated on p. 242 of AEM (without proof, which is tricky). For specific cases, e.g. for all m and n up to 100 (or more), you can verify these statements, using the following command (which does it for 0, 1, ..., 10).

```
> with(orthopoly):
> seq(seq(int(P(m, x)*P(n, x), x = -1..1), m = 0..n), n = 0..10);
```

$$2, 0, \frac{2}{3}, 0, 0, \frac{2}{5}, 0, 0, 0, \frac{2}{7}, 0, 0, 0, 0, \frac{2}{9}, 0, 0, 0, 0, 0, \frac{2}{11}, 0, 0, 0, 0, 0, 0, \frac{2}{13}, 0, 0, 0,$$

$$0, 0, 0, 0, \frac{2}{15}, 0, 0, 0, 0, 0, 0, 0, 0, \frac{2}{17}, 0, 0, 0, 0, 0, 0, 0, 0, 0, \frac{2}{19}, 0, 0, 0, 0, 0, 0, 0,$$

$$0, 0, 0, 0, \frac{2}{21}$$

You see the values $2/(2m + 1) = 2, 2/3, 2/5, \ldots$ for $m = n = 0, 1, 2, \ldots$ as well as the zeros for the index pairs $(0,1)$, $(0,2)$, $(1,2)$, $(0,3)$, $(1,3)$, $(2,3)$, and so on.

Orthogonal expansions are series for a given function f in terms of orthogonal functions, in which orthogonality helps determining the coefficients in a simple fashion – this is a main point of orthogonality. For the Legendre polynomials this gives **Fourier-Legendre series**, as follows. Take, for instance $f(x) = \sin \pi x$,

```
> f := sin(Pi*x);                          # Resp. f := sin (π x)
```

The series will be $f = a_0 \, P_0(x) + a_1 \, P_1(x) + \ldots$ with coefficients a_m equal $(2m + 1)/2$ times the integral of $f P_m$ from $x = -1$ to 1. Recall the simplest procedure in Example 4.2 in this Guide and type (note that the orthopoly package has been loaded before)

```
> c := proc(m)
>    (2*m + 1)/2*int(f*P(m, x), x = -1..1);
> end:
```

Then you get single coefficients or, better, several at once, say, a_0, a_1, a_2, a_3, by the command seq applied to c(m) – don't forget the m,

```
> seq(c(m), m = 0..3);
```
$$\text{\# Resp. } 0, \, 3\frac{1}{\pi}, \, 0, \, 7\frac{-15 + \pi^2}{\pi^3}$$

```
> evalf(%, 6);
```
$$\text{\# Resp. } 0., \, .954930, \, 0, \, -1.15825$$

The even-numbered coefficients are zero since f is odd. You should now design a procedure that gives you the *terms* of your series, not just the coefficients,

```
> term := proc(m)
>    (2*m + 1)/2*int(f*P(m, x), x = -1..1)*P(m, x);
> end:
> term(3);
```

$$7\frac{\left(-15 + \pi^2\right)\left(\frac{5}{2}x^3 - \frac{3}{2}x\right)}{\pi^3}$$

This is not what you want and need. The Legendre polynomials should remain untouched, as shown above in the series. To accomplish this, type '$P_n(x)$' in primes as shown,

```
> term := proc(m)
>    (2*m + 1)/2*int(f*P(m, x), x = -1..1)*'P(m, x)';
> end:
> term(3);
```

$$7\frac{\left(-15 + \pi^2\right)P(3, x)}{\pi^3}$$

```
> evalf(%);
```
$$\text{\# Resp. } -2.895604778\,x^3 + 1.737362867\,x$$

Similarly, type your series (it is short, to show that often a small number of terms gives good approximations)

```
> sum('term(m)', m = 0..3);
```
$$\text{\# Resp. } 3\frac{P(1, x)}{\pi} + 7\frac{\left(-15 + \pi^2\right)P(3, x)}{\pi^3}$$

To have the coefficients as decimal fractions, use evalf. Then plot.

```
> S := sum('evalf(term(m), 4)', m = 0..3);
```

$$S := .9549\,P\left(1, x\right) - 1.157\,P\left(3, x\right)$$

```
> plot({f, S}, x = -1..1, xtickmarks = [-1, -0.5, 0, 0.5, 1],
   ytickmarks = [-1, -0.5, 0, 0.5, 1]);
```

The accuracy is surprising. The areas under the curves seem almost equal, and so are the locations and size of the extrema. The zeros of $\sin \pi x$ are not too well approximated. In terms of powers of x this approximation of $\sin \pi x$ is

```
> evalf(expand(S, x), 6);
```
$$\text{\# Resp. } 2.69040\,x - 2.89250\,x^3$$

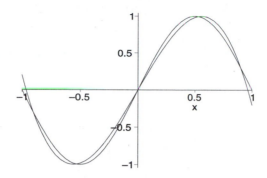

Example 4.4. Sine function and approximation by $a_1 P_1(x) + a_3 P_3(x)$

Similar Material in AEM: p. 242

EXAMPLE 4.5   **FROBENIUS METHOD**

In this example we solve some ODE's by `dsolve` for which the Frobenius method is needed, because the coefficients of the equations are singular when the equations are written in standard form with 1 as the coefficient of y''. You will see in what form the computer will give a general solution.

```
> ode1 := x^2*diff(y(x), x, x) + b*x*diff(y(x), x) + c*y(x) = 0;
```

$$ode1 := x^2 \left(\frac{\partial^2}{\partial x^2} \, y(x) \right) + b\,x \left(\frac{\partial}{\partial x} \, y(x) \right) + c\,y(x) = 0$$

For this Euler-Cauchy equation $x^2 y'' + bxy' + cy = 0$ with arbitrary constant b and c you obtain solutions x^μ, where μ is a solution of the auxiliary equation as given by the usual formula for a quadratic equation; that is,

```
> dsolve(ode1);
```

$$y(x) = \_C1\, x^{\left(-\frac{1}{2} b + \frac{1}{2} + \frac{1}{2} \sqrt{b^2 - 2b + 1 - 4c}\right)} + \_C2\, x^{\left(-\frac{1}{2} b + \frac{1}{2} - \frac{1}{2} \sqrt{b^2 - 2b + 1 - 4c}\right)}$$

For the next equation, $x(x-1)y'' + (3x-1)y' + y = 0$, type

```
> ode2 := x*(x - 1)*diff(y(x), x, x) + (3*x - 1)*diff(y(x), x) + y(x)
   = 0;
```

$$ode2 := x\,(x-1) \left(\frac{\partial^2}{\partial x^2} \, y(x) \right) + (3\,x - 1) \left(\frac{\partial}{\partial x} \, y(x) \right) + y(x) = 0$$

and obtain by `dsolve` a general solution

```
> dsolve(ode2);
```

$$y(x) = \frac{\_C1 \ln(x) + \_C2}{x - 1}$$

```
> expand(%);
```

$$y(x) = \frac{\ln(x)\,\_C1}{x - 1} + \frac{\_C2}{x - 1}$$

This response involving a logarithm shows that the **indicial equation** (see AEM, p. 213) has a double root, which has the value zero. You may try a power series solution, expecting a geometric series with sum $1/(1 - x)$. To accomplish this, proceed as in the previous example. Type

```
> ser := sum(a[m]*x^m, m = s..s + 2);
```

$$ser := a_s\, x^s + a_{1+s}\, x^{(1+s)} + a_{s+2}\, x^{(s+2)}$$

```
> eval(subs(y(x) = ser, ode2));        # A long response
> simplify(%);                         # Another long response
> coeff(lhs(%), x^s);                  # Terms may appear in different order.
```

$$-2\, a_{1+s} s - a_{1+s} + a_s\, s^2 + a_s - a_{1+s}\, s^2 + 2\, a_s\, s$$

```
> a[s+1] := solve(%, a[s+1]);          # Resp. a_{1+s} := a_s
```

This is the recursion for the coefficients of a geometric series, as expected.

As a third ODE, consider $(x^2 - x)\, y'' - x\, y' + y = 0$. Type and solve it.

```
> ode3 := (x^2 - x)*diff(y(x), x, x) - x*diff(y(x), x) + y(x) = 0;
```

$$ode3 := (x^2 - x)\left(\frac{\partial^2}{\partial x^2}\, y(x)\right) - x\left(\frac{\partial}{\partial x}\, y(x)\right) + y(x) = 0$$

```
> dsolve(ode3);
```

$$y(x) = \_C1\, x + \_C2\, (1 + \ln(x)\, x)$$

This indicates that the indicial equation has roots differing by an integer (these are 1 and 0) and the second solution does involve a logarithm.

Similar Material in AEM: p. 215

EXAMPLE 4.6   **BESSEL'S EQUATION. BESSEL FUNCTIONS**

Bessel's differential equation

$$x^2\, y'' + xy' + (x^2 - \nu^2)y = 0$$

appears in various practical problems, in particular in those that exhibit cylindrical symmetry. ν (Greek nu) is a given real number (a real parameter suggested by the kind of problem). Type the equation as

```
> BES := x^2*diff(y(x), x, x) + x*diff(y(x), x) + (x^2 - nu^2)*y(x) = 0;
```

$$BES := x^2\left(\frac{\partial^2}{\partial x^2}\, y(x)\right) + x\left(\frac{\partial}{\partial x}\, y(x)\right) + (x^2 - \nu^2)\, y(x) = 0$$

Obtain a general solution by the command

```
> sol := dsolve(BES);
```

$$sol := y(x) = \_C1\, \text{BesselJ}(\nu,\, x) + \_C2\, \text{BesselY}(\nu,\, x)$$

The first term involves the **Bessel function of the first kind** $J_\nu(x)$ and the second term the **Bessel function of the second kind** $Y_\nu(x)$. When ν is an integer, one writes n instead of ν.

The functions J_0 and J_1 are particularly important. J_0 satisfies Bessel's equation with parameter 0, that is,

```
> subs(nu = 0, BES)/x;        # Resp.
```
$$\frac{x^2\left(\frac{\partial^2}{\partial x^2}\,y(x)\right) + x\left(\frac{\partial}{\partial x}\,y(x)\right) + y(x)\,x^2}{x} = 0$$

```
> BES0 := simplify(%);
```

$$BES0 := x\left(\frac{\partial^2}{\partial x^2}\,y(x)\right) + \left(\frac{\partial}{\partial x}\,y(x)\right) + y(x)\,x = 0$$

Maple did not drop the factor x automatically, hence our trick. From the general solution (or equally well from the Bessel equation with $n = 0$) you obtain J_0, its Maclaurin series, and the conversion of the response to a polynomial by the commands

```
> subs(_C1 = 1, _C2 = 0, nu = 0, sol);        # Resp. y(x) = BesselJ(0, x)
> ser0 := series(rhs(%), x, 12);
```

$$ser0 := 1 - \frac{1}{4}x^2 + \frac{1}{64}x^4 - \frac{1}{2304}x^6 + \frac{1}{147456}x^8 - \frac{1}{14745600}x^{10} + O(x^{12})$$

```
> p0 := convert(%, polynom);
```

$$p0 := 1 - \frac{1}{4}x^2 + \frac{1}{64}x^4 - \frac{1}{2304}x^6 + \frac{1}{147456}x^8 - \frac{1}{14745600}x^{10}$$

Similarly for the Bessel function $J_1(x)$

```
> subs(_C1 = 1, _C2 = 0, nu = 1, sol);        # Resp. y(x) = BesselJ(1, x)
> ser1 := series(rhs(%), x, 12);
```

$$ser1 := \frac{1}{2}x - \frac{1}{16}x^3 + \frac{1}{384}x^5 - \frac{1}{18432}x^7 + \frac{1}{1474560}x^9 - \frac{1}{176947200}x^{11} + O(x^{12})$$

```
> p1 := convert(%, polynom);
```

$$p1 := \frac{1}{2}x - \frac{1}{16}x^3 + \frac{1}{384}x^5 - \frac{1}{18432}x^7 + \frac{1}{1474560}x^9 - \frac{1}{176947200}x^{11}$$

Plot the two polynomials and the two functions jointly, so that you see that the approximation seems good for x to about 3 or 4. For larger x you would need more terms. The other figure shows that the curves of the two functions look similar to those of cosine and sine, due to the similarity of the respective series.

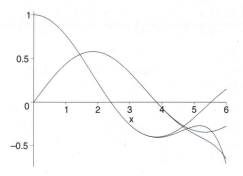

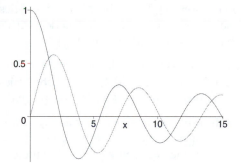

Example 4.6. Partial sum approximations of the Bessel functions J_0 and J_1

Example 4.6. Bessel functions J_0 (starting from 1) and J_1 (starting from 0)

```
> plot({p0, p1, BesselJ(0, x), BesselJ(1, x)}, x = 0..6, y = -0.8..1,
  xtickmarks = [1, 2, 3, 4, 5, 6], ytickmarks = [-0.5, 0, 0.5, 1]);
> plot({BesselJ(0, x), BesselJ(1, x)}, x = 0..15, y = -0.8..1,
  xtickmarks = [5, 10, 15], ytickmarks = [-0.5, 0, 0.5, 1]);
```

Zeros of Bessel functions are often needed in applications (vibration of membranes, celestial mechanics, etc.) and have been extensively tabulated, as have the functions themselves, see Ref. [1] in Appendix 1 of this Guide. You obtain the zeros by the command `fsolve`. For instance, for the first positive zero of J_0 you obtain as an approximation (exact to 5 digits) and as an exact 10-digit value

```
> x01appr := fsolve(p0 = 0, x, 2..3);        # Resp. x01appr := 2.404792604
```

```
> x01 := fsolve(BesselJ(0, x) = 0, x, 2..3);   # Resp. x01 := 2.404825558
```

For the second zero (near 5.5) you see from the first figure that you cannot expect great accuracy. Try it.

Finally, remember that `dsolve(.., series)` gives series directly. In the present case,

```
> Order := 8:
> dsolve(BES0, y(x), series);
```

$$y(x) = \_C1 \left(1 - \frac{1}{4} x^2 + \frac{1}{64} x^4 - \frac{1}{2304} x^6 + \mathrm{O}(x^8) \right) + \_C2$$

$$\left(\ln(x) \left(1 - \frac{1}{4} x^2 + \frac{1}{64} x^4 - \frac{1}{2304} x^6 + \mathrm{O}(x^8) \right) + \left(\frac{1}{4} x^2 - \frac{3}{128} x^4 + \frac{11}{13824} x^6 + \mathrm{O}(x^8) \right) \right)$$

The second line results from the Bessel function of the second kind $Y_0(x)$ (as explained on p. 229 of AEM).

   Similar Material in AEM: pp. 218-220, 228, 229

Problem Set for Chapter 4

Pr.4.1 (**Power series**) Find the Maclaurin series of $\tan x$ (5 nonzero terms). Find the coefficient of x^7 (as a decimal fraction with 6 digits) by applying a suitable command to that series. (*AEM Ref.* p. 195)

Pr.4.2 (**Power series**) Find the Maclaurin series of $f = \cos \pi x$ on the computer (powers up to x^{18}, inclusively). Plot f and the series jointly to see for what x the series gives useful approximations. (*AEM Ref.* p. 195)

Pr.4.3 (**Coefficient recursion. Do-loop**) Obtain the coefficient recursion for the power series solution of $y' = 2xy$ with $a_0 = 1$ and write the series thus obtained. (For another coefficient recursion, see Example 4.3 in this Guide.)

Pr.4.4 (**Legendre polynomials**) Find and solve an initial value problem for which $P_4(x)$ is a solution. (*AEM Ref.* p. 208)

Pr.4.5 (**Legendre's equation**) Solve Legendre's equation with $n = 0$ by `dsolve` both as a series and in closed form. (*AEM Ref.* p. 209 (#1))

Pr.4.6 (Boundary value problem) Solve the boundary value problem consisting of the Legendre equation with parameter $n = 5$ and boundary values $y(-1) = -1$ and $y(1) = 1$. Plot the solution. (*AEM Ref.* pp. 205-210)

Pr.4.7 (Experiment on Fourier-Legendre series) Represent e^x on the interval from -3 to 3 by partial sums S_n of a Fourier-Legendre series. Plot e^x and the approximating partial sum on common axes, beginning with 2 terms and taking more and more terms stepwise until the two curves practically coincide in the plot. Describe what you can see regarding the increase of accuracy (a) on the subinterval from -1 to 1, (b) on the entire interval from -3 to 3. Calculate e approximately from S_n with increasing n. Find an empirical formula for the (approximate) size of the error as a function of n. (*AEM Ref.* pp. 242, 246 (#6))

Pr.4.8 (Fourier-Legendre series) Develop x^6 in a Fourier-Legendre series. Why will this series reduce to a polynomial? Plot x^6 and the corresponding Fourier-Legendre series of powers up to x^4 and comment on the quality of approximation. Show graphically that the error (as a function of x) is oscillating. (*AEM Ref.* pp. 242, 246)

Pr.4.9 (Frobenius method) Using `dsolve` , solve the initial value problem

$$xy'' + (1 - 2x)y' + (x - 1)y = 0, \qquad y(1) = 1, \qquad y'(1) = 1,$$

involving an ODE for which the Frobenius method is needed. Also obtain a general solution and find out whether you could prescribe initial values at $x = 0$.
(*AEM Ref.* p. 216 (#7))

Pr.4.10 (Frobenius method) Find a general solution of $(x - 1)^2 y'' + (x - 1)y' - 4y = 0$. Set up two initial value problems that give the two solutions of the basis separately, and find series for these solutions. How can you check your results by a transformation of the given ODE? (*AEM Ref.* p. 216 (#10))

Pr.4.11 (Hypergeometric equation) Apply the command `dsolve` to the hypergeometric equation
$$x(1 - x)y'' + [c - (a + b + 1)x]\, y' - aby = 0.$$

In many cases, the solutions of this equation become elementary functions. Show that this is the case when $a = 1$, $b = 1$, $c = 2$, $x = -t$. As another case, consider $a = -n$, $c = b$, $x = -t$. (*AEM Ref.* pp. 216, 217)

Pr.4.12 (Frobenius method) The equation $xy'' + 2y' + xy = 0$ requires the Frobenius method (why?). Solve it by `dsolve(..., series)` . Try to identify the result in terms of known series. Then apply `dsolve` . (*AEM Ref.* p. 216 (#5))

Pr.4.13 (Hypergeometric equation) Solve the initial value problem

$$x(1 - x)y'' + (2 - 4x)y' - 2y = 0, \qquad y(0) = 1, \qquad y'(0) = 1.$$

Give the solution also in the form of a series. To what a, b, c in the general hypergeometric equation does the present equation correspond? (*AEM Ref.* p. 216)

Pr.4.14 (Bessel functions) The Bessel functions J_ν satisfy the basic relations

$$[x^\nu J_\nu(x)]' = x^\nu J_{\nu-1}(x), \qquad [x^{-\nu} J_\nu(x)]' = -x^{-\nu} J_{\nu+1}(x),$$

where $' = d/dx$. Try to obtain these relations on the computer, either directly or, if this will not work, in integrated form on both sides. (*AEM Ref.* p. 223)

Pr.4.15 (Zeros of Bessel functions) Find approximations of the first two positive zeros of $J_1(x)$ from the Maclaurin series, using powers up to x^{19}, and compare with the values obtained directly from $J_1(x) = 0$. (*AEM Ref.* p. 220)

Pr.4.16 (Experiment on asymptotics of Bessel functions) For large x the Bessel function $J_n(x)$ is approximately equal to

$$\sqrt{\frac{2}{\pi x}} \, \cos\left(x - \frac{n\pi}{2} - \frac{\pi}{4}\right).$$

Experiment with this formula for various n; find out empirically from the graphs how accurate the formula is for small integers n and for larger ones.
(*AEM Ref.* pp. 220, 228 (#30))

Pr.4.17 (Bessel functions Y_0 and Y_1) Solve the general Bessel equation

$$x^2 \, y'' + xy' + (x^2 - \nu^2)y = 0$$

by `dsolve`. Obtain from the response the particular solutions Y_0 and Y_1 and plot them on common axes. (*AEM Ref.* p. 231)

Pr.4.18 (Bessel's equation) A large number of ODE's can be reduced to Bessel's equation. Show that $x^2 y'' + xy' + (4x^4 - 1/4)y = 0$ is of this kind, by solving it by `dsolve`, which will give a general solution in terms of Bessel functions.
(*AEM Ref.* p. 226 (#3))

Pr.4.19 (Reduction to Bessel's equation) Find out from the solution by `dsolve` what transformation will reduce $y'' + x^2 \, y = 0$ to Bessel's equation.
(*AEM Ref.* p. 232 (#5))

Pr.4.20 (Orthogonality) The functions $\sin nx$ with positive integer are orthogonal on any interval of length 2π and have the norm $\sqrt{\pi}$. Verify this for $n = 1, ..., 10$ by integration.
(*AEM Ref.* pp. 234-236)

Chapter 5

Laplace Transform Method for Solving ODE's

Content. Transforms, inverse transforms (Ex. 5.1, Prs. 5.1-5.5)
Solving ODE's (Exs. 5.2-5.4, Prs. 5.6, 5.7, 5.11-5.14)
Forced oscillations, resonance (Ex. 5.3)
Unit step, Dirac delta (Ex. 5.4, Prs. 5.9-5.15)
Solving systems (Ex. 5.5)
Differentiation, s-shifting, t-shifting, convolution, etc. (Ex. 5.6, Pr. 5.8)
Electric circuits, rectifiers (Prs. 5.11, 5.12, 5.14, 5.15)

inttrans[laplace] package (suggesting 'integral transforms'). Load it by typing

> `with(inttrans):`   # Ignore the warning.

Laplace transforms. Type `laplace(f, t, s)`, f the given function, t its variable, s the variable of the transform F. For instance,

> `laplace(1 + t + t^2 + exp(a*t) + cos(omega*t) + sin(Pi*t), t, s);`

$$\frac{1}{s} + \frac{1}{s^2} + \frac{2}{s^3} + \frac{1}{s-a} + \frac{s}{s^2+\omega^2} + \frac{\pi}{s^2+\pi^2}$$

Inverse transforms. Type `invlaplace(F, s, t)`. For instance,

> `invlaplace(1/s^4 + s/(s^2 + 25), s, t);`   # Resp. $\dfrac{1}{6}t^3 + \cos(5\,t)$

Transforms of the derivatives $\mathcal{L}(f') = s\mathcal{L}(f) - f(0)$ and $\mathcal{L}(f'') = s^2\mathcal{L}(f) - sf(0) - f'(0)$ (where $f' = df/dt$, etc.) are typed as

> `laplace(diff(f(t), t), t, s);`   # Resp. $s\,\text{laplace}(f(t),\ t,\ s)\ -\ f(0)$

> `laplace(diff(f(t), t, t), t, s);`

$$s\,(s\,\text{laplace}(f(t),\ t,\ s)\ -\ f(0))\ -\ \text{D}(f)(0)$$

Solving ODE's by `dsolve(ode, y(t), method = laplace)` see Ex. 5.2, etc.

Examples for Chapter 5

EXAMPLE 5.1   **FURTHER TRANSFORMS AND INVERSE TRANSFORMS**

> `with(inttrans):`   # Ignore the warning.

For e^{at}, $\cos\omega t$, and $\sin\omega t$ see before. Further frequently used functions $\cosh at$, $\sinh at$, $e^{at}\cos(\omega t)$, $e^{at}\sin(\omega t)$ have the transforms

```
> laplace(cosh(a*t), t, s);
```

$$\frac{s}{s^2 - a^2}$$

```
> laplace(sinh(a*t), t, s);
```

$$\frac{a}{s^2 - a^2}$$

```
> laplace(exp(a*t)*cos(omega*t), t, s);
```

$$\frac{s - a}{(s - a)^2 + \omega^2}$$

```
> laplace(exp(a*t)*sin(omega*t), t, s);
```

$$\frac{\omega}{(s - a)^2 + \omega^2}$$

Powers of t are transformed as follows.

```
> laplace(1, t, s);
```
     # Resp. $\dfrac{1}{s}$

```
> laplace(t, t, s);
```
     # Resp. $\dfrac{1}{s^2}$

```
> assume(n, positive);
> laplace(t^n, t, s);
```
     # Resp. $s^{(-n^\sim - 1)}\,\Gamma(n^\sim + 1)$

Hence t^n with $n = 0, 1, 2, \ldots$ has the transform $n!/s^{(n+1)}$ because $\Gamma(n + 1) = n!$ when n is a positive integer. (The tilde is just a reminder that n is assumed to be positive.)

Inverse transforms are more difficult to obtain than transforms. Corresponding tables are available (e.g. in AEM, pp. 297-299), and you can use them just as you would use a table of integrals in integration. The Maple command `invlaplace(F, s, t)` gives help. (Note that now, `s` comes first in the command!) For example,

```
> invlaplace((s + 2)/(s^2 + 4*s + 5), s, t);
```
     # Resp. $\mathrm{e}^{(-2\,t)}\cos(t)$

```
> invlaplace(1/s^4, s, t);
```
     # Resp. $\dfrac{1}{6}t^3$

```
> invlaplace((s - 4)/(s^2 - 4), s, t);
```
     # Resp. $-\dfrac{1}{2}\,\mathrm{e}^{(2\,t)} + \dfrac{3}{2}\,\mathrm{e}^{(-2\,t)}$

```
> invlaplace(laplace(f, t, s), s, t);
```
     # Resp. f

```
> laplace(invlaplace(F, s, t), t, s);
```
     # Resp. F

Thus the two commands are inverses of each other, as had to be expected.

      Similar Material in AEM: pp. 251-255

| **EXAMPLE 5.2** | **DIFFERENTIAL EQUATIONS** |

Consider the differential equation and initial conditions

$$y'' + 2y' + 2y = 0, \qquad y(0) = 1, \quad y'(0) = -1.$$

I. Find a general solution $y(t)$ in three ways, first by `dsolve`, then by `dsolve(..., method = Laplace)`, and finally by using the subsidiary equation and transforming its solution $Y(s)$ back.

II. Apply the same three methods to the initial value problem.

Solution. **I.** Type the equation in the following form, and then apply `dsolve`.

```
> ode := diff(y(t), t, t) + 2*diff(y(t), t) + 2*y(t) = 0;
```

$$ode := \left(\frac{\partial^2}{\partial t^2} y(t) \right) + 2 \left(\frac{\partial}{\partial t} y(t) \right) + 2 y(t) = 0$$

```
> dsolve(ode);          # Terms may appear with cos and sin interchanged.
```

$$y(t) = \_C1 \, e^{(-t)} \, \sin(t) + \_C2 \, e^{(-t)} \, \cos(t)$$

```
> factor(%);            # Resp. y(t) = e^{(-t)} (_C1 sin (t) + _C2 cos (t))
```

This is a general solution in the form expected. (The arbitrary constants may appear interchanged.) Now turn to the Laplace transform method. Type

```
> with(inttrans);                          # Ignore the warning.
```

and

```
> dsolve(ode, y(t), method = laplace);
```

$$y(t) = e^{(-t)} \, y(0) \, \cos(t) + e^{(-t)} \, y(0) \, \sin(t) + e^{(-t)} \, D(y)(0) \, \sin(t)$$

```
> factor(%);
```

$$y(t) = e^{(-t)} \, (y(0) \, \cos(t) + y(0) \, \sin(t)) + D(y)(0) \, \sin(t))$$

This form of the solution should not surprise you. It is needed to make the solution equal to $y(0)$ when $t = 0$ (so that the sine terms are zero) and its derivative equal to $y'(0)$.

In the third method you get the **subsidiary equation** by applying the command `laplace` to the differential equation,

```
> t := 't':    s := 's':
> subsid := laplace(ode, t, s);
```

$$subsid := s \, (s \, \text{laplace} \, (y(t), \, t, \, s) - y(0)) - D(y)(0) + 2 \, s \, \text{laplace} \, (y(t), \, t, \, s) - 2 \, y(0)$$

$$+ 2 \, \text{laplace} \, (y(t), \, t, \, s) = 0$$

As the next step, solve the subsidiary equation *algebraically* for the transform, call it Y, of the unknown solution; thus,

```
> Y := solve(subsid, laplace(y(t), t, s));
```

$$Y := \frac{s \, y \, (0) + D \, (y) \, (0) + 2 \, y \, (0)}{s^2 + 2 \, s + 2}$$

Finally, obtain $y(t)$ itself by taking the inverse of the expression for Y. This gives the general solution

```
> sol := invlaplace(Y, s, t);
```

$$sol := e^{(-t)} \, y(0) \, \cos(t) + e^{(-t)} \, y(0) \, \sin(t) + e^{(-t)} \, D \, (y) \, (0) \, \sin(t)$$

in agreement with the previous result.

II. Now apply the same three methods to the initial value problem, remembering from Chap. 2 what will change when initial conditions are given. Basically, not much; put braces {} around as shown.

```
> dsolve({ode, y(0) = 1, D(y)(0) = -1});
```

$$y(t) = e^{(-t)} \cos(t)$$

```
> dsolve({ode, y(0) = 1, D(y)(0) = -1}, y(t), method = laplace);
```

$$y(t) = e^{(-t)} \cos(t)$$

```
> subsid2 := subs(y(0) = 1, D(y)(0) = -1, subsid);
```

$subsid2 := s\,(s\,\mathrm{laplace}\,(\mathrm{y}(t),\,t,\,s) - 1) - 1 + 2\,s\,\mathrm{laplace}\,(\mathrm{y}(t),\,t,\,s)$
$\quad + 2\,\mathrm{laplace}\,(\mathrm{y}(t),\,t,\,s) = 0$

```
> Y2 := solve(subsid2, laplace(y(t), t, s));
```

$$Y2 := \frac{s+1}{s^2 + 2\,s + 2}$$

```
> y2 := invlaplace(Y2, s, t);
```

$$y2 := e^{(-t)} \cos(t)$$

Similar Material in AEM: pp. 260-262

$\boxed{\textbf{EXAMPLE 5.3}}$ **FORCED VIBRATIONS. RESONANCE**

Resonance occurs in undamped mechanical or electrical systems if the driving force (the electromotive force, respectively) has the frequency equal to the natural frequency of the free vibrations of the system. This is the case for the differential equation

$$y'' + \omega_0{}^2\, y = K \sin \omega_0\, t.$$

Using the Laplace transform method, find the solution of this equation satisfying the initial conditions $y(0) = 0$, $y'(0) = 0$. Specifying K and ω_0 to be 1, plot that solution.

Solution. Type the equation; then obtain the subsidiary equation.

```
> ode := diff(y(t), t, t) + omega[0]^2*y(t) = K*sin(omega[0]*t);
```

$$ode := \left(\frac{\partial^2}{\partial t^2}\, \mathrm{y}(t)\right) + \omega_0{}^2\, \mathrm{y}(t) = K \sin(\omega_0\, t)$$

```
> with(inttrans):                                          # Ignore the warning.
> subsid := laplace(ode, t, s);
```

$subsid := s\,(s\,\mathrm{laplace}\,(\mathrm{y}(t),\,t,\,s) - \mathrm{y}\,(0)) - \mathrm{D}\,(y)\,(0) + \omega_0{}^2\,\mathrm{laplace}\,(\mathrm{y}(t),\,t,\,s)$

$$= \frac{K\,\omega_0}{s^2 + \omega_0{}^2}$$

Now include the initial conditions in the subsidiary equation,

```
> subsid2 := subs(y(0) = 0, D(y)(0) = 0, subsid);
```

$$subsid2 := s^2\,\mathrm{laplace}(\mathrm{y}(t),\,t,\,s) + \omega_0{}^2\,\mathrm{laplace}(\mathrm{y}(t),\,t,\,s) = \frac{K\,\omega_0}{s^2 + \omega_0{}^2}$$

Next solve the subsidiary equation *algebraically* for the Laplace transform, call it Y, of the unknown solution y:

```
> Y := solve(subsid2, laplace(y(t), t, s));
```

$$Y := \frac{K\,\omega_0}{s^4 + 2\,s^2\,\omega_0{}^2 + \omega_0{}^4}$$

The inverse Laplace transform then gives the solution

```
> yp := invlaplace(Y, s, t);
```

$$yp := K\,\omega_0 \left(\frac{1}{2}\,\frac{\sin(\omega_0\,t)}{\omega_0{}^3} - \frac{1}{2}\,\frac{t\,\cos(\omega_0\,t)}{\omega_0{}^2} \right)$$

For plotting choose $K = 1$ and $\omega_0 = 1$ by the command

```
> yp0 := subs(K = 1, omega[0] = 1, yp);
```

$$yp0 := \frac{1}{2}\,\sin(t) - \frac{1}{2}\,t\,\cos(t)$$

The maximum amplitude of the solution grows beyond bound as t increases indefinitely.

```
> plot(yp0, t = 0..100);
```

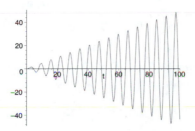

Example 5.3. Particular solution exhibiting resonance

Similar Material in AEM: p. 288

| **EXAMPLE 5.4** | **UNIT STEP FUNCTION (HEAVISIDE FUNCTION), DIRAC'S DELTA** |

The **unit step function** $u(t-a)$ (or **Heaviside function**) is 0 for $t < a$, has a jump of size 1 at $t = a$ (where we need not assign a value to it) and is 1 for $t > a$. Here, $a \geq 0$. The Maple notation is `Heaviside(t - a)`. Type `?Heaviside` for information. The transform of $u(t-a)$ is obtained by typing

```
> with(inttrans):                          # Ignore the warning.
> assume(a >= 0);                          # Essential. Try without!
> laplace(Heaviside(t - a), t, s);
```

$$\frac{e^{(-s\,a^{\sim})}}{s}$$

($-s\,a^{\sim}$ means $-as$. The tilde after the a is a reminder that we have made an assumption about a; don't be disturbed by that symbol.) The unit step function is the basic building block for representing "piecewise functions". By this, Maple means piecewise continuous functions that are given by different expressions over different intervals. (Type `?piecewise`.) For instance, type

```
> f := piecewise(t < Pi, 2, t < 2*Pi, 0, t > 2*Pi, sin(t));
```

$$f := \begin{cases} 2 & t < \pi \\ 0 & t < 2\pi \\ \sin t & 2\pi < t \end{cases}$$

To explain: for $t < \pi$ the function equals 2. For π to 2π it equals zero. For $t > 2\pi$ it equals $\sin t$. See the figure. To represent `f` in terms of unit step functions, use the command `convert(..., Heaviside)`.

```
> f := convert(f, Heaviside);
```

$$f := 2 - 2\,\text{Heaviside}(-\pi + t) + \sin(t)\,\text{Heaviside}(-2\pi + t)$$

that is, $f = 2 - 2u(t - \pi) + u(t - 2\pi)\sin t$. To plot f, type the following, where the (optional) command `scaling = constrained` gives equal scales on both axes.

```
> plot(f, t = 0..6*Pi, labels = [t, y], scaling = constrained);
```

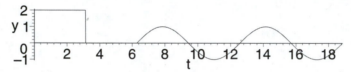

Example 5.4. "Piecewise function"

The transform of f is

```
> F := laplace(f, t, s);
```

$$F := 2\,\frac{1}{s} - 2\,\frac{e^{(-s\pi)}}{s} + \frac{e^{(-2s\pi)}}{s^2 + 1}$$

  Functions such as f may occur as driving forces in mechanics or as electromotive forces in circuits. For instance, the response of an undamped mass-spring system of mass 1 and spring constant 1 (initially at rest at equilibrium $y = 0$) to a single rectangular wave, say, $g(t) = 4$ if $2 < t < 10$ and 0 otherwise, is the solution of the ODE

```
> ode := diff(y(t), t, t) + y(t) = 4*Heaviside(t - 2)
   - 4*Heaviside(t - 10);
```

$$ode := \left(\frac{\partial^2}{\partial t^2}\,y(t)\right) + y(t) = 4\,\text{Heaviside}(t - 2) - 4\,\text{Heaviside}(t - 10)$$

```
> yp := dsolve({ode, y(0) = 0, D(y)(0) = 0}, y(t), method = laplace);
```

$$yp := y(t) = 8\,\text{Heaviside}\,(t - 2)\,\sin\left(\frac{1}{2}t - 1\right)^2 - 8\,\text{Heaviside}\,(t - 10)\,\sin\left(\frac{1}{2}t - 5\right)^2$$

```
> plot(rhs(yp), t = 0..20);
```

You see that the motion begins at $t = 2$. Until $t = 10$ there is a transition period (sinusoidal), followed by the steady-state harmonic oscillation that begins at $t = 10$

when the driving force shuts off.

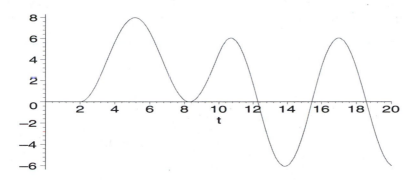

Example 5.4. Response of a mass-spring system
to a single rectangular wave

The **Dirac delta function** $\delta(t-a)$ (also known as the **unit impulse function**) models an impulse at $t = a$ (practically, for instance, a hammer blow). The command is `Dirac(t - a)` (type `?Dirac`). Recall that we have assumed $a \geq 0$.

> `Dirac(t - a);`   `# Resp.` $\mathrm{Dirac}(t - a\tilde{})$

> `laplace(Dirac(t - a), t, s);`   `# Resp.` $\mathrm{e}^{(-s\,a\tilde{})}$

For instance, if at $t = 1$ a hammer blow is imposed on an undamped mass-spring system of mass 1 and spring constant 1, the model is as follows and the solution (the displacement $y(t)$) is obtained by `dsolve(.., method = laplace)`, as before.

> `ode := diff(y(t), t, t) + y(t) = Dirac(t - 1);`

$$ode := \left(\frac{\partial^2}{\partial t^2}\, \mathrm{y}(t) \right) + \mathrm{y}(t) = \mathrm{Dirac}\,(t - 1)$$

> `sol := dsolve({ode, y(0) = 0, D(y)(0) = 0}, y(t), method = laplace);`

$$sol := \mathrm{y}(t) = \mathrm{Heaviside}\,(t - 1)\,\sin\,(t - 1)$$

> `plot(rhs(sol), t = 0..10, xtickmarks = [0, 1, 5, 10]);`

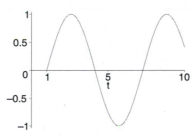

Example 5.4. Response of a mass-spring system to a hammer blow at $t = 1$

You see that the system remains at rest until the blow happens, and then immediately begins its steady-state harmonic oscillation.

Similar Material in AEM: pp. 266-273

EXAMPLE 5.5   SOLUTION OF SYSTEMS BY LAPLACE TRANSFORM

The process of solving systems by `dsolve(..., method = laplace)` is quite similar to that for single differential equations. For instance, type (where $D(y1)(t) = y_1'(t)$, etc.)

```
> y1 := 'y1': y2 := 'y2': y3 := 'y3':
> sys := D(y1)(t) + D(y2)(t) = 2*sinh(t),
>          D(y2)(t) + D(y3)(t) = exp(t),
>          D(y3)(t) + D(y1)(t) = 2 + exp(t);
```

$$sys := D(y1)(t) + D(y2)(t) = 2\sinh(t), \; D(y2)(t) + D(y3)(t) = e^t,$$
$$D(y3)(t) + D(y1)(t) = 2 + e^t$$

Then type initial conditions, so that you will obtain particular solutions; say,

```
> inits := y1(0) = 1, y2(0) = 0, y3(0) = 0;
```

$$inits := y1(0) = 1, \; y2(0) = 0, \; y3(0) = 0$$

Now apply the Laplace solution command. (`y1`, `y2`, `y3` may appear in a different order.)

```
> with(inttrans):                              # Ignore the warning.
> dsolve({sys, inits},  {y1(t), y2(t), y3(t)}, method = laplace);
```

$$\{y2(t) = -t - 1 + \tfrac{1}{2}e^t + \tfrac{1}{2}e^{(-t)}, \; y3(t) = t + \tfrac{1}{2}e^t - \tfrac{1}{2}e^{(-t)}, \; y1(t) = t + \tfrac{1}{2}e^t + \tfrac{1}{2}e^{(-t)}\}$$

Another system with unit step functions on the right, representing constant driving forces acting from $t = 0$ to 1 only, is the following.

```
> sys2 := D(y1)(t) = -y2(t) + 1 - Heaviside(t - 1),
>          D(y2)(t) = y1(t) + 1 - Heaviside(t - 1);
```

$$sys2 := D(y1)(t) = -y2(t) + 1 - \text{Heaviside}(t - 1),$$
$$D(y2)(t) = y1(t) + 1 - \text{Heaviside}(t - 1)$$

Applying `dsolve(..., method = laplace)` and prescribing initial conditions, say, $y_1(0) = 0, y_2(0) = 0$, you obtain particular solutions, first in terms of unit step functions,

```
> y := dsolve({sys2, y1(0) = 0, y2(0) = 0},  {y1(t), y2(t)}, method
  = laplace);
```

$$y := \{y2(t) = \sin(t) - \text{Heaviside}(t-1)\sin(t-1) + 2\sin\left(\tfrac{1}{2}t\right)^2 - \text{Heaviside}(t-1)$$

$$+\text{Heaviside}(t-1)\cos(t-1)\}, \; y1(t) = -2\sin\left(\tfrac{1}{2}t\right)^2 - \text{Heaviside}(t-1)\sin(t-1)$$

$$+\text{Heaviside}(t-1) - \text{Heaviside}(t-1)\cos(t-1) + \sin(t)\}$$

Note that the two components have come out in the wrong order. You can convert these solutions componentwise to the usual form

```
> p1 := convert(y[2], piecewise);
```

$$p1 := \text{y1}(t) = \begin{cases} -2\sin\left(\tfrac{1}{2}t\right)^2 + \sin(t) & t \le 1 \\ -\sin(t-1) + 1 - \cos(t-1) - 2\sin\left(\tfrac{1}{2}t\right)^2 + \sin(t) & 1 < t \end{cases}$$

```
> p2 := convert(y[1], piecewise);
```

$$p2 := \text{y2}(t) = \begin{cases} \sin(t) + 2\sin(\tfrac{1}{2}t)^2 & t \le 1 \\ -\sin(t-1) - 1 + \cos(t-1) + \sin(t) + 2\sin(\tfrac{1}{2}t)^2 & 1 < t \end{cases}$$

```
> plot({rhs(p1), rhs(p2)}, t = 0..5);
```

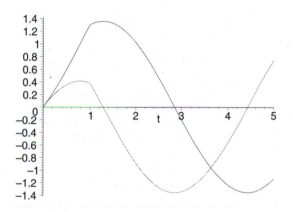

Example 5.5. Solutions $y_1(t)$ (lower curve for t from 0 to 3) and $y_2(t)$ of the second system sys2

At $t = 1$, where the driving force jumps from 1 down to 0, the curves have a cusp, where the derivative (the slope) suddenly decreases, as you can see.

    Similar Material in AEM: pp. 291-294

| EXAMPLE 5.6 | **FORMULAS ON GENERAL PROPERTIES OF THE LAPLACE TRANSFORM** |

The Laplace transform has various general properties that are essential for its practical usefulness. In this example we collect some of the most important corresponding formulas.

```
> with(inttrans):                                    # Ignore the warning.
> f := 'f':    g := 'g':    a := 'a':
> laplace(a*f(t) + b*g(t), t, s);                    # Linearity
```

$$a\,\text{laplace}\,(\text{f}(t),\,t,\,s) + b\,\text{laplace}\,(\text{g}(t),\,t,\,s)$$

```
> laplace(diff(f(t), t), t, s);                      # Differentiation
```

$$s\,\text{laplace}\,(\text{f}(t),\,t,\,s) - \text{f}\,(0)$$

```
> laplace(D(f)(t), t, s);                          # Use of the D-notation
```

$$s \, \text{laplace} \, (\text{f}(t), \, t, \, s) - \text{f}\,(0)$$

```
> laplace((D@@2)(f)(t), t, s);                     # Second derivative
```

$$s \, (s \, \text{laplace} \, (\text{f}(t), \, t, \, s) - \text{f}\,(0)) - \text{D}\,(f)\,(0)$$

Thus, roughly speaking, differentiation of a function $f(t)$ corresponds to the multiplication of its transform by s. And integration of $f(t)$ from 0 to t (with the variable of integration denoted by τ) corresponds to division of the transform by s, namely,

```
> laplace(int(f(tau), tau = 0..t), t, s);
```

$$\frac{\text{laplace} \, (\text{f}(t), \, t, \, s)}{s}$$

Multiplying a function by e^{at} corresponds to replacing s by $s - a$ in the transform. This is called *s*-**shifting**. For instance,

```
> laplace(exp(a*t)*cos(t), t, s);
```

$$\frac{s - a}{(s - a)^2 + 1}$$

Multiplying a transform by e^{-as} corresponds to replacing t by $t - a$ in the function and to equating it to 0 for $t < a$. This is called *t*-**shifting**. For instance,

```
> invlaplace(exp(-a*s)*laplace(cos(t), t, s), s, t);
```

$$\text{Heaviside} \, (t - a) \, \cos \, (t - a)$$

Multiplying a function by t corresponds to differentiating its transform (and multiplying it by -1),

```
> laplace(t*f(t), t, s);
```

$$- \left(\frac{\partial}{\partial s} \, \text{laplace} \, (\text{f}\,(t), \, t, \, s) \right)$$

Thus, $\mathcal{L}(t \, f(t)) = -F'(s)$. Taking the inverse transform on both sides (and interchanging the two sides), you thus have $\mathcal{L}^{-1}(F'(s)) = -t \, f(t)$; indeed,

```
> invlaplace(diff(laplace(f(t), t, s), s), s, t);
```

$$-t \, \text{f}(t)$$

Finally, taking the product of two transforms corresponds to taking the transform of the **convolution** (the integral shown) of the two corresponding functions,

```
> laplace(int(f(tau)*g(t - tau), tau = 0..t), t, s);
```

$$\text{laplace} \, (\text{f}(t), \, t, \, s) \, \text{laplace} \, (\text{g}(t), \, t, \, s)$$

For instance, if $f(t) = \cos t$ and $g(t) = e^t$, their transforms are $s/(s^2 + 1)$ and $1/(s - 1)$, and you should get the product of these two transforms on the right; indeed,

```
> laplace(int(cos(tau)*exp(t - tau), tau = 0..t), t, s);
```

$$-\frac{1}{2} \, \frac{s}{s^2 + 1} + \frac{\frac{1}{2}}{s^2 + 1} + \frac{\frac{1}{2}}{s - 1}$$

> `simplify(%);`

$$\frac{s}{(s^2 + 1)(s - 1)}$$

Similar Material in AEM: p. 296

Problem Set for Chapter 5

Pr.5.1 (**Transform**) Find $\mathcal{L}(\sin \pi t)$ by evaluating the defining integral.
(*AEM Ref.* p. 257 (#3))

Pr.5.2 (**Transform by integration**) Find $\mathcal{L}(\cos^2 \omega t)$ by evaluating the defining integral of the transform. (*AEM Ref.* p. 257 (#4))

Pr.5.3 (**Transform by integration**) Find the transform of $f(t) = k$ if $0 < t < c$, $f(t) = 0$ otherwise, by integration. (*AEM Ref.* p. 257 (#11))

Pr.5.4 (**Inverse transform**) Using `invlaplace`, find the function whose transform is $(s - 4)/(s^2 - 4)$. (*AEM Ref.* p. 257 (#20))

Pr.5.5 (**Inverse transform**) Find the inverse transform of $\dfrac{s}{L^2 s^2 + n^2 \pi^2}$.
(*AEM Ref.* p. 257 (#23))

Pr.5.6 (**Initial value problem, subsidary equation**) Find the solution of the initial value problem $y' - 5y = 1.5\,e^{-4t}$, $y(0) = 1$ by obtaining the subsidiary equation, solving it, and transforming the solution back. (*AEM Ref.* p. 264 (#2))

Pr.5.7 (**Initial value problem, subsidiary equation**) Solve the initial value problem $y'' + 2y' - 3y = 6e^{-2t}$, $y(0) = 2$, $y'(0) = -14$ by using the subsidiary equation.
(*AEM Ref.* p. 264 (#9))

Pr.5.8 (***t*-shifting**) Plot the function $4u(t - \pi)\cos t$ and find its transform.
(*AEM Ref.* p. 273 (#7))

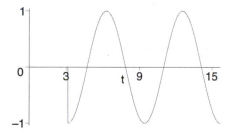

Problem 5.8. Given function $f(t)$

Pr.5.9 (**Unit step function**) Find and plot the inverse of $3(1 - e^{-\pi s})/(s^2 + 9)$.
(*AEM Ref.* p. 273 (#17))

Pr.5.10 (**Unit step function**) Find and plot the inverse transform of $e^{-2\pi s}/(s^2 + 2s + 2)$.
(*AEM Ref.* p. 273 (#18))

Pr.5.11 (*RC*-circuit, unit step) Find the subsidiary equation of

$$R\,i(t) + \frac{1}{C}\int_0^t i(\tau)\,d\tau = v(t), \qquad i(0) = 0, \qquad i'(0) = 0$$

determining the current $i(t)$ in an *RC*-circuit, assuming that $v(t) = K = const$ for t from 1 to 3 and $v(t) = 0$ otherwise. Find $i(t)$. Plot $i(t)$ when $R = 1$ ohm, $C = 1$ farad, and $K = 110$ volts. (*AEM Ref.* p. 274)

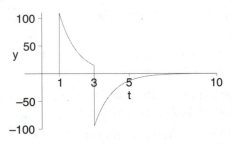

Problem 5.11. Current in an *RC*-circuit with a single rectangular wave as electromotive force

Pr.5.12 (*RC*-circuit, Dirac's delta) Solve Pr.5.11 when $v(t) = K\,\delta(t-1)$, the other data being as before. (*AEM Ref.* p. 271)

Pr.5.13 (Experiment on repeated Dirac's delta) Solve $y'' + y = \delta(t - \pi) - \delta(t - 2\pi)$, $y(0) = 0$, $y'(0) = 1$. First guess what the solution may look like. Then solve and plot. What will happen if you add further terms $\delta(t - 3\pi) - \delta(t - 4\pi) \pm \ldots$? (*AEM Ref.* p. 274 (#27)).

Pr.5.14 (*RL*-circuit) Using the command `dsolve(..., method = laplace)`, solve $L i' + R i = v(t)$, $i(0) = 0$, where $v(t) = \sin t$ if $0 < t < 2\pi$ and 0 otherwise. Plot the solution with $R = 1$, $L = 1$ and comment. (*AEM Ref.* p. 274 (#32))

Pr.5.15 (Full-wave rectifier) Plot the function obtained by the full-wave rectification of $\sin t$, that is, by multiplying the negative half-waves by -1. Let t go from 0 to 5π. (*AEM Ref.* p. 290 (#16))

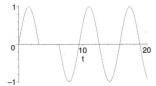

Problem 5.13. Current in an *RC*-circuit with $\delta(t - \pi) - \delta(t - 2\pi)$ as electromotive force

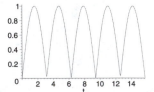

Problem 5.15. Full-wave rectification of the sine curve

PART B. LINEAR ALGEBRA, VECTOR CALCULUS

Content. Matrices, vectors, determinants, linear systems of equations (Chap. 6)
Matrix eigenvalue problems (Chap. 7)
Vectors in R^2 and R^3, dot and cross products, grad, div, curl (Chap. 8)
Vector integral calculus, integral theorems (Chap. 9)

linalg package. Load it by typing `with(linalg)`. Type ?linalg, ?matrix, ?vector, ?determinant, ?gausselim. You will need the package practically all the time.

Chapter 6

Matrices, Vectors, Determinants.
Linear Systems of Equations

Content. Addition, scalar multiplication, matrix multiplication, determinants
(Ex. 6.1, Prs. 6.1-6.6)
Special matrices, composing matrices (Exs. 6.2, 6.3, Prs. 6.7, 6.9, 6.17)
Transpose, inverse, linear transformations (Prs. 6.8, 6.10-6.12)
Orthogonality, norm (Prs. 6.13-6.15)
Solution of linear systems of equations (Exs. 6.4, 6.5, Pr. 6.14)
Rank of a matrix, row space, linear independence (Ex. 6.6,
Prs. 6.16, 6.18-6.20)

Numerical methods for matrices see Chap 18.

Examples for Chapter 6

EXAMPLE 6.1 **MATRIX ADDITION, SCALAR MULTIPLICATION,
MATRIX MULTIPLICATION. VECTORS**

```
> with(linalg):                                      # Ignore the warning.
> A := matrix([[5, -3, 0], [6, 1, -4]]);
```

$$A := \begin{bmatrix} 5 & -3 & 0 \\ 6 & 1 & -4 \end{bmatrix}$$

This shows how to type a matrix. If you prefer to arrange the entries as they will occur, you can type, receiving the same response,

67

```
> A := matrix([[5, -3,  0],
>                [6,  1, -4]]);                          # Response as before
```

As another way of writing matrices, you can type

```
> A := matrix(2, 3, [5, -3, 0,   6, 1, -4]);
```

$$A := \begin{bmatrix} 5 & -3 & 0 \\ 6 & 1 & -4 \end{bmatrix}$$

2 is the number of rows and 3 the number of columns. For larger matrices the other way is perhaps better because you can see the positions of the individual entries more easily.

Addition of matrices. To define addition of matrices, let

```
> B := matrix([[-2, 4, -1], [1, 1, 5]]);
```

$$B := \begin{bmatrix} -2 & 4 & -1 \\ 1 & 1 & 5 \end{bmatrix}$$

The sum is typed with a + sign,

```
> C := A + B;                                           # Resp. A + B
> C := evalm(A + B);                       # evalm suggests "evaluate matrix".
```

$$C := \begin{bmatrix} 3 & 1 & -1 \\ 7 & 2 & 1 \end{bmatrix}$$

Scalar multiplication. Transposition. Multiplication of matrices. Try

```
> D := evalm(4*C);
```

`Error, attempting to assign to 'D' which is protected`

```
> evalm(A*B);
```

`Error, (in evalm/evaluate) use the &* operator for matrix/vector multiplication`

```
> E := evalm(A&*B);
```

`Error, (in linalg[multiply]) non matching dimensions for vector/matrix product`

So the second message told you the right sign &* for multiplication, and the third that the product $\mathbf{AB}$ is not defined. But $\mathbf{AF}$ ($\mathbf{F} = \mathbf{B}^T$ the **transpose** of $\mathbf{B}$) should work, by the definition of matrix multiplication.

```
> F := transpose(B);                                    # Transpose of a matrix
```

$$F := \begin{bmatrix} -2 & 1 \\ 4 & 1 \\ -1 & 5 \end{bmatrix}$$

```
> G := evalm(A&*F);                                     # Product of two matrices
```

$$G := \begin{bmatrix} -22 & 2 \\ -4 & -13 \end{bmatrix}$$

```
> evalm(G^2);
```
                                                        # Square of a matrix

$$\begin{bmatrix} 476 & -70 \\ 140 & 161 \end{bmatrix}$$

Inverse of a matrix. Determinant. Trace

```
> inverse(G);
```
                                                        # Inverse of a matrix

$$\begin{vmatrix} \dfrac{-13}{294} & \dfrac{-1}{147} \\[2mm] \dfrac{2}{147} & \dfrac{-11}{147} \end{vmatrix}$$

```
> det(G);
```
                                        # Resp. 294 # Determinant of a matrix
```
> trace(G);
```
                                        # Resp. -35 # trace $-22 - 13 = -35$

Vectors occur together with matrices, particularly in connection with linear systems. Let

```
> v := [3, 5];
```
                                                        # Resp. $v := [3, 5]$
```
> evalm(v&*A);
```
                                                        # Resp. $[45, -4, -20]$
```
> evalm(A&*v);
```
`Error, (in linalg[multiply]) non matching dimensions for vector/matrix product`
```
> evalm(F&*v);
```
                                                        # Resp. $[-1, 17, 22]$

Note that for vectors entered in the form $v = [v_1, v_2, ...]$ Maple does not distinguish between row and column vectors, leaving the interpretation to you. Thus the last response is a column vector. If $\mathbf{C}$ is a 3×3 matrix and $\mathbf{x} = [x_1, x_2, x_3]$, you can compute $\mathbf{C}\&*\mathbf{x}$ or $\mathbf{x}\&*\mathbf{C}$ obtaining a row vector $\mathbf{r} = [r_1, r_2, r_3]$ in both cases, but in the first case $\mathbf{x}$ and $\mathbf{r}$ are column vectors and in the second case they are row vectors.

Further let

```
> u := [4, -2];
```
                                                        # Resp. $u = [4, -2]$
```
> u*v;
```
                                                        # Resp. $[4, -2][3, 5]$
```
> evalm(u&*v);
```
                                        # Resp. 2 # Inner product (dot product)
```
> dotprod(u, v);
```
                                                # Resp. 2 # Inner product
```
> innerprod(u, v);
```
                                                # Resp. 2 # Inner product

So here you have three different commands for the **inner product** (**dot product**) $\mathbf{u} \bullet \mathbf{v}$ or, regarding the vectors as column vectors, $\mathbf{u}^T\mathbf{v}$. Similarly, you obtain $\mathbf{v}^T\mathbf{G}$ ($\mathbf{v}$ regarded as a column vector).

```
> evalm(v&*G);
```
                                                        # Resp. $[-86, -59]$

On rare occasions you may need the $n \times n$ matrix $\mathbf{vu}^T$ (where $n = 2$ for our vectors).

```
> v := matrix([[3, 5]]);
```
                                                        # Resp. $v := [3 \quad 5]$
```
> u := matrix([[4], [-2]]);
```
                                        # Resp. $u := \begin{bmatrix} 4 \\ -2 \end{bmatrix}$
```
> evalm(u&*v);
```
                                        # Resp. $\begin{bmatrix} 12 & 20 \\ -6 & -10 \end{bmatrix}$

Similar Material in AEM: pp. 307-309, 311-313

EXAMPLE 6.2   SPECIAL MATRICES

Matrices of the subsequent form will be needed quite frequently.

```
> with(linalg):                      # Load the linalg package. Ignore the warning.
> matrix(3, 5, 0);                        # Zero matrix with 3 rows and 5 columns
```

$$\begin{bmatrix} 0 & 0 & 0 & 0 & 0 \\ 0 & 0 & 0 & 0 & 0 \\ 0 & 0 & 0 & 0 & 0 \end{bmatrix}$$

```
> diag(a, b, c);                                              # A diagonal matrix
```

$$\begin{bmatrix} a & 0 & 0 \\ 0 & b & 0 \\ 0 & 0 & c \end{bmatrix}$$

```
> diag(1, 1, 1);                                       # The 3 × 3 unit matrix
```

$$\begin{bmatrix} 1 & 0 & 0 \\ 0 & 1 & 0 \\ 0 & 0 & 1 \end{bmatrix}$$

Matrices whose entries are given by a formula of the subscripts of the entries can be obtained as illustrated by a famous example (the 3×3 **Hilbert matrix**):

```
> matrix(3, 3, (j, k) -> 1/(j + k - 1));
```

$$\begin{bmatrix} 1 & \dfrac{1}{2} & \dfrac{1}{3} \\ \dfrac{1}{2} & \dfrac{1}{3} & \dfrac{1}{4} \\ \dfrac{1}{3} & \dfrac{1}{4} & \dfrac{1}{5} \end{bmatrix}$$

Similar Material in AEM: p. 314

EXAMPLE 6.3   CHANGING AND COMPOSING MATRICES,
               ACCESSING ENTRIES. SUBMATRICES

Having discussed the basic operations with matrices and vectors in Example 6.1, we now turn to operations of accessing entries and of accessing, interchanging or changing rows or columns of a matrix.

```
> with(linalg):                                           # Load the linalg package.
> A := matrix([[0, 1, -1, 2], [1, -2, 4, -3], [3, -4, 0, 0]]); # Given matrix
```

$$A := \begin{bmatrix} 0 & 1 & -1 & 2 \\ 1 & -2 & 4 & -3 \\ 3 & -4 & 0 & 0 \end{bmatrix}$$

```
> A[2,3];                                                          # Resp. 4
```

This shows how to extract an entry of $\mathbf{A} = [a_{jk}]$, namely, $a_{23} = 4$. A whole row or column is obtained by

```
> row(A, 3);                                          # Resp. [3, −4, 0, 0]
> col(A, 1);                                          # Resp. [0, 1, 3]
> submatrix(A, 2..3, 2..4);                           # Command for submatrix
```

$$\begin{bmatrix} -2 & 4 & -3 \\ -4 & 0 & 0 \end{bmatrix}$$

The first range 2..3 indicates the rows to be included (Rows 2 and 3), and the second range 2..4 indicates the columns (the last three columns). Instead of ranges you can list which rows (1 and 3, e.g.) and which columns (2 and 4, e.g.) you want to include, so that you do get any desired submatrix.

```
> submatrix(A, [1, 3], [2, 4]);                       # Resp.
```
$$\begin{bmatrix} 1 & 2 \\ -4 & 0 \end{bmatrix}$$

Changing rows and columns. Rows are interchanged by swaprow . For instance,

```
> B := swaprow(A, 1, 2);
```

$$B := \begin{bmatrix} 1 & -2 & 4 & -3 \\ 0 & 1 & -1 & 2 \\ 3 & -4 & 0 & 0 \end{bmatrix}$$

You will need this in the Gauss elimination. Similarly for interchanging columns, say, 2 and 4, type swapcol(A, 2, 4) . The next command, basic in the Gauss elimination, adds −3 times Row 1 to Row 3, creating a 0 in the left lower corner.

```
> C := addrow(B, 1, 3, -3);
```

$$C := \begin{bmatrix} 1 & -2 & 4 & -3 \\ 0 & 1 & -1 & 2 \\ 0 & 2 & -12 & 9 \end{bmatrix}$$

Composition of matrices from vectors. The *augmented matrix* [**A**, **b**] of **A** and a vector **b**, say,

```
> b := [4, 5, 2];                                     # Resp. b := [4, 5, 2]
```

is obtained by typing

```
> augment(A, b);                           # Here, Maple takes b as a column vector.
```

$$\begin{bmatrix} 0 & 1 & -1 & 2 & 4 \\ 1 & -2 & 4 & -3 & 5 \\ 3 & -4 & 0 & 0 & 2 \end{bmatrix}$$

More generally, the command augment composes a matrix from given vectors as columns, and by taking the transpose you get the matrix with these vectors as rows. For instance, let

```
> a := [2, 1]:  b := [3, 8]:  c := [-1, 2]:
```

Then

```
> augment(a, b, c);                                   # Resp.
```
$$\begin{bmatrix} 2 & 3 & -1 \\ 1 & 8 & 2 \end{bmatrix}$$

```
> transpose(%);
```

$$\begin{bmatrix} 2 & 1 \\ 3 & 8 \\ -1 & 2 \end{bmatrix}$$

Similar Material in AEM: pp. 324-326

EXAMPLE 6.4   SOLUTION OF A LINEAR SYSTEM

Solve the linear system

$$\begin{aligned} -x_1 &+ x_2 + 2x_3 = 2 \\ 3x_1 &- x_2 + x_3 = 6 \\ -x_1 &+ 3x_2 + 4x_3 = 4 \end{aligned}$$

Solution. First method. Write the system in matrix form $\mathbf{Ax} = \mathbf{b}$ and type the coefficient matrix $\mathbf{A}$ and the vector $\mathbf{b}$ as shown.

```
> with(linalg):                              # Ignore the warning.
> A := matrix([[-1, 1, 2], [3, -1, 1], [-1, 3, 4]]);
```

$$A := \begin{bmatrix} -1 & 1 & 2 \\ 3 & -1 & 1 \\ -1 & 3 & 4 \end{bmatrix}$$

```
> b := [2, 6, 4];                            # Resp. b := [2, 6, 4]
> x := linsolve(A, b);                       # Resp. x := [1, -1, 2]
```

    Second method. In the method just discussed you get the solution without seeing what is going on. The second method does elimination and back substitution separately. Accordingly, first type the augmented matrix by the command augment (see the previous example) and then apply gausselim , which does the Gauss elimination. As Step 2 of the method then follows the back substitution by the command backsub .

```
> A1 := augment(A, b);
```

$$A1 := \begin{bmatrix} -1 & 1 & 2 & 2 \\ 3 & -1 & 1 & 6 \\ -1 & 3 & 4 & 4 \end{bmatrix}$$

```
> B := gausselim(A1);
```

$$B := \begin{bmatrix} -1 & 1 & 2 & 2 \\ 0 & 2 & 7 & 12 \\ 0 & 0 & -5 & -10 \end{bmatrix}$$

```
> x := 'x':                                  # Unassign x (used just before).
> x := backsub(B);
```

$$x := [1, -1, 2]$$

```
> rank(A);                                   # Resp. 3
```

This implies that the solution is unique.

    Similar Material in AEM: p. 327

EXAMPLE 6.5   GAUSS ELIMINATION; FURTHER CASES

Case 1. Infinitely many solutions. The system is

$$3.0\,x_1 + 2.0\,x_2 + 2.0\,x_3 - 5.0\,x_4 = 8.0$$
$$0.6\,x_1 + 1.5\,x_2 + 1.5\,x_3 - 5.4\,x_4 = 2.7.$$
$$1.2\,x_1 - 0.3\,x_2 - 0.3\,x_3 + 2.4\,x_4 = 2.1$$

The process of solution is the same as in the first method in the previous example.

```
> with(linalg):                              # Ignore the warning.
```

```
> Digits := 5:                   # This restricts floating numbers to 5 digits.
```

```
> A := matrix([[3.0,   2.0,   2.0, -5.0],
>              [0:6,   1.5,   1.5, -5.4],
>              [1.2, -0.3, -0.3,   2.4]]);
```

$$A := \begin{bmatrix} 3.0 & 2.0 & 2.0 & -5.0 \\ .6 & 1.5 & 1.5 & -5.4 \\ 1.2 & -.3 & -.3 & 2.4 \end{bmatrix}$$

```
> b := [8.0,   2.7,   2.1];
```

$$b := [8.0, 2.7, 2.1]$$

```
> x := linsolve(A, b);
```
$$x := [-.2500, \_t_1, 10. - 1.\_t_1, 2.2500]$$

Hence $x_1 = -0.25$, x_2 arbitrary (here denoted by t_1), $x_3 = 10 - x_2$, $x_4 = 2.25$. But these are not all solutions. We claim that x_4 can also be arbitrary, and then, for instance, you can take $x_1 = 2 - x_4$, $x_2 = 1 - x_3 + 4\,x_4$, x_3 and x_4 arbitrary. Indeed, substituting these expressions into the equations on the left, you obtain the values on the right,

```
> x1 := 2 - x4;   x2 := 1 - x3 + 4*x4;
```

$$x1 := 2 - x4$$
$$x2 := 1 - x3 + 4\,x4$$

```
> 3*x1 + 2*x2 + 2*x3 - 5*x4;                    # Resp. 8
> 0.6*x1 + 1.5*x2 + 1.5*x3 - 5.4*x4;           # Resp. 2.7
> 1.2*x1 - 0.3*x2 - 0.3*x3 + 2.4*x4;           # Resp. 2.1
```

Furthermore, if in the general solution you choose $x_4 = 2.25$, you obtain the partial solution set first obtained. Hence be prepared that ***your software may not always give you all the solutions.*** Thus it may often be worthwhile to do all the steps of the Gauss elimination.

     Try out whether the second method in the previous example will give better results. Type

```
> A1 := augment(A, b);
```

$$A1 := \begin{bmatrix} 3.0 & 2.0 & 2.0 & -5.0 & 8.0 \\ .6 & 1.5 & 1.5 & -5.4 & 2.7 \\ 1.2 & -.3 & -.3 & 2.4 & 2.1 \end{bmatrix}$$

```
> B := gausselim(A1);
```

$$B := \begin{bmatrix} 3.0 & 2.0 & 2.0 & -5.0 & 8.0 \\ 0 & 1.1000 & 1.1000 & -4.4000 & 1.1000 \\ 0 & 0 & 0 & 0 & 0 \end{bmatrix}$$

```
> x := 'x':                                    # Unassign x used just before.

> x := backsub(B);
```

$$x := [\, 2.0000 - .99999\_t_1,\ 1.0000 - 1.0000\_t_2 + 4.0000\_t_1,\ \_t_2,\ \_t_1\,]$$

With this method you have obtained the full solution set (except for the small round-off error; -0.99999 should be -1), $x_1 = 2 - t_1$, $x_2 = 1 - t_2 + 4\,t_1$, $x_3 = t_2$ *arbitrary*, $x_4 = t_1$ *arbitrary*.

Case 2. A unique solution. See the previous example.

Case 3. No solutions. For instance, $x_1 + x_2 = 3$, $2\,x_1 + 2\,x_2 = 5$ has no solutions. To see what happens, type

```
> A := matrix([[1, 1], [2, 2]]);    b := [3, 5];
```

$$A := \begin{bmatrix} 1 & 1 \\ 2 & 2 \end{bmatrix}$$

$$b := [3,\ 5]$$

```
> linsolve(A, b);                   # No response. Hence you get no solutions.
```

Similar Material in AEM: p. 327

EXAMPLE 6.6   **RANK. ROW SPACE. LINEAR INDEPENDENCE**

The **rank of a matrix** is the key concept in connection with the existence and uniqueness of solutions of a linear system of equations. Illustrate by the following matrix **A** that rank **A** is invariant under transposition.

```
> with(linalg):                                # Ignore the warning.

> A := matrix([[3, 0, 2, 2], [-6, 42, 24, 54], [21, -21, 0, -15]]);
```

$$A := \begin{bmatrix} 3 & 0 & 2 & 2 \\ -6 & 42 & 24 & 54 \\ 21 & -21 & 0 & -15 \end{bmatrix}$$

```
> rank(A);   rank(transpose(A));                       # Resp. 2   2
```

Show the invariance of rank **A** under elementary row operations. Interchange, for instance, Rows 1 and 3:

```
> rank(swaprow(A, 1, 3));                               # Resp. 2
```

The second row operation is the addition of a constant multiple of a row to another row, for instance, add -5 times Row 2 to Row 3:

```
> B := addrow(A, 2, 3, -5);                          # Type ?addrow, ?addcol
```

$$B := \begin{bmatrix} 3 & 0 & 2 & 2 \\ -6 & 42 & 24 & 54 \\ 51 & -231 & -120 & -285 \end{bmatrix}$$

```
> rank(B);                                                          # Resp. 2
```

The third row operation is the multiplication of a row by a *nonzero* constant c. For instance, multiply Row 3 of **A** by $-1/3$ (by adding $-2/3$ of Row 3 to Row 3):

```
> addrow(A, 3, 3, -2/3);
```

$$\begin{bmatrix} 3 & 0 & 2 & 2 \\ -6 & 42 & 24 & 54 \\ 7 & -7 & 0 & -5 \end{bmatrix}$$

```
> rank(%);                                                          # Resp. 2
```

Methods of determining rank. *I.* From the "triangularized form" of the matrix (see Example 6.5 in this Guide) you see directly that rank $\mathbf{A} = 2$:

```
> gausselim(A);
```

$$\begin{bmatrix} 3 & 0 & 2 & 2 \\ 0 & 42 & 28 & 58 \\ 0 & 0 & 0 & 0 \end{bmatrix}$$

II. By the command `rowspace(A)` or `colspace(A)`, which compute a **basis** of the **row space** or the **column space** of **A**, respectively, and the fact that rank **A** equals the dimension 2 of these spaces:

```
> rowspace(A);   colspace(A);
```

$$\left\{ \left[0, 1, \frac{2}{3}, \frac{29}{21}\right], \left[1, 0, \frac{2}{3}, \frac{2}{3}\right] \right\}$$

$$\left\{ [1, 0, 6], \left[0, 1, \frac{-1}{2}\right] \right\}$$

Linear independence and dependence of vectors can also be tested by the use of a rank, namely, by the rank of the matrix whose rows or columns are the given vectors. For instance, let the vectors be

```
> a := [-6, 42, 24, 54];   b := [21, -21, 0, -15];   c := [3, 0, 2, 2];
```

$$a := [-6, 42, 24, 54]$$

$$b := [21, -21, 0, -15]$$

$$c := [3, 0, 2, 2]$$

Obtain the matrix with these vectors as columns by the command `augment`,

```
> M := augment(a, b, c);
```

$$M := \begin{bmatrix} -6 & 21 & 3 \\ 42 & -21 & 0 \\ 24 & 0 & 2 \\ 54 & -15 & 2 \end{bmatrix}$$

```
> rank(M);                                        # Resp. 2
```

Hence the given vectors are linearly dependent. Indeed, $\mathbf{M}$ is obtained from $\mathbf{A}$ by interchanging the rows and then taking the transpose.

      Similar Material in AEM: pp. 331-336

Problem Set for Chapter 6

Pr.6.1 (Addition, scalar multiplication) In Example 6.1 of this Guide, compute $3\mathbf{A}$, $\mathbf{A} - \mathbf{B}$, and $(2\mathbf{A} - (1/2)\mathbf{B})^T$. (*AEM Ref.* p. 308)

Pr.6.2 (Matrix multiplication) In Example 6.1 compute $\mathbf{A}\mathbf{A}^T$, $\mathbf{A}^T\mathbf{A}$, $(\mathbf{A}^T\mathbf{A})^2$, $(\mathbf{A} + \mathbf{B})(\mathbf{A} - \mathbf{B})^T$. (*AEM Ref.* pp. 311-313)

Pr.6.3 (Matrix multiplication) Let

$$\mathbf{A} = \begin{bmatrix} 1 & 3 & 2 \\ 3 & 5 & 0 \\ 2 & 0 & 4 \end{bmatrix}, \quad \mathbf{B} = \begin{bmatrix} 0 & 2 & 1 \\ -2 & 0 & -3 \\ -1 & 3 & 0 \end{bmatrix}, \quad \mathbf{c} = \begin{bmatrix} 1 \\ 0 \\ -2 \end{bmatrix}, \quad \mathbf{d} = \begin{bmatrix} 3 \\ 1 \\ 2 \end{bmatrix}.$$

Compute $\mathbf{A}^2$, $\mathbf{A}^4$, $\mathbf{B}^2$, $\mathbf{B}^4$, $\mathbf{A}\mathbf{B} - \mathbf{B}\mathbf{A}$, $\mathbf{A} - \mathbf{A}^T$, $\mathbf{B} + \mathbf{B}^T$, $\det \mathbf{A}$, $\det \mathbf{B}$, $\mathbf{A}\mathbf{c}$, $\mathbf{B}(\mathbf{c} - 3\mathbf{d})$, $\mathbf{c} \bullet \mathbf{d}$. (*AEM Ref.* pp. 311-316)

Pr.6.4 (Matrices and vectors) Using the data given in Pr.6.3, compute $\mathbf{B}\mathbf{c}$, $\mathbf{c}^T\mathbf{A}\mathbf{c}$, $\mathbf{A}\mathbf{B}\mathbf{c}$, $(\mathbf{A}\mathbf{c}) \bullet (\mathbf{B}\mathbf{d})$, $\mathbf{d}^T\mathbf{B}\mathbf{d}$. Why is the last result 0?

Pr.6.5 (Experiments on multiplication) By experimenting with 3×3 or 4×4 ***symmetric*** and ***skew-symmetric*** matrices whose entries are numbers or general letters try to answer the following questions. Are sums, products, and powers of symmetric matrices symmetric? Study the same questions for skew-symmetric matrices. What can you say about products of a symmetric matrix times a skew-symmetric one? Is $\det \mathbf{A} = 0$ for every skew-symmetric matrix? For some symmetric matrices? (*AEM Ref.* p. 307)

Pr.6.6 (Associativity, distributivity) Verify the associativity of matrix multiplication and the distributivity by 3×3 matrices of your own choice. (*AEM Ref.* p. 313)

Pr.6.7 (Rotation, the command `map`**)** If

$$\mathbf{A} = \begin{bmatrix} \cos\theta & -\sin\theta \\ \sin\theta & \cos\theta \end{bmatrix}, \quad \text{verify that} \quad \mathbf{A}^n = \begin{bmatrix} \cos n\theta & -\sin n\theta \\ \sin n\theta & \cos n\theta \end{bmatrix}$$

for $n = 2, 3, 4$. What does this mean in terms of rotations through an angle θ? (Use the command `map(combine, A^2)`, etc., which operates on each entry separately. Type `?map` for information.) (*AEM Ref.* p. 320)

Pr.6.8 (Transposition rule for products) Prove $(\mathbf{A}\,\mathbf{B})^T = \mathbf{B}^T\mathbf{A}^T$ for general 2×2 matrices on the computer. (*AEM Ref.* p. 315)

Pr.6.9 **(Experiment on Hankel matrices)** Find empirically a law for the smallest n as a function of m (> 0, integer) such that $\det \mathbf{A} = 0$, where the $n \times n$ matrix $\mathbf{A} = [a_{jk}]$ has the entries $a_{jk} = (j+k)^m$. (Enjoy these special Hankel matrices, whose determinants have very fast growing values, but all of sudden become 0 from some n on. This is of practical interest in connection with the so-called Padé approximation.)

Pr.6.10 **(Inverse)** Using the computer, find the formula for the inverse of a 2×2 matrix $\mathbf{A} = [a_{jk}]$ in terms of a_{jk} and $\det \mathbf{A}$. (*AEM Ref.* p. 353)

Pr.6.11 **(Inverse of a product)** Verify the basic relation $(\mathbf{A}\,\mathbf{B})^{-1} = \mathbf{B}^{-1}\mathbf{A}^{-1}$ for the matrices

$$\mathbf{A} = \begin{bmatrix} 0 & -2 & -1 \\ -2 & 3 & 2 \\ -1 & 2 & 1 \end{bmatrix} \quad \text{and} \quad \mathbf{B} = \begin{bmatrix} 1 & 2 & 3 \\ 2 & 3 & 4 \\ 3 & 4 & 6 \end{bmatrix}$$

(*AEM Ref.* p. 355)

Pr.6.12 **(Linear transformations)** With respect to Cartesian coordinates in space, let $\mathbf{y} = \mathbf{A}\mathbf{x}$ and $\mathbf{x} = \mathbf{B}\mathbf{w}$ with $\mathbf{A}$ and $\mathbf{B}$ as in the previous problem. Find the transformation $\mathbf{y} = \mathbf{C}\mathbf{w}$ which transforms $\mathbf{w}$ directly into $\mathbf{y}$. Find the inverse of this transformation. (*AEM Ref.* pp. 316-318)

Pr.6.13 **(Orthogonal vectors)** Show that the following vectors are orthogonal. (*AEM Ref.* p. 365)

$$\mathbf{c} = \begin{bmatrix} 3 & 2 & -2 & 1 & 0 \end{bmatrix}, \quad \mathbf{d} = \begin{bmatrix} 2 & 0 & 3 & 0 & 4 \end{bmatrix}, \quad \mathbf{e} = \begin{bmatrix} 1 & -3 & -2 & -1 & 1 \end{bmatrix}$$

Pr.6.14 **(Extension of an orthogonal system)** Find a vector $\mathbf{x}$ orthogonal to the three vectors in Pr.6.13. (*AEM Ref.* p. 365)

Pr.6.15 **(Norm)** Find the Euclidean norm $(c_1^2 + c_2^2 + \ldots + c_n^2)^{1/2}$, etc., of the vectors in Pr.6.13. (*AEM Ref.* p. 362)

Pr.6.16 **(Linear independence)** Check the set of vectors $\begin{bmatrix} -1 & 5 & 0 \end{bmatrix}$, $\begin{bmatrix} 16 & 8 & -3 \end{bmatrix}$, $\begin{bmatrix} -64 & 56 & 9 \end{bmatrix}$ for linear independence. (*AEM Ref.* p. 336 (#3))

Pr.6.17 **(Hilbert matrices)** Find the determinant and the inverse of the Hilbert matrix $\mathbf{H} = [h_{jk}]$ with $n = 3$ rows and columns, where $h_{jk} = 1/(j+k-1)$. Comment on the size of $\det \mathbf{H}$ and of the entries of the inverse. Do the same tasks when $n = 4$ and 5. (*AEM Ref.* p. 358. See also Example 6.2 in this Guide.)

Pr.6.18 **(Experiment on rank)** Find the rank of the $n \times n$ matrix $\mathbf{A} = [a_{jk}]$ with entries $a_{jk} = j + k - 1$ and any n. To understand the reason for the perhaps somewhat surprising result, use the command `gausselim` and then write a program for the steps of the Gauss elimination in the case of the present matrix. (*AEM Ref.* pp. 331-336)

Pr.6.19 **(Linear independence)** Are the vectors $\begin{bmatrix} 0 & 16 & 0 & -24 & 0 \end{bmatrix}$, $\begin{bmatrix} 1 & 0 & -1 & 0 & 2 \end{bmatrix}$, $\begin{bmatrix} 0 & -14 & 0 & 21 & 0 \end{bmatrix}$ linearly independent or dependent? (*AEM Ref.* p. 336)

Pr.6.20 **(Row and column space)** Find a basis of the row space and of the column space of the matrix with the rows $\begin{bmatrix} 3 & 1 & 4 \end{bmatrix}$, $\begin{bmatrix} 0 & 5 & 8 \end{bmatrix}$, $\begin{bmatrix} -3 & 4 & 4 \end{bmatrix}$, $\begin{bmatrix} 1 & 2 & 4 \end{bmatrix}$. (Type `?rowspace`, `?columnspace`. *AEM Ref.* p. 337 (#29))

Chapter 7

Matrix Eigenvalue Problems

Content. Basic commands (Ex. 7.1)

Complex eigenvalues, orthogonal matrices (Exs. 7.2, 7.3, Pr. 7.3)

Complex matrices (Ex. 7.4, Prs. 7.6-7.9)

Similar matrices, diagonalization (Ex. 7.5, Prs. 7.10-7.12)

All matrices in this chapter are square. The **linalg package** (see Part opening) is needed again. `eigenvectors` gives eigenvalues and eigenvectors. `eigenvalues` gives eigenvalues only. Further commands (for characteristic matrix, accessing parts of the spectrum, etc.) see in Example 7.1.

Examples for Chapter 7

EXAMPLE 7.1 **EIGENVALUES, EIGENVECTORS, ACCESSING SPECTRUM**

Find the characteristic determinant, the characteristic polynomial, the eigenvalues, and eigenvectors of the matrix

$$\mathbf{A} = \begin{bmatrix} -2 & 2 & -3 \\ 2 & 1 & -6 \\ -1 & -2 & 0 \end{bmatrix}.$$

Solution. You can obtain all the information about eigenvalues and eigenvectors from the single command `eigenvectors`. That is, type

```
> with(linalg):                                    # Ignore the warning.
```

```
> A := matrix([[-2, 2, -3],  [2, 1, -6],  [-1, -2, 0]]);
```

$$A = \begin{bmatrix} -2 & 2 & -3 \\ 2 & 1 & -6 \\ -1 & -2 & 0 \end{bmatrix}$$

```
> eig := eigenvectors(A);
```

$$eig := [-3, 2, \{[3, 0, 1], [-2, 1, 0]\}], [5, 1, \{[-1, -2, 1]\}]$$

You see that **A** has two eigenvalues, $\lambda_2 = -3$, of multiplicity 2, and $\lambda_1 = 5$, of multiplicity 1. An eigenvector corresponding to 5 is $[-1 \quad -2 \quad 1]$. Two linearly independent eigenvectors corresponding to -3 are $[3 \quad 0 \quad 1]$ and $[-2 \quad 1 \quad 0]$. Keep in mind that ***the order of the eigenvalues and eigenvectors may be different from that appearing in this Guide.*** You could terminate your work here. But we shall continue with a few further useful commands.

Accessing parts of the spectrum. This can be useful, for instance, in connection with systems of ODE's (Chap. 3). The commands are self-explanatory.

```
> eig[1];                               # Resp. [−3, 2, {[3, 0, 1], [−2, 1, 0]}]
> eig[1][1];                                       # Resp. −3 ( Eigenvalue λ_2)
> eig[1][2];                                        # Resp. 2  (Its multiplicity)
> eig[1][3];             # Resp. {[3, 0, 1], [−2, 1, 0]}  (Independent eigenvectors)
> eig[1][3][2];                           # Resp. [−2, 1, 0]  (Second eigenvector)
> eig[2][1];                                        # Resp. 5 ( Eigenvalue λ_1)
> eig[2][3];                               # Resp. {[−1, −2, 1]}  (Its eigenvector)
> eig[2][3][1];                            # Resp. [−1, −2, 1]  (No more braces{})
```

Characteristic matrix, determinant, and polynomial; eigenvalues

```
> charmat(A, lambda);                               # Characteristic matrix
```

$$\begin{bmatrix} \lambda + 2 & -2 & 3 \\ -2 & \lambda - 1 & 6 \\ 1 & 2 & \lambda \end{bmatrix}$$

Maple here gives $\lambda\mathbf{I} - \mathbf{A}$. More common (and used throughout AEM and many other books) is $\mathbf{A} - \lambda\mathbf{I}$. The corresponding characteristic polynomial is $(-1)^n$ times charpoly. Of course, the eigenvalues are the same for the usual and Maple definitions.

```
> charpoly(A, lambda);          # Characteristic polynomial (Maple definition)
```

$$\lambda^3 + \lambda^2 - 21\lambda - 45$$

```
> roots(%, lambda);                    # Eigenvalues and multiplicities
```

$$[[5, 1], [-3, 2]]$$

Similar Material in AEM: p. 374

| EXAMPLE 7.2 | **REAL MATRICES WITH COMPLEX EIGENVALUES**

A *real* matrix may have *complex* eigenvalues. If $a + ib$ (a, b real) is one of them, the conjugate $a - ib$ must also be an eigenvalue because if a polynomial with *real* coefficients has complex roots, the latter must occur in conjugate pairs. A simple case is

```
> with(linalg):                                 # Ignore the warning.
> A := matrix([[0, 1],  [-1, 0]]);
```

$$A := \begin{bmatrix} 0 & 1 \\ -1 & 0 \end{bmatrix}$$

```
> eigenvectors(A);      # Eigenvectors may appear multiplied by some factor.
```

$$[I, 1, \{[1, I]\}], [-I, 1, \{[1, -I]\}]$$

You see that the eigenvalues are i and $-i$, where $\mathtt{I}$ is the Maple notation for $i = \sqrt{-1}$. Each eigenvalue has multiplicity 1 because for an $n \times n$ matrix the sum of the algebraic multiplicities of all the eigenvalues must equal n. Eigenvectors are $[1 \quad i\,]$ and $[1 \quad -i\,]$.

More generally, consider the matrix

```
> A := matrix([[a, b],  [-b, a]]);
```

$$A := \begin{bmatrix} a & b \\ -b & a \end{bmatrix}$$

You obtain the eigenvalue $a + ib$ and its conjugate

```
> eigenvalues(A);                        # Resp. a + Ib, a - Ib
```

For $a = 0$ and $b = 1$ you get the previous matrix and its eigenvalues. If you want to see the eigenvectors, too, type

```
> eigenvectors(A);          # Resp. [a + Ib, 1, {[-I, 1]}], [a - Ib, 1, {[I, 1]}]
```

Hence you get $[-i \quad 1]$ as an eigenvector for the first eigenvalue and $[i \quad 1]$ for the second. Now an eigenvector is determined up to a nonzero constant only. Multiplying the first eigenvector by i and the second by $-i$, you will obtain the previous eigenvectors for the special case $a = 0$, $b = 1$.

This example, solvable almost by inspection, serves to explain relevant Maple commands and to show how the computer will handle them.

Similar Material in AEM: p. 375

| EXAMPLE 7.3 | **ORTHOGONAL MATRICES AND TRANSFORMATIONS**

The main reason for the importance of orthogonal matrices and transformations results from the fact that these matrices preserve the inner product (the dot product), hence also the length of a vector and the angle between two vectors. Show this for vectors in three-space R^3 in the case of the matrix

$$\mathbf{A} = \begin{bmatrix} \dfrac{2}{3} & \dfrac{1}{3} & \dfrac{2}{3} \\[2mm] \dfrac{-2}{3} & \dfrac{2}{3} & \dfrac{1}{3} \\[2mm] \dfrac{1}{3} & \dfrac{2}{3} & \dfrac{-2}{3} \end{bmatrix}.$$

Show that this matrix is orthogonal.

Solution. Type the matrix

```
> with(linalg):                          # Ignore the warning.
> A := matrix([[2/3,  1/3,  2/3],
>              [-2/3, 2/3,  1/3],
>              [1/3,  2/3, -2/3]]);
```

$$A := \begin{bmatrix} \dfrac{2}{3} & \dfrac{1}{3} & \dfrac{2}{3} \\[2mm] \dfrac{-2}{3} & \dfrac{2}{3} & \dfrac{1}{3} \\[2mm] \dfrac{1}{3} & \dfrac{2}{3} & \dfrac{-2}{3} \end{bmatrix}$$

```
> B := inverse(A);
```

$$B := \begin{bmatrix} \dfrac{2}{3} & \dfrac{-2}{3} & \dfrac{1}{3} \\ \dfrac{1}{3} & \dfrac{2}{3} & \dfrac{2}{3} \\ \dfrac{2}{3} & \dfrac{1}{3} & \dfrac{-2}{3} \end{bmatrix}$$

You see that the inverse of **A** is the transpose of **A**. To confirm this, type

```
> evalm(B - transpose(A));                    # The 3 × 3 zero matrix
```

Hence **A** is orthogonal, by definition.

Now type two vectors and their inner product (dot product)

```
> v := [v1, v2, v3]:
> w := [w1, w2, w3]:

> innerprod(v, w);                    # Resp. v1 w1 + v2 w2 + v3 w3
```

Multiply each of the vectors by **A** and take the inner product of the resulting vectors **Av** and **Aw**,

```
> innerprod(A&*v, A&*w);              # Resp. v1 w1 + v2 w2 + v3 w3
```

This shows the invariance of the inner product under our (special) orthogonal transformation.

As a special case this includes the invariance of the **length** of a vector under orthogonal transformations; just set $\mathbf{w} = \mathbf{v}$, then you have the square of the Euclidean norm (the square of the usual length of **v**) in the previous formula, and of **Av** in the present formula.

Finally, show that the **angle** between vectors is preserved under an orthogonal transformation.

```
> angle(v, w);
```

$$\arccos\left(\frac{v1\ w1 + v2\ w2 + v3\ w3}{\sqrt{v1^2 + v2^2 + v3^2}\ \sqrt{w1^2 + w2^2 + w3^2}}\right)$$

```
> angle(A&*v, A&*w):                  # Terrible response not shown

> simplify(%);
```

$$\arccos\left(\frac{v1\ w1 + v2\ w2 + v3\ w3}{\sqrt{v1^2 + v2^2 + v3^2}\ \sqrt{w1^2 + w2^2 + w3^2}}\right)$$

Similar Material in AEM: pp. 382-384

EXAMPLE 7.4   COMPLEX MATRICES

Three classes of complex matrices are of practical interest (in physics, for instance), mainly because of their spectrum. They are complex generalizations of symmetric, skew-symmetric, and orthogonal real matrices. They are defined as follows.

A. A matrix is **Hermitian** if it is equal to its conjugate transpose. Its eigenvalues are real.

B. A matrix is **skew-Hermitian** if it is equal to minus its conjugate transpose. Its eigenvalues are pure imaginary or zero.

C. A matrix is **unitary** if its inverse is equal to its conjugate transpose. Its eigenvalues have absolute value 1.

A. Show that the following matrix **A** is Hermitian and find its spectrum.

```
> with(linalg):                              # Ignore the warning.
> A := matrix([[4,   1-3*I],   [1+3*I,   7]]);
```

$$A := \begin{bmatrix} 4 & 1 - 3I \\ 1 + 3I & 7 \end{bmatrix}$$

```
> AT := transpose(A);
```

$$AT := \begin{bmatrix} 4 & 1 + 3I \\ 1 - 3I & 7 \end{bmatrix}$$

```
> ACT := map(conjugate, AT);
```

$$ACT := \begin{bmatrix} 4 & 1 - 3I \\ 1 + 3I & 7 \end{bmatrix}$$

You see that `map(conjugate, AT)` acts (operates) on each entry of the transpose and transforms it into its complex conjugate entry (of course, with no effect on the real entries).

```
> evalm(A - ACT);          # The 2 × 2 zero matrix shows that A is Hermitian.

> eigenvalues(A);                       # Resp. 9, 2  #  Real eigenvalues
```

B. Show that the following matrix **B** is skew-Hermitian and its eigenvalues are pure imaginary.

```
> B := matrix([[3*I,   2+I],   [-2+I,   -I]]);
```

$$B := \begin{bmatrix} 3I & 2 + I \\ -2 + I & -I \end{bmatrix}$$

```
> BCT := map(conjugate, transpose(B));
```

$$BCT := \begin{bmatrix} -3I & -2 - I \\ 2 - I & I \end{bmatrix}$$

```
> evalm(B + BCT);                              # The 2 × 2 zero matrix
```

This shows that the conjugate transpose BCT of **B** equals minus **B**. Hence **B** is skew-Hermitian, by definition.

```
> eigenvalues(B);                       # Resp. 4I, −2I   #  Imaginary eigenvalues
```

C. Show that the following matrix **C** is unitary and its eigenvalues have absolute value 1.

```
> C := matrix([[I/2,   sqrt(3)/2],   [sqrt(3)/2,   I/2]]);
```

$$C := \begin{bmatrix} \dfrac{1}{2} I & \dfrac{1}{2} \sqrt{3} \\ \dfrac{1}{2} \sqrt{3} & \dfrac{1}{2} I \end{bmatrix}$$

> evalm(C&*map(conjugate, transpose(C)));　　　　　# Resp. $\begin{bmatrix} 1 & 0 \\ 0 & 1 \end{bmatrix}$

So in this last command you have formed the transpose of **C** and then used map to get the conjugate transpose of **C**. For a unitary matrix **C** that conjugate transpose should equal the inverse of **C**, so that multiplication by **C** should produce the unit matrix. This is precisely what has happened.

> eig := eigenvalues(C);

$$eig := \frac{1}{2} I + \frac{1}{2} \sqrt{3}, \ \frac{1}{2} I - \frac{1}{2} \sqrt{3}$$

> abs(eig[1]);　　　　　　　　　　　　　　　　# Resp. 1

eig[1] gives the first eigenvalue, and abs its absolute value. Type abs(eig[2]) to do the same for the second eigenvalue of **C**.

Similar Material in AEM: pp. 385-387

| **EXAMPLE 7.5** | **SIMILARITY OF MATRICES. DIAGONALIZATION** |

A real $n \times n$ matrix **B** is called **similar** to a real $n \times n$ matrix **A** if $\mathbf{B} = \mathbf{P}^{-1} \mathbf{A} \mathbf{P}$ for some nonsingular matrix **P**. Then **A** is similar to **B**, and **A** and **B** are called *similar matrices*. Similar matrices have the same eigenvalues. This accounts for the practical importance of similarity, for instance, in the design of numerical methods for eigenvalues. Furthermore, if $\mathbf{x_1}$ is an eigenvector of **A**, then $\mathbf{y_1} = \mathbf{P}^{-1} \mathbf{x_1}$ is an eigenvector of the similar matrix **B** corresponding to the same eigenvalue. Verify these statements for the matrices **A** and **P** given by

> with(linalg):　　　　　　　　　　　　# Ignore the warning.
> A := matrix([[10, -3, 5],　[0, 1, 0],　[-15, 9, -10]]);

$$A := \begin{bmatrix} 10 & -3 & 5 \\ 0 & 1 & 0 \\ -15 & 9 & -10 \end{bmatrix}$$

> P := matrix([[2, 0, 3],　[0, 1, 0],　[3, 0, 5]]);

$$P := \begin{bmatrix} 2 & 0 & 3 \\ 0 & 1 & 0 \\ 3 & 0 & 5 \end{bmatrix}$$

Solution. In addition to **A** and **P** you need the inverse of **P**, call it **Q**. Then type $\mathbf{B} = \mathbf{P}^{-1} \mathbf{A} \mathbf{P}$.

> Q := inverse(P);

$$Q := \begin{bmatrix} 5 & 0 & -3 \\ 0 & 1 & 0 \\ -3 & 0 & 2 \end{bmatrix}$$

```
> B := evalm(Q&*A&*P);
```

$$B := \begin{bmatrix} 355 & -42 & 560 \\ 0 & 1 & 0 \\ -225 & 27 & -355 \end{bmatrix}$$

Now compare the eigenvalues. Also obtain eigenvectors, to be used afterwards.

```
> eigA := eigenvectors(A);    eigB := eigenvectors(B);
```

$$eigA := [5, 1, \{[-1, 0, 1]\}], [-5, 1, \{[1, 0, -3]\}], [1, 1, \{[1, -2, -3]\}]$$

$$eigB := \left[1, 1, \{\left[\frac{-14}{9}, \frac{2}{9}, 1\right]\}\right], \left[-5, 1, \{\left[\frac{-14}{9}, 0, 1\right]\}\right], \left[5, 1, \{\left[\frac{-8}{5}, 0, 1\right]\}\right]$$

You see that the eigenvalues are the same. They may come out in a different order, so you have to watch a little when you compare eigenvectors. You get the latter for **A**, call them $\mathbf{x_1}$, $\mathbf{x_2}$, $\mathbf{x_3}$, and those for **B**, call them $\mathbf{y_1}$, $\mathbf{y_2}$, $\mathbf{y_3}$, by typing

```
> x1 := eigA[1][3][1];    x2 := eigA[2][3][1];    x3 := eigA[3][3][1];
```

$$x1 := [-1, 0, 1]$$
$$x2 := [1, 0, -3]$$
$$x3 := [1, -2, -3]$$

```
> y1 := evalm(Q&*x1);    y2 := evalm(Q&*x2);    y3 := evalm(Q&*x3);
```

$$y1 := [-8, 0, 5]$$
$$y2 := [14, 0, -9]$$
$$y3 := [14, -2, -9]$$

This agrees with the vectors accessed from eigB , in the corresponding order (third vector, second vector, first vector) and multiplied by a (nonzero) constant factor (up to which an eigenvector is only determined).

Diagonalization of A. If **X** is the matrix with a basis of eigenvectors of **A** as column vectors, then $\mathbf{D} = \mathbf{X}^{-1}\mathbf{A}\mathbf{X}$ is diagonal with the eigenvalues as diagonal entries. Indeed, write **C** for **D** (which is protected, that is, it must not be used here) and type

```
> X := augment(x1, x2, x3);
```

$$X := \begin{bmatrix} -1 & 1 & 1 \\ 0 & 0 & -2 \\ 1 & -3 & -3 \end{bmatrix}$$

```
> C := evalm(inverse(X)&*A&*X);
```

$$C := \begin{bmatrix} 5 & 0 & 0 \\ 0 & -5 & 0 \\ 0 & 0 & 1 \end{bmatrix}$$

Similar Material in AEM: pp. 392-394, 397 (#5)

Problem Set for Chapter 7

Pr.7.1 (Symmetric and skew-symmetric matrices) Write a program for representing a square matrix as the sum of a symmetric and a skew-symmetric matrix and apply the program to the matrix in Example 7.1 in this Guide. (*AEM Ref.* p. 381)

Pr.7.2 (Polynomial matrix) A *polynomial matrix* is a matrix of the form

$$q(\mathbf{A}) = a_0\mathbf{I} + a_1\,\mathbf{A} + a_2\,\mathbf{A}^2 + \ldots + a_m\,\mathbf{A}^m,$$

where $\mathbf{A}$ is any $n \times n$ matrix and $\mathbf{I}$ is the $n \times n$ unit matrix. If λ is an eigenvalue of $\mathbf{A}$, then $q(\lambda)$ is an eigenvalue of $q(\mathbf{A})$. This is called the **spectral mapping theorem**. Verify it for the matrix $\mathbf{A}$ in Example 7.5 in this Guide and the polynomial $q(x) = x^3 + 4\,x^2 - 10\,x + 8$. (*AEM Ref.* p. 381)

Pr.7.3 (Orthogonal transformation, rotation) Show that the matrix $\mathbf{A}$ is orthogonal and the orthogonal transformation $\mathbf{y} = \mathbf{Ax}$ leaves the inner product (the dot product) invariant, where $\mathbf{A}$ is as follows. (*AEM Ref.* pp. 381-383)

$$\mathbf{A} = \begin{bmatrix} \cos\theta & -\sin\theta \\ \sin\theta & \cos\theta \end{bmatrix}$$

Pr.7.4 (Cayley's theorem) Cayley's theorem states that every square matrix $\mathbf{A}$ satisfies its characteristic equation, that is, $p(\mathbf{A}) = \mathbf{0}$, where $p(\lambda)$ is the characteristic polynomial of $\mathbf{A}$. Verify this for the matrix $\mathbf{A}$ in Example 7.1 in this Guide. (*AEM Ref.* p. 373)

Pr.7.5 (Experiment on Perron's theorem) Perron's famous theorem states that a square matrix $\mathbf{A}$ all of whose entries are positive has a real eigenvalue of multiplicity 1 which is greater than the absolute value of any other eigenvalue of $\mathbf{A}$, and a corresponding real eigenvector whose components are all positive. Verify this theorem for matrices of your own choice. (*AEM Ref.* p. 381. A proof of the theorem is not easy.)

Pr.7.6 (Complex matrix) Show that the matrix with rows $[2 \quad 1 - i]$ and $[1 + i \quad 3]$ is Hermitian and find its spectrum. (*AEM Ref.* p. 385)

Pr.7.7 (Hermitian and skew-Hermitian matrices) Represent the matrix $\mathbf{A}$ as the sum of a Hermitian and a skew-Hermitian matrix and find the eigenvalues and eigenvectors of $\mathbf{A}$, where $\mathbf{A}$ is as follows. (*AEM Ref.* p. 385)

$$\mathbf{A} = \begin{bmatrix} 4 + 12i & -12 - 12i & 12 - 12i \\ -6 + 6i & 10 - 6i & 6 + 6i \\ 6 + 6i & -6 + 6i & -2 - 6i \end{bmatrix}$$

Pr.7.8 (Complex matrix) Let $\mathbf{A}$ be the 3×3 matrix with main diagonal entries 0 and all other entries i. Show that $\mathbf{A}$ is skew-Hermitian and find its spectrum. (Note the multiplicities!) Find the inverse and show that $\mathbf{A} = \mathbf{I} - 2\mathbf{A}^{-1}$, where $\mathbf{I}$ is the 3×3 unit matrix. (*AEM Ref.* p. 385)

Pr.7.9 (Experiment on unitary matrices) Consider the powers $\mathbf{C}^m$ (m a positive integer) of the matrix $\mathbf{C}$ in Example 7.4 in this Guide, their eigenvalues, and their inverses. Show that the eigenvalues have absolute value 1. Verify that the inverses are unitary. For what values of m will $\mathbf{C}^m$ be real? For what m will the eigenvalues be 1, 1? Verify that $\mathbf{C}$ satisfies Cayley's theorem (see Pr.7.4). Find other unitary matrices and consider them in a similar fashion. (*AEM Ref.* pp. 385-387)

Pr.7.10 (Similar matrices) Find and discuss the relations between the eigenvalues and eigenvectors of $\mathbf{A}$ and $\mathbf{B} = \mathbf{P}^{-1}\mathbf{A}\mathbf{P}$, where $\mathbf{A}$ and $\mathbf{P}$ are as follows.
(*AEM Ref.* p. 397 (#3))

$$\mathbf{A} = \begin{bmatrix} 3 & 4 \\ 4 & -3 \end{bmatrix} \qquad \text{and} \qquad \mathbf{P} = \begin{bmatrix} -4 & 2 \\ 3 & -1 \end{bmatrix}$$

Pr.7.11 (Similarity transformation) Show that $\mathbf{A}$ and $\mathbf{B} = \mathbf{P}^{-1}\mathbf{A}\mathbf{P}$ have the same eigenvalues and establish the relation between their eigenvectors, where $\mathbf{A}$ and $\mathbf{P}$ are as follows. (*AEM Ref.* p. 392)

$$\mathbf{A} = \begin{bmatrix} -1 & -3 & 3 \\ -6 & 2 & 6 \\ -3 & 3 & 5 \end{bmatrix} \qquad \text{and} \qquad \mathbf{P} = \begin{bmatrix} 3 & -1 & 1 \\ -15 & 6 & -5 \\ 5 & -3 & 2 \end{bmatrix}$$

Pr.7.12 (Transformation to diagonal form) Show that $\mathbf{D} = \mathbf{X}^{-1}\mathbf{A}\mathbf{X}$ is diagonal with the eigenvalues of $\mathbf{A}$ as diagonal entries. Here $\mathbf{X}$ is a matrix with eigenvectors of $\mathbf{A}$ as column vectors, and $\mathbf{A}$ is as follows. (*AEM Ref.* pp. 392-394)

$$\mathbf{A} = \begin{bmatrix} 5 & 4 \\ 1 & 2 \end{bmatrix}$$

Chapter 8

Vectors in R^2 and R^3.
Dot and Cross Products.
Grad, Div, Curl

Content. Addition of vectors, scalar multiplication (Ex. 8.1, Prs. 8.1-8.4)
Inner product, cross product, triple product (Ex. 8.2, Prs. 8.5-8.13)
Vector fields (Pr. 8.14)
Derivatives, curves (Ex. 8.3, Prs. 8.15-8.18)
Gradient, divergence, curl (Exs. 8.4, 8.5, Prs. 8.19-8.25)

linalg package. Load it by typing `with(linalg):` You will need it.
For vectors in R^2 and R^3 various commands are those for vectors in R^n in Chap. 6, for instance, `innerprod(u,v)`. New are `angle(u, v)`, `crossprod(u, v)`, and

`grad(f(x,y,z), [x, y, z])`	(Ex. 8.4)
`diverge(v, [x, y, z])`	(Ex. 8.5)
`curl(v, [x, y, z])`	(Ex. 8.5)
`potential(v, [x, y, z], 'f')`	(Ex. 8.4)
`laplacian(f, [x, y, z])`	(Ex. 8.5)

Examples for Chapter 8

EXAMPLE 8.1	VECTORS, LENGTH, ADDITION, SCALAR MULTIPLICATION

In the applications in this chapter, vectors **v** are often given as arrows from an initial point P to a terminal point Q, where, for instance,

```
> P := [3, 1, 4];   Q := [1, -2, 4];
```

$$P := [3, 1, 4]$$
$$Q := [1, -2, 4]$$

Use brackets [..], not parentheses (..), not braces {..}. (Try the latter; they are for sets; order may change and numbers may be omitted. Try $\{1, 1, 1\}$.)
The vector **v** has the components

```
> v := Q - P;                          # Resp. v := [−2, −3, 0]
```

and the length

```
> with(linalg):                        # Ignore the warning.
> lenghtv := sqrt(innerprod(v, v));    # Resp. lenghtv := √13
```

Vector addition and scalar multiplication proceed as in Chap. 6. For instance, let (with **v** as before)

```
> w := [5, -1, 3];                          # Resp. w := [5, −1, 3]
> v + w;   -4*v;   0.5*w;
```

$$[3, -4, 3]$$
$$[8, 12, 0]$$
$$[2.5, -.5, 1.5]$$

Similar Material in AEM: pp. 401-406

EXAMPLE 8.2   **INNER PRODUCT. CROSS PRODUCT**

Inner products (dot products) have already occurred in Chap. 6 (see Example 6.1 in this Guide). In 2- and 3-space they can be motivated by the **work done by a force** in a displacement. For instance, find the work W done by the force $\mathbf{p} = [2, 1, 4]$ in the displacement from $A : (1, 1, 0)$ to $B : (2, 1, 10)$. Also find the angle between $\mathbf{p}$ and a vector in the direction of the displacement.

Solution. W is defined by $W = \mathbf{p} \bullet \mathbf{d}$, where $\mathbf{d}$ is the "displacement vector" from A to B; thus,

```
> with(linalg):                             # Ignore the warning.
> A := [1, 1, 0];    B := [2, 1, 10];    p := [2, 1, 4];
```

$$A := [1, 1, 0]$$
$$B := [2, 1, 10]$$
$$p := [2, 1, 4]$$

```
> d := B - A;                               # Resp. d := [1, 0, 10]
> W := innerprod(p, d);                     # Resp. W := 42
```

For the **angle** you have $\mathbf{p} \bullet \mathbf{d} = \|\mathbf{p}\|\|\mathbf{d}\| \cos \theta$, hence $\theta = \arccos(\mathbf{p} \bullet \mathbf{d}/(\|\mathbf{p}\|\|\mathbf{d}\|))$. Hence type

```
> theta := arccos(innerprod(p, d)/(norm(p, 2)*norm(d, 2)));
```

$$\theta := \arccos\left(\frac{2}{101}\sqrt{21}\sqrt{101}\right)$$

```
> angle(p, d);                              # For confirmation of the result
```

$$\arccos\left(\frac{2}{101}\sqrt{21}\sqrt{101}\right)$$

```
> evalf(convert(%, degrees), 4);            # "4" gives 4 digits.
```

$$24.21\ degrees$$

Cross products $\mathbf{v} = \mathbf{a} \times \mathbf{b}$ are vectors $\mathbf{v}$ perpendicular to $\mathbf{a}$ and $\mathbf{b}$ and of length $\|\mathbf{v}\|$ equal to the area of the parallelogram with $\mathbf{a}$ and $\mathbf{b}$ as adjacent sides. (If $\mathbf{a}$ and $\mathbf{b}$ are parallel or at least one of them is $\mathbf{0}$, then $\mathbf{v} = \mathbf{0}$ by definition.) Hence, among other applications, cross products can be used to calculate the area of a triangle when its vertices are given, say $A : (1, 1, 1)$, $B : (5, 2, 3)$, $C : (-1, 4, 5)$.

Solution. . You need two vectors **a** and **b** representing two sides of the triangle, say, AB and AC. Accordingly, type

```
> A := [1, 1, 1];   B := [5, 2, 3];   C := [-1, 4, 5];
> a := B - A;       b := C - A;
```

$$a := [4, 1, 2]$$
$$b := [\ 2, 3, 4]$$

The command for the cross product is `crossprod(a, b)` . To obtain its length, type `norm(..., 2)` , and the area of the triangle is $1/2$ of this length. Thus,

```
> v := crossprod(a, b);                    # Resp. v := [-2, -20, 14]
> answer := evalf(norm(%, 2)/2, 4);        # Area of the triangle; 4 digits
```

$$answer := 12.25$$

Recalling that a cross product can be written as a symbolical determinant whose first row is **i, j, k** (unit vectors in the positive directions of the coordinate axes) and whose second and third rows are the two vectors, you can check your cross product by typing

```
> r := [i, j, k]:
> C := transpose(augment(r, a, b));
```

$$C := \begin{bmatrix} i & j & k \\ 4 & 1 & 2 \\ -2 & 3 & 4 \end{bmatrix}$$

```
> det(C);                 # Resp. -2 i - 20 j + 14 k       Cross product [-2, -20, 14]
```

Similar Material in AEM: p. 410

EXAMPLE 8.3 | **DIFFERENTIATION OF VECTORS.**
CURVES AND THEIR PROPERTIES

Investigate the main geometrical properties of the **helix**

$$\mathbf{r}(t) = [a\cos t,\ a\sin t,\ ct],$$

which lies on a cylinder of radius a, is right-handed if $c > 0$, and has pitch $2\pi c$.

Solution. You will need the first two derivatives and their inner products (dot products). Type

```
> with(linalg):                            # Ignore the warning.
> r := [a*cos(t), a*sin(t), c*t];          # Gives nonsense result.
```

$$r := [[4,\ 1,\ 2]\cos(t),\ [4,\ 1,\ 2]\sin(t),\ ct]$$

```
> a := 'a':                  # Unassign a , used as a vector in Example 8.2.
> r := [a*cos(t), a*sin(t), c*t];          # Resp. r := [a\cos(t), a\sin(t), ct]
> r1 := diff(r, t);   r2 := diff(r1, t);
```

$$r1 := [-a\sin(t),\ a\cos(t),\ c]$$
$$r2 := [-a\cos(t),\ -a\sin(t),\ 0]$$

```
> i11 := innerprod(r1, r1);    i12 := innerprod(r1, r2);
  i22 := innerprod(r2, r2);
```

$$i11 := a^2 \sin(t)^2 + a^2 \cos(t)^2 + c^2$$
$$i12 := 0$$
$$i22 := a^2 \cos(t)^2 + a^2 \sin(t)^2$$

```
> i22 := simplify(%);                                    # Resp. i22 := a²
```

Arc length. The arc length s is the integral of the square root of `i11`,

```
> s := int(sqrt(simplify(i11)), t);            # Resp. s := √(c² + a²) t
```

Curvature. The curvature κ of a curve represented by $\mathbf{r} = \mathbf{r}(t)$ is given by

```
> kappa := (i11*i22 - i12^2)^(1/2)/i11^(3/2);
```

$$\kappa := \frac{\sqrt{(a^2 \sin(t)^2 + a^2 \cos(t)^2 + c^2)\, a^2}}{(a^2 \sin(t)^2 + a^2 \cos(t)^2 + c^2)^{(3/2)}}$$

```
> assume(a, positive);    assume(c, positive);
> simplify(kappa);
```

$$\frac{a^\sim}{a^{\sim 2} + c^{\sim 2}}$$

Type `?sqrt` to understand what Maple is doing here; taking the square root was conditional on the radicand's being positive. Tildes indicate that a and c are conditioned to be positive. (Without this condition, Maple would not reduce $\sqrt{x^2}$ to x!)

Torsion. You will need the third derivative of $\mathbf{r}(t)$. The numerator of the torsion τ is the determinant of the matrix with $\mathbf{r}', \mathbf{r}'', \mathbf{r}'''$ as rows (or columns), and the denominator is $i11 * i22 - i12^2$. Hence type

```
> r3 := diff(r2, t);
```

$$r3 := [a^\sim \sin(t),\, -a^\sim \cos(t),\, 0]$$

```
> tau := det(augment(r1, r2, r3))/(i11*i22 - i12^2);
```

$$\tau := \frac{c^\sim(a^{\sim 2} \cos(t)^2 + a^{\sim 2} \sin(t)^2)}{(a^{\sim 2} \sin(t)^2 + a^{\sim 2} \cos(t)^2 + c^{\sim 2})\, a^{\sim 2}}$$

```
> simplify(%);
```

$$\frac{c^\sim}{a^{\sim 2} + c^{\sim 2}}$$

The command `augment` has occurred before; here it is again called for composing a matrix from given vectors to be used as column vectors.

    Note that the helix has the remarkable property that both its curvature and its torsion are constant.

    If you want to plot the helix, you have to choose specific values for a and c, for instance, $a = 2$ and $c = 1$. The optional `axes = NORMAL` introduces coordinate axes in space. The optional `numpoints = 400` makes the curve smooth. To see this, try without. Move the curve in space by clicking on the figure and 'dragging' it.

```
> r0 := [2*cos(t), 2*sin(t), t]:
> with(plots):                                    # Ignore the warning.
```

```
> spacecurve([r0[1], r0[2], r0[3], t = 0..6*Pi], axes = NORMAL,
  numpoints = 400, orientation = [45, 70]);
```

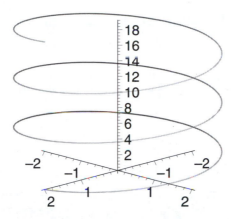

Example 8.3. Helix on a cylinder of radius 2

Similar Material in AEM: pp. 428-435, 440-442

| EXAMPLE 8.4 | **GRADIENT. DIRECTIONAL DERIVATIVE. POTENTIAL** |

Gradient. Type `?grad` for information. The command to compute the gradient is `grad(f(x,y,z), [x, y, z])`. For instance, type

```
> with(linalg):                            # Ignore the warning.
> f := 2*x^2 + 3*y^2 + z^2;                # Resp. f := 2x² + 3y² + z²
> v := grad(f, [x, y, z]);                 # Resp. v := [4x, 6y, 2z]
```

$$\text{\# Resp. } f := 2\,x^2 + 3\,y^2 + z^2$$
$$\text{\# Resp. } v := [4\,x,\, 6\,y,\, 2\,z]$$

Main applications of the gradient occur in connection with directional derivatives, surface normals, and potentials, as we shall now see.

Directional derivative. The directional derivative of a function f at a point P in the direction of a vector **a** can be expressed in terms of the gradient by

$$\mathrm{D}_a f = (1/\|\mathbf{a}\|)\mathbf{a} \bullet \operatorname{grad} f.$$

For instance, find the directional derivative of the above f at the point (2, 1, 3) in the direction of the vector [1, 0, -2].

Solution. Type **a** and a corresponding unit vector **b**, then the directional derivative, call it `deriv`. Finally take the value of the derivative at the given point.

```
> a := [1, 0, -2]; b := a/norm(a, 2);
```

$$a := [1,\, 0,\, -2]$$

$$b := \frac{1}{5}\,[1,\, 0,\, -2]\,\sqrt{5}$$

```
> deriv := innerprod(b, v);                # v = grad f; see above.
```

$$deriv := \frac{4}{5}\,\sqrt{5}\,x - \frac{4}{5}\,\sqrt{5}\,z$$

```
> subs(x = 2, y = 1, z = 3, deriv);
```
$\qquad$ # Resp. $-\dfrac{4}{5}\sqrt{5}$

```
> evalf(%, 4);
```
$\qquad$ # Resp. -1.789

Hence f is decreasing in the direction of **a**.

Surface normal. The gradient grad f is a normal vector to a level surface $f = const$ of a function $f(x, y, z)$, that is, if grad f is not the zero vector, it is perpendicular to the surface $f = const$ passing through the point considered. For instance, find a normal vector to the cone $z^2 = 4\,(x^2 + y^2)$ at the point $(1, 0, 2)$.

Solution. The cone S is the level surface $f = 0$ of the function

```
> f := 4*x^2 + 4*y^2 - z^2;
```
$\qquad$ # Resp. $f := 4\,x^2 + 4\,y^2 - z^2$

Thus, a normal vector of S is $\mathbf{N} = \mathrm{grad}\,f$,

```
> N := grad(f, [x, y, z]);
```
$\qquad$ # Resp. $N := [8\,x,\ 8\,y,\ -2\,z]$

```
> NP := subs(x = 1, y = 0, z = 2, [N[1], N[2], N[3]]);
```

$$NP := [8,\ 0,\ -4]$$

Hence a unit normal vector of S at the point $(1, 0, 2)$ is (multiply by the reciprocal length of `NP`)

```
> n := 1/norm(NP, 2)*NP;
```

$$n := \frac{1}{20}\,\sqrt{5}\,[8,\ 0,\ -4]$$

Potential. A potential is a scalar function f associated with a vector function **v** such that $\mathbf{v} = \mathrm{grad}\,f$. Not every **v** has a potential. If a **v** does, it is generally more convenient to work with a single scalar function (the potential) than with the triple of component functions of **v**. Consider the following basic example. Begin by typing the distance r of a point (x, y, z) from the origin. Then type **v**. Note that $\|\mathbf{v}\| = c/r^2$, as it appears in Newton's law of gravitation or in Coulomb's law of attraction or repulsion of electrically charged particles, where c is a constant. Then ask whether **v** has a potential. It does, and you will obtain it.

```
> r := sqrt(x^2 + y^2 + z^2);
```
$\qquad$ # Distance of (x, y, z) from the origin

$$r := \sqrt{x^2 + y^2 + z^2}$$

```
> v := [-c*x/r^3, -c*y/r^3, -c*z/r^3];
```

$$v := \left[-\frac{c\,x}{(x^2 + y^2 + z^2)^{(3/2)}},\ -\frac{c\,y}{(x^2 + y^2 + z^2)^{(3/2)}},\ -\frac{c\,z}{(x^2 + y^2 + z^2)^{(3/2)}} \right]$$

```
> potential(v, [x, y, z], 'f');
```
$\qquad$ # Resp. $true$

```
> f;
```

$$\frac{c}{\sqrt{x^2 + y^2 + z^2}}$$

This is the potential at (x, y, z) of a point mass or electrical point charge located at the origin.

     Similar Material in AEM: pp. 446-452

EXAMPLE 8.5   **DIVERGENCE, LAPLACIAN, CURL**

The command `diverge(v, [x, y, z])` is used to obtain the **divergence** div **v** of a vector function $\mathbf{v}(x, y, z) = [v_1(x, y, z),\ v_2(x, y, z),\ v_3(x, y, z)]$.

```
> with(linalg):                          # Ignore the warning.
> v := [v1, v2, v3];                     # Resp. v := [v1, v2, v3]
> div := diverge(v(x,y,z), [x, y, z]);
```

$$div := \left(\frac{\partial}{\partial x}\, v1(x,\, y,\, z)\right) + \left(\frac{\partial}{\partial y}\, v2(x,\, y,\, z)\right) + \left(\frac{\partial}{\partial z}\, v3(x,\, y,\, z)\right)$$

For example, find the divergence of the following vector function **w**:

```
> x*y*z*[x, y, z];                       # Resp. x y z [x, y, z]
> w := expand(%);                        # Resp. w := [x² y z, x y² z, x y z²]
> divw := diverge(w, [x, y, z]);         # Resp. divw := 6 x y z
```

Laplacian. The Laplacian of a function f referred to Cartesian coordinates is

$$\nabla^2 f = \mathrm{div}\,(\mathrm{grad}\, f) = f_{xx} + f_{yy} + f_{zz}.$$

For instance,

```
> f := x*y/z:
> lapf := diff(f,x,x) + diff(f,y,y) + diff(f,z,z);   # Resp. lapf := 2 (x y)/z³
```

or more quickly,

```
> laplacian(f, [x, y, z]);               # Resp. 2 (x y)/z³
```

The Laplacian of a function f equals the divergence of grad f. Indeed,

```
> f := 'f':                              # Remove the special assignment from f.
> diverge(grad(f(x,y,z), [x, y, z]), [x, y, z]);
```

$$\left(\frac{\partial^2}{\partial x^2}\, f(x,\, y,\, z)\right) + \left(\frac{\partial^2}{\partial y^2}\, f(x,\, y,\, z)\right) + \left(\frac{\partial^2}{\partial z^2}\, f(x,\, y,\, z)\right)$$

Curl. The command `curl(v, [x, y, z])` is used to obtain the curl of a vector function **v**. The curl plays a role in connection with rotations. A **rotation** can be described by a rotation vector **w**, whose direction is that of the axis of rotation and whose length equals the angular velocity. Then the velocity **v** at a point P with position vector **r** (whose initial point lies on the axis of rotation) is

$$\mathbf{v} = \mathbf{w} \times \mathbf{r}.$$

```
> w := [w1, w2, w3]:   r := [x, y, z]:
> v := crossprod(w, r);
```

$$v := [w2\, z - w3\, y,\ w3\, x - w1\, z,\ w1\, y - w2\, x]$$

If you take the curl of **v**, you obtain

```
> curl(v, [x, y, z]);                    # Resp. [2 w1, 2 w2, 2 w3]
```

This proves that the curl of the velocity vector of the rotation equals twice the rotation vector. This is a basic relation on rotations which characterizes the nature of the curl in this connection.

If a vector field **v** has a potential f so that $\mathbf{v} = \operatorname{grad} f$, then **v** is **irrotational**, that is, its curl is the zero vector. To prove this, type

```
> f := 'f':                          # Unassign f , i.e., make it arbitrary again.

> v := grad(f, [x, y, z]);              # Resp. v := [0, 0, 0]   # Nonsense
> v := grad(f(x,y,z), [x, y, z]);
```

$$v := \left[\frac{\partial}{\partial x} f(x, y, z), \frac{\partial}{\partial y} f(x, y, z), \frac{\partial}{\partial z} f(x, y, z) \right]$$

```
> curl(v, [x, y, z]);                                    # Resp. [0, 0, 0]
```

Similarly, the divergence of the curl of **v** is zero. Indeed,

```
> v := 'v':

> diverge(curl([v1, v2, v3], [x, y, z]), [x, y, z]);            # Resp. 0
```

Similar Material in AEM: pp. 453-458

Problem Set for Chapter 8

Pr.8.1 (Components, length) Find the components and length of the vector **v** with initial point (1, 2, 3) and terminal point (2, 4, 6). (*AEM Ref.* p. 407 (#3))

Pr.8.2 (Addition, scalar multiplication) Find $4\mathbf{a} + 8\mathbf{c}$ and $4(\mathbf{a} + 2\mathbf{c})$, where **a** = [3, -2, 1] and **c** = [4, 1, -1]. (*AEM Ref.* p. 407 (#19))

Pr.8.3 (Resultant force) Find the resultant of the forces **p** = [4, -2, -3], **q** = [8, 8, 1], **u** = [-12, -6, 2]. (*AEM Ref.* p. 407 (#25))

Pr.8.4 (Equilibrium) Find **p** such that **p**, **q** = [3, 2, 0], and **u** = [-2, 4, 0] are in equilibrium. Sketch these forces. (*AEM Ref.* p. 407 (#29))

Pr.8.5 (Inner product) Find $2\mathbf{b} \bullet 5\mathbf{c}$ and $10\mathbf{b} \bullet \mathbf{c}$, where **b** = [2, 0, -5] and **c** = [4, -2, 1]. (*AEM Ref.* p. 413 (#5))

Pr.8.6 (Angle) Find the angle between **b** and **c** in Pr.8.5 in two ways, (a) by using inner products, (b) by using the command `angle`. (*AEM Ref.* p. 409)

Pr.8.7 (Work) Find the work done by the force **p** = [2, 6, 6] in the displacement from the point $A : (3, 4, 0)$ to the point $B : (5, 8, 0)$. Sketch the vectors involved. (*AEM Ref.* p. 413 (#13))

Pr.8.8 (Vector product) Find $\mathbf{a} \times \mathbf{c}$, $\mathbf{c} \times \mathbf{a}$, $|\mathbf{a} \times \mathbf{c}|$, $|\mathbf{c} \times \mathbf{a}|$, $\mathbf{a} \bullet \mathbf{c}$, where **a** = [1, 2, 0] and **c** = [2, 3, 4]. (*AEM Ref.* p. 421 (#3))

Pr.8.9 (Scalar triple product) Find $(\mathbf{a} \times \mathbf{b}) \bullet \mathbf{c}$ and $\mathbf{a} \bullet (\mathbf{b} \times \mathbf{c})$, where **b** = [−3, 2, 0] and **a** and **c** are as in Pr.8.8. (*AEM Ref.* p. 421 (#11))

Pr.8.10 (Tetrahedron) Find the volume of the tetrahedron with vertices (1,3,6), (3,7,12), (8,8,9), (2,2,8). (*AEM Ref.* p. 422 (#34))

Pr.8.11 **(Linear independence)** Are the vectors [3, 5, 9], [73, −56, 76], [−4, 7, −1] linearly independent? (*AEM Ref.* p. 422 (#37))

Pr.8.12 **(Plane)** Find the plane through the points (1, 2, 1/4), (4, 2, -2), (0, 8. 4). (*AEM Ref.* p. 422 (#31))

Pr.8.13 **(Length of cross product)** Show that for any vectors **a** and **b** the length of **a** × **b** equals the square root of $(\mathbf{a} \bullet \mathbf{a})(\mathbf{b} \bullet \mathbf{b}) - (\mathbf{a} \bullet \mathbf{b})^2$ (*AEM Ref.* p. 422 (#38))

Pr.8.14 **(Vector field)** Plot the vector field $\mathbf{v} = [y^2, 1]$. (Type `?fieldplot`.) (*AEM Ref.* p. 427 (#19))

Pr.8.15 **(Length of a curve)** Find the length of the **catenary** $\mathbf{r} = [t, \cosh t]$ from $t = 0$ to $t = 1$. Plot this portion of the curve. (*AEM Ref.* p. 434 (#27))

Pr.8.16 **(Lissajous curves)** Plot the special Lissajous curve $[\cos t, \sin 3t]$ for $t = 0 \ldots 2\pi$ (named after the French physicist J. A. Lissajous, 1822-1880). First guess what the curve might look like. (*AEM Ref.* Sec. 8.5)

Pr.8.17 **(Torsion)** Find the torsion $\tau = (\mathbf{r}' \ \mathbf{r}'' \ \mathbf{r}''')/[(\mathbf{r}' \bullet \mathbf{r}')(\mathbf{r}'' \bullet \mathbf{r}'') - (\mathbf{r}' \bullet \mathbf{r}'')^2]$ of the curve $\mathbf{r}(t) = [t, \ t^2, \ t^3]$. (*AEM Ref.* p. 443 (#15). See also Example 8.3 in this Guide.)

Pr.8.18 **(Tangential acceleration)** Find the tangential acceleration of the curve $\mathbf{r}(t) = [\cos t, \ \sin 2t, \ \cos 2t]$. Plot the curve. (*AEM Ref.* p. 439 (#8c))

Pr.8.19 **(Gradient)** Find the gradient of the function $f(x, y) = \ln(x^2 + y^2)$ and its value at the point (2, 0). Plot the gradient field in the first quadrant near the origin, say, for x and y from 0.1 to 0.5. (*AEM Ref.* p. 452 (#3))

Pr.8.20 **(Directional derivative)** Find the directional derivative of the function in Pr.8.19 at the point (4, 0) in the direction of the vector $\mathbf{a} = [1, -1]$. (*AEM Ref.* p. 453 (#32))

Pr.8.21 **(Surface normal)** Find a unit normal vector of the cone $z^2 = x^2 + y^2$ at the point (3, 4, 5). (*AEM Ref.* p. 450)

Pr.8.22 **(Potential)** Does $\mathbf{v} = [2xy \cos(yz), \ x^2 \cos(yz) - x^2 yz \sin(yz), -x^2 y^2 \sin(yz)]$ have a potential? If so, find it. (*AEM Ref.* p. 450)

Pr.8.23 **(Divergence)** Find the divergence of $(x^2 + y^2)^{-1}[-y, x]$. (*AEM Ref.* p. 456 (#5))

Pr.8.24 **(Divergence)** Prove $\text{div}(f\mathbf{v}) = f \text{div} \mathbf{v} + \mathbf{v} \bullet \text{grad} f$. (*AEM Ref.* p. 456 (#13))

Pr.8.25 **(Curl)** Illustrate the important relations $\text{curl}(\text{grad} f) = \mathbf{0}$ and $\text{div}(\text{curl} \mathbf{v}) = 0$ by examples of your own and list some typical physical applications of these relations. (*AEM Ref.* p. 458)

Chapter 9

Vector Integral Calculus. Integral Theorems

Content. Line integrals (Exs. 9.1, 9.2, Prs. 9.1-9.4)
Double integrals, Green's theorem (Exs. 9.3, 9.4, Prs. 9.5-9.9)
Surface integrals, Gauss and Stokes theorems
(Exs. 9.5-9.7, Prs. 9.10-9.13, 9.16-9.20)
Triple integrals (Prs. 9.14, 9.15)

`int(fC, t = a..b)` evaluates line integrals converted to integrals over the parameter interval `t = a..b` of the path of integration C (see Ex. 9.1). `int(int(...))` evaluates double integrals (Ex. 9.3). `diff(F, x)`, `diff(F, y)` give partial derivatives (Ex. 9.4).

Examples for Chapter 9

EXAMPLE 9.1 | LINE INTEGRALS

A **line integral** of a function f over a curve $C : \mathbf{r}(t)$, $a \le t \le b$, in space or in the plane (called the **path of integration**) can be defined by

$$\int_C f(\mathbf{r})\, d\mathbf{r} = \int_a^b f(\mathbf{r}(t))\, dt.$$

For example, let C be the helix $\mathbf{r}(t) = [\cos t,\ \sin t,\ 3t]$, $a = 0$, $b = 2\pi$, and

$$f(\mathbf{r}) = f(x, y, z) = (x^2 + y^2 + z^2)^2.$$

Then type

```
> r := [cos(t), sin(t), 3*t];
> f := (x^2 + y^2 + z^2)^2;           # Resp. f := (x² + y² + z²)²
```

Substituting $x = \cos t$, $y = \sin t$, $z = 3t$ gives $f(\mathbf{r}(t))$, that is, f on C. Denote this by `fC`. Accordingly, type (`r[1]` being the first component of `r`, etc.)

```
> subs(x = r[1], y = r[2], z = r[3], f);   # Resp. (cos(t)² + sin(t)² + 9 t²)²
> fC := simplify(%);                        # Use cos² t + sin² t = 1.
```

$$fC := 81\, t^4 + 18\, t^2 + 1$$

```
> answer := int(fC, t = 0..2*Pi);   # Resp. answer := 2592/5 π⁵ + 48 π³ + 2 π
> evalf(%,7);                         # Resp. 160135.3
```

Work integrals. This is a practical (nonstandard) name for another very useful kind of line integral in which a vector function $\mathbf{F}(\mathbf{r})$ is given and one takes and integrates

96

the tangential component of $\mathbf{F}(\mathbf{r})$ in the tangent direction of the path of integration C. Hence this is the work done by a force $\mathbf{F}$ in a displacement of a body along C. In terms of a formula,

$$\int_C \mathbf{F}(\mathbf{r}) \bullet d\mathbf{r} = \int_a^b \mathbf{F}(\mathbf{r}(t)) \bullet \frac{d\mathbf{r}}{dt}\,dt.$$

For instance, let $\mathbf{F}(\mathbf{r}) = [z,\ x,\ y]$, and let C and $\mathbf{r}(t)$ be as before. Denote $\mathbf{F}(\mathbf{r}(t))$ by FC and $\mathbf{r}' = d\mathbf{r}/dt$ by r1. Then

```
> F := [z, x, y];                              # Resp. F := [z, x, y]
```

```
> r1 := diff(r, t);                            # Resp. r1 := [−sin(t), cos(t), 3]
```

```
> FC := eval(subs(x = r[1], y = r[2], z = r[3], [F[1], F[2], F[3]]));
```

$$FC := [3\,t,\ \cos(t),\ \sin(t)]$$

In this command, x = r[1] is the first component $\cos t$ of the position vector $\mathbf{r}$, etc. Similarly, F[1] is the first component of $\mathbf{F}$, namely, z, etc. To compute the inner product (the dot product), you need the linalg package.

```
> with(linalg):                                # Ignore the warning.
```

```
> integrand := innerprod(FC, r1);
```

$$integrand := -3\,t\,\sin(t) + \cos(t)^2 + 3\,\sin(t)$$

```
> answer := int(integrand, t = 0..2*Pi);       # Resp. answer := 7 π
```

```
> with(plots):                                 # Ignore the warning.
```

```
> spacecurve(r, t = 0..2*Pi, axes = NORMAL);
```

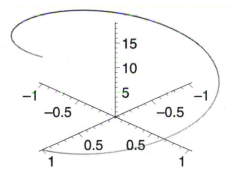

Example 9.1. Path of integration C (circular helix)

Similar Material in AEM: pp. 465-469

| EXAMPLE 9.2 | **INDEPENDENCE OF PATH** |

A line integral from a given point A to a given point B will generally depend on the path along which you integrate from A to B. A line integral is called **independent of path in a domain** D in space if for every pair of endpoints A, B in D the integral has the same value for all paths in D that begin at A and end at B.

For instance, show that the work integral of $\mathbf{F} = [2x,\ 2y,\ 4z]$ is independent of path in any domain D in space and find its value if you integrate from (0,0,0) to (2,2,2).

Solution. Necessary and sufficient for path independence in D is that $\mathbf{F}$ (with continuous coefficients) have a potential f in D, that is, that $\mathbf{F}$ be the gradient of a function f in D. Thus type

```
> with(linalg):                                    # Ignore the warning.
> F := [2*x, 2*y, 4*z];
> potential(F, [x, y, z], 'f');                    # Resp. true
```

This proves the path independence claimed and gives the potential

```
> f;                                               # Resp. x² + y² + 2z²
```

Furthermore, if the components of $\mathbf{F}$ are continuous and have continuous first partial derivatives and if $\mathbf{F}$ is path independent in a domain D, then its curl is the zero vector. In our case,

```
> curl(F, [x, y, z]);                              # Resp. [0, 0, 0]
```

This condition is also sufficient for path independence in D, provided D is simply connected.

Path independence being guaranteed, you may now choose the most convenient path, namely, the straight-line segment C from the origin $A : (0,0,0)$ to $B : (2,2,2)$, and perform the integration as in the previous example. Let $\mathbf{r}$ be the position vector of C and $\mathbf{r1}$ its derivative. Let $\mathbf{FC}$ denote $\mathbf{F}$ on C, that is, $\mathbf{F}(\mathbf{r}(t))$.

```
> r := [t, t, t];   r1 := diff(r, t);
```

$$r := [t,\, t,\, t]$$
$$r1 := [1,\, 1,\, 1]$$

```
> FC := eval(subs(x = r[1], y = r[2], z = r[3], [F[1], F[2], F[3]]));
```

$$FC := [2\,t,\, 2\,t,\, 4\,t]$$

```
> answer := int(innerprod(FC, r1), t = 0..2);      # Resp. answer := 16
```

This result is much more quickly obtained by using the potential f and noting that the integral equals $f(B) - f(A)$ (the analog of a well-known formula from calculus) valid in the case of path independence.

```
> subs(x = 2, y = 2, z = 2, f) - subs(x = 0, y = 0, z = 0, f)# Resp. 16
```

Of course, this simple example merely serves to explain the basic facts and techniques, and a computer or even a calculator would hardly be needed here.

    Similar Material in AEM: pp. 471-475

| EXAMPLE 9.3 | **DOUBLE INTEGRALS. MOMENTS OF INERTIA**

Find the moments of inertia I_x and I_y of a mass of density $\sigma = 1$ in the triangle with vertices $(0,0)$, $(b,0)$, (b,h) about the coordinate axes, as well as the polar moment $I_0 = I_x + I_y$. (Sketch the triangle.)

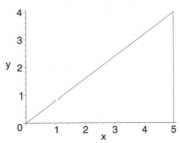

Example 9.3. Region of integration when $b = 5$ and $h = 4$

Solution. Since $\sigma = 1$, the integrand of I_x is y^2. If you integrate stepwise first over y from 0 to xh/b and then over x from 0 to b, you obtain

```
> int(y^2, y = 0..h*x/b);
```
　　　　　　　　　　　　　　　　　　# Resp. $\dfrac{1}{3}\dfrac{h^3 x^3}{b^3}$

```
> Ix := int(%, x = 0..b);
```
　　　　　　　　　　　　　　　　　　# Resp. $Ix := \dfrac{1}{12} h^3 b$

If you combine the two steps into one, you obtain the same result,

```
> int(int(y^2, y = 0..h*x/b), x = 0..b);
```
　　　　　　　　　　　　　　　　　　# Resp. $\dfrac{1}{12} h^3 b$

If you integrate y^2 first over x, you must integrate from by/h to b and then over y from 0 to h, obtaining

```
> int(int(y^2, x = b*y/h..b), y = 0..h);
```
　　　　　　　　　　　　　　　　　　# Resp. $\dfrac{1}{12} h^3 b$

Similarly, the integrand of I_y is x^2 and integration gives

```
> Iy := int(int(x^2, y = 0..h*x/b), x = 0..b);
```
　　　　　　　　　　　# Resp. $Iy := \dfrac{1}{4} h b^3$

The **polar moment of inertia** I_0 is the sum $I_x + I_y$,

```
> I0 = Ix + Iy;
```
　　　　　　　　　　　　　　# Resp. $I0 = \dfrac{1}{12} h^3 b + \dfrac{1}{4} h b^3$

```
> factor(%);
```
　　　　　　　　　　　　　　# Resp. $I0 = \dfrac{1}{12} h b (h^2 + 3 b^2)$

Similar Material in AEM: pp. 481, 484 (#17)

EXAMPLE 9.4　　**GREEN'S THEOREM IN THE PLANE**

Green's theorem in the plane transforms a double integral over a region R in the xy-plane into a line integral over the boundary C of R and conversely. The formula is

$$\iint\limits_{R} \left[\left(\frac{\partial}{\partial x} F_2 \right) - \left(\frac{\partial}{\partial y} F_1 \right) \right] dx\, dy = \int_C (F_1\, dx + F_2\, dy).$$

It is valid under suitable regularity assumptions on R, C, F_1, and F_2, usually satisfied in applications (see AEM, p. 485).

　　For instance, proceeding first in terms of components, (and later repeating the calculation in terms of vectors), let $F_1 = y^2 - 7\,y$, $F_2 = 2\,x\,y + 2\,x$. Type

```
> F1 := y^2 - 7*y;   F2 := 2*x*y + 2*x;
```

$$F1 := y^2 - 7\,y$$
$$F2 := 2\,x\,y + 2\,x$$

The integrand of the double integral is

```
> diff(F2, x) - diff(F1, y);                                    # Resp. 9
```

Hence the integrand is constant, so that the integral equals 9 times the area of the region of integration R. For instance, if R is a circular disk of radius 1, the integral equals 9π.

We verify the formula of Green's theorem for this case, assuming that the disk has the center at the origin of the xy-plane. Then its boundary curve C is the circle $x^2 + y^2 = 1$. In polar coordinates, the position vector $\mathbf{r}$ of C (oriented counterclockwise!) and its derivative $\mathbf{r}'$ have the components

```
> x := cos(t);   y := sin(t);    x1 := diff(x, t);    y1 := diff(y, t);
```

$$x := \cos(t)$$
$$y := \sin(t)$$
$$x1 := -\sin(t)$$
$$y1 := \cos(t)$$

Hence the line integral on the right-hand side of the formula is

```
> int(F1*x1 + F2*y1, t = 0..2*Pi);                             # Resp. 9 π
```

This verifies Green's theorem for the present example.

The same example in vectorial notation. Let

```
> x := 'x':  y := 'y':                                 # Unassign x and y.
> r := [cos(t), sin(t), 0];   r1 := diff(r, t);
```

$$r := [\cos(t), \sin(t), 0]$$
$$r1 := [-\sin(t), \cos(t), 0]$$

```
> F := [y^2 - 7*y, 2*x*y + 2*x, 0];
```

$$F := [y^2 - 7\,y, \, 2\,x\,y + 2\,x, \, 0]$$

```
> with(linalg):                                        # Ignore the warning.
> CU := curl(F, [x, y, z]);                        # Resp. CU := [0, 0, 9]
```

Hence the integrand (curl $\mathbf{F}$) $\bullet\, \mathbf{k}$ of the double integral (with $\mathbf{k}$ the unit vector in the z-direction) equals 9, as before,

```
> innerprod(CU, [0, 0, 1]);                                    # Resp. 9
```

For the line integral you get (with r[1] and r[2] the x and y components of $\mathbf{r}$), as before,

```
> FC := subs(x = r[1], y = r[2], F);
```

$$FC := [\sin(t)^2 - 7\sin(t), \, 2\cos(t)\sin(t) + 2\cos(t), \, 0]$$

```
> int(innerprod(FC, r1), t = 0..2*Pi);                         # Resp. 9 π
```

Note that in using the curl, Maple requires that you carry along a third component 0 in $\mathbf{r}$. (Can you find the reason?)

    Similar Material in AEM: pp. 485, 486

| EXAMPLE 9.5 | **SURFACE INTEGRALS. FLUX**

These are integrals taken over a surface S. We assume S to be given parametrically in the form $\mathbf{r}(u, v)$, where (u, v) varies over a region R in the uv-plane. The integrand is a scalar function given in the form $\mathbf{F} \cdot \mathbf{n}$, where the vector function $\mathbf{F}$ is given and $\mathbf{n}$ is a unit normal vector of S ($-\mathbf{n}$ being the other one). This is very practical in flow problems, where $\mathbf{F} = \rho\mathbf{v}$ (ρ the density, $\mathbf{v}$ the velocity) and $\mathbf{F} \cdot \mathbf{n}$ is the **flux** across S, that is, the mass of fluid crossing S per unit time. The integral is evaluated by reducing it to a double integral over R, namely,

$$\iint_S \mathbf{F} \cdot \mathbf{n} \, dA = \iint_R \mathbf{F}(\mathbf{r}(u, v)) \cdot \mathbf{N}(u, v) \, du \, dv,$$

where dA is the element of area, $\mathbf{N} = \mathbf{r}_u \times \mathbf{r}_v$ is a normal vector of S, and $\mathbf{r}_u$ and $\mathbf{r}_v$ are the partial derivatives of $\mathbf{r}(u, v)$.

    For instance, let $\mathbf{r} = [u, \ u^2, \ v]$ with u varying from 0 to 2 and v from 0 to 3. You see that $y = x^2$, so that this surface intersects the xy-plane along the parabola $y = x^2$. See the figure. To understand the plot command, type `?plot3d` and `?plot3d[options]`. `orientation = [50, 70]` gives the spherical coordinates θ and ϕ of the point from which the plot is viewed. Try other angles.

    Let $\mathbf{v} = \mathbf{F} = [3z^2, \ 6, \ 6xz]$ and $\rho = 1$ gram/cm^3 = 1 ton/meter^3, as for water, speed being measured in meters/sec. Type

```
> r := [u, u^2, v];        # Resp. r := [u, u², v]  # 0 ≤ u ≤ 2, 0 ≤ v ≤ 3
> ru := diff(r, u);   rv := diff(r, v);
```

$$ru := [1, \, 2\,u, \, 0]$$
$$rv := [0, \, 0, \, 1]$$

```
> with(linalg):                          # Ignore the warning.
> N := crossprod(ru, rv);                 # Resp. N := [2 u, −1, 0]
> F := [3*z^2, 6, 6*x*z];                 # Resp. F := [3 z², 6, 6 x z]
```

On S the vector function $\mathbf{F}$ takes the form $\mathbf{F}(\mathbf{r}(u, v))$. Call it FS and type it and the integrand on the right-hand side of the general formula

```
> FS := subs(x=r[1], y=r[2], z=r[3], F);  # Resp. FS := [3 v², 6, 6 u v]
> integrand := innerprod(FS, N);          # Resp. integrand := 6 v² u − 6
```

Now integrate over u and v,

```
> flux := int(int(integrand, u = 0..2), v = 0..3);   # Resp. flux := 72
```

Since speed is measured in meters/sec, the flux is measured in tons/sec; this gives the answer 72 tons/sec or 72000 liters/sec.

```
> plot3d([r[1], r[2], r[3]], u = 0..2, v = 0..3, axes = NORMAL,
  labels = [x, y, z], orientation = [50, 70]);
```

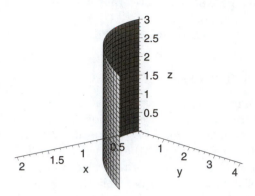

Example 9.5. Surface over which the integral is extended

Similar Material in AEM: pp. 497, 498

EXAMPLE 9.6   **DIVERGENCE THEOREM OF GAUSS**

The **divergence theorem** transforms a triple integral over a region T in space into a surface integral over the boundary surface S of T and conversely. The formula is

$$\iiint\limits_{T} \operatorname{div} \mathbf{F}\, dV = \iint\limits_{S} \mathbf{F} \bullet \mathbf{n}\, dA$$

where $\mathbf{F}$ is a given vector function, dV is the volume element, $\mathbf{n}$ is the outer unit normal vector of S, and dA is the element of area of S. The formula is valid under suitable regularity assumptions on T, S, and $\mathbf{F}$ usually satisfied in applications (see AEM, p. 506). Note that if in a flow problem, $\mathbf{F} = \rho \mathbf{v}$ (ρ the density, $\mathbf{v}$ the velocity), then the integral over S gives the flux through S. This is similar to the previous example.

For instance, let T be the solid circular cylinder of radius a with the z-axis as the axis, extending from $z = 0$ to $z = b$ (> 0), and let $\mathbf{F} = [x^3, x^2 y, x^2 z]$. Using the divergence theorem, evaluate the surface integral of $\mathbf{F} \bullet \mathbf{n}$ over the surface S of T (consisting of the upper and lower disks and the lateral cylindrical part).

Solution. Type

```
> F := [x^3, x^2*y, x^2*z];          # Resp. F := [x³, y x², x² z]
> with(linalg):                      # Ignore the warning.
> divF := diverge(F, [x, y, z]);     # Resp. divF := 5 x²
```

The form of T suggests the use of cylindrical coordinates $x = r \cos \theta$, $y = r \sin \theta$, $z = z$. Accordingly, make the substitution (note that div $\mathbf{F}$ depends only on x)

```
> r := 'r': theta := 'theta':  a := 'a':  b := 'b':     # Unassign
> f := subs(x = r*cos(theta), divF);   # Resp. f := 5 r² cos(θ)²
```

This is the integrand of the triple integral in the divergence theorem, which must be multiplied by the volume element $r\, dr\, d\theta\, dz$ and integrated over r from 0 to a, over θ from 0 to 2π, and over z from 0 to b. Type

```
> answer := int(int(int(f*r, r = 0..a), theta = 0..2*Pi), z = 0..b);
```

$$answer := \frac{5}{4}\,\pi\,a^4\,b$$

The direct evaluation of the surface integral should give the same result. It is more elaborate and thus illustrates the usefulness of the divergence theorem. Begin with the disk on the top, whose outer unit normal vector is pointing vertically upward,

```
> n := [0, 0, 1];                          # Resp. n := [0, 0, 1]
```

You need the dot product $\mathbf{F} \bullet \mathbf{n}$,

```
> ip := innerprod(F, n);                    # Resp. ip := x^2 z
```

Introduce polar coordinates by setting $x = r\cos\theta$ and then integrate over r from 0 to a and over θ from 0 to 2π. Use $dA = r\,dr\,d\theta$. Call the integral J1 .

```
> g := subs(x = r*cos(theta), z = b, ip);   # Resp. g := r^2 cos(θ)^2 b
> J1 := int(int(g*r, r = 0..a), theta = 0..2*Pi);
```

$$J1 := \frac{1}{4}\,\pi\,a^4\,b$$

For the bottom $z = 0$ the inner product is $-x^2z$, hence it is zero. (The outer normal vector points downward; this gives the minus sign.)

Now turn to the vertical cylinder surface. Represent it by

```
> r := [a*cos(theta), a*sin(theta), z];
```

$$r := [a\,\cos(\theta),\,a\,\sin(\theta),\,z]$$

Obviously, its normal is horizontal and at each point has the direction of the position vector. Hence the outer unit normal vector n2 of the cylinder is

```
> n2 := [cos(theta), sin(theta), 0];        # Resp. n2 := [cos(θ), sin(θ), 0]
```

Let FC denote $\mathbf{F}$ on the cylinder. Recall that r[1] , r[2] , r[3] are the components of $\mathbf{r}$. Hence type

```
> FC := subs(x = r[1], y = r[2], z = r[3], F);
```

$$FC := [\,a^3\,\cos(\theta)^3,\,a^3\,\cos(\theta)^2\,\sin(\theta),\,a^2\,\cos(\theta)^2\,z\,]$$

```
> ip2 := innerprod(FC, n2);
```

$$ip2 := a^3\,\cos(\theta)^4 + a^3\,\cos(\theta)^2\,\sin(\theta)^2$$

```
> simplify(%);                              # Resp. a^3 cos(θ)^2
```

This is the integrand. Multiply it by the element of area $a\,d\theta\,dz$ and integrate over θ from 0 to 2π and over z from 0 to b. Call the integral J2 . Add it to J1 to get the answer, which agrees with the previous one.

```
> J2 := int(int(ip2*a, theta = 0..2*Pi), z = 0..b);;
```

$$J2 := \pi\,a^4\,b$$

```
> answer2 := J1 + J2;                       # Resp. answer2 := \frac{5}{4}\,\pi\,a^4\,b
```

Similar Material in AEM: pp. 506, 507

| EXAMPLE 9.7 | STOKES'S THEOREM

Stokes's theorem transforms a surface integral over a surface S into a line integral over the boundary curve C of S and conversely. The formula is

$$\iint\limits_{S} (\text{curl } \mathbf{F}) \bullet \mathbf{n} \, dA = \oint_{C} \mathbf{F} \bullet \mathbf{r}'(s) \, ds$$

where $\mathbf{n}$ is a unit surface normal vector of S and $\mathbf{r}'(s)$ is a unit tangent vector of C and the direction of integration around C appears counterclockwise if one looks from the terminal point of $\mathbf{n}$ onto the surface. s is the arc length of C. The formula holds under suitable regularity conditions on S, C, and $\mathbf{F}$ usually satisfied in applications (see AEM, p. 516).

For instance, using Stokes's theorem, evaluate the line integral in the formula when $\mathbf{F} = [2\,y^2, \, x, \, -z^3]$ and C is the circle $x^2 + y^2 = a^2$, $z = b$ (> 0).

Solution. Type $\mathbf{F}$ as given. Then obtain the curl.

```
> F := [2*y^2, x, -z^3];                    # Resp. F := [2 y², x, −z³]
```

```
> with(linalg):                             # Ignore the warning.
```

```
> curlF := curl(F, [x, y, z]);              # Resp. curlF := [0, 0, 1 − 4 y]
```

As surface S choose the disk bounded by C. Obtain the curl on S; denote it by `curlFS`.

```
> r := [u*cos(v), u*sin(v), b];
```

$$r := [u \cos(v), \, u \sin(v), \, b]$$

```
> curlFS := [0, 0, subs(y = r[2], curlF[3])];
```

$$curlFS := [0, \, 0, \, 1 - 4\,u \sin(v)]$$

A unit normal vector of the disk is

```
> n := [0, 0, 1];                           # Resp. n := [0, 0, 1]
```

You can now type the integrand `ip` and then integrate over u from 0 to a and over the angle v from 0 to 2π.

```
> ip := innerprod(curlFS, n);               # Resp. ip := 1 − 4 u sin (v)
> answer := int(int(ip*u, u = 0..a), v = 0..2*Pi);
```

$$answer := a^2\,\pi$$

Confirmation by direct evaluation of the line integral around the boundary circle $C: x^2 + y^2 = a^2$, $z = b$. A representation of C is the following, in which t goes from 0 to 2π.

```
> r := [a*cos(t), a*sin(t), b];             # Resp. r := [a cos(t), a sin(t), b]
```

You can use t as a variable of integration because in the integral, by calculus, $\mathbf{r}'(s)\,ds = (d\mathbf{r}/ds)\,ds = (d\mathbf{r}/dt)\,dt$. Denote $\mathbf{F}$ on C by `FC` and obtain it by typing

```
> FC := subs(x = r[1], y = r[2], z = r[3], F);
```

$$FC := [2\,a^2\,\sin(t)^2, \, a \cos(t), \, -b^3]$$

Denote $d\mathbf{r}/dt$ by `r1` and type

```
> r1 := diff(r, t);                          # Resp. r1 := [−a sin(t), a cos(t), 0]
```

Now obtain the integrand `ip` and finally the integral, which will agree with the previous result.

```
> ip := innerprod(FC, r1);                    # Resp. ip := −2 a³ sin(t)³ + a² cos(t)²
> int(ip, t = 0..2*Pi);                       # Resp. a² π
```

Similar Material in AEM: pp. 516, 521 (#8)

Problem Set for Chapter 9

Pr.9.1 **(Line integral. Work)** Evaluate the work integral (see Example 9.1 in this Guide) of the force $\mathbf{F} = [2z,\ x,\ -y]$ from $(1,0,0)$ to $(1,0,4\pi)$ along the helix C: $\mathbf{r} = [\cos t,\ \sin t,\ 2t]$. (*AEM Ref.* p. 470 (#7))

Pr.9.2 **(Path dependence, same endpoints)** Show that the integral in Pr.9.1 changes its value if you integrate from $(1,0,0)$ to $(1,0,4\pi)$ along the straight-line segment with these endpoints. (*AEM Ref.* p. 470 (#7))

Pr.9.3 **(Independence of path. Potential)** Using a suitable curl, show that the integral $\int_C (3x^2\,dx + 2yz\,dy + y^2\,dz)$ is independent of path in any domain in space. Find a potential and use it to obtain the value of the integral from $A\colon (0,1,2)$ to $B\colon (1,-1,7)$. (*AEM Ref.* p. 473)

Pr.9.4 **(Independence of path)** Is the integral of $e^z\,dx + 2y\,dy + xe^z\,dz$ independent of path in space? If this is the case, find a potential by integration (not by using the command `potential`). Using the potential, integrate the given form from the origin to the point (a,b,c). (*AEM Ref.* p. 478 (#19))

Pr.9.5 **(Double integral. Center of gravity)** Find the center of gravity of a mass of density $\sigma = 1$ in the portion of the disk $x^2 + y^2 \le 1$ in the first quadrant of the xy-plane. (*AEM Ref.* pp. 481, 482)

Pr.9.6 **(Double integral. Moment of inertia)** Consider a mass of density $\sigma = 1$ in the portion of the disk $x^2 + y^2 \le a^2$ in the upper half-plane. Find the polar moment of inertia I_0 of this mass about the origin. (*AEM Ref.* pp. 481-483)

Pr.9.7 **(Green's theorem in the plane)** Using the formula of Green's theorem (see Example 9.4 in this Guide), integrate $\mathbf{F}(\mathbf{r}(t)) \bullet \mathbf{r}'(t)$ counterclockwise around the boundary of the region $R : 1 + x^4 \le y \le 2$; here, $\mathbf{F} = [e^y/x,\ e^y \ln x + 2x]$. (*AEM Ref.* p. 490 (#9))

Pr.9.8 **(Green's theorem in the plane)** Using Green's theorem in the plane (see Example 9.4 in this Guide), integrate $\mathbf{F}(\mathbf{r}(t)) \bullet \mathbf{r}'(t)$ with $\mathbf{F} = [x \cosh 2y,\ 2x^2 \sinh 2y]$ counterclockwise around the boundary of the region $R : x^2 \le y \le x$. (*AEM Ref.* p. 490 (#10))

Pr.9.9 **(Area)** Choosing $F_1 = 0$, $F_2 = x$ in the formula of Green's theorem in Example 9.4 of this Guide gives $\iint_R dx\,dy = \oint_C x\,dy$. Similarly, $\iint_R dx\,dy = -\oint_C y\,dx$ by choosing $F_1 = -y$, $F_2 = 0$. The double integral is the area A of R. Together,

$$A = \frac{1}{2} \oint_C (x\,dy - y\,dx).$$

Obtain from this the area of an ellipse $x^2/a^2 + y^2/b^2 = 1$. (*AEM Ref.* p. 488)

Pr.9.10 (**Experiment on surface normal**) Find a representation of the **ellipsoid**

$$S : \mathbf{r}(u, v) = [a \cos v \cos u, \ b \cos v \sin u, \ c \sin v]$$

in terms of Cartesian coordinates. Find a normal vector of S. Plot S for some triples a, b, c, for instance, $a = 3$, $b = 2$, $c = 1$. Choose other triples and observe how the surface and its normal change. (*AEM Ref.* p. 495 (#6))

Pr.9.11 (**Surface integral**) Find the flux integral (see Example 9.5 in this Guide for the definition) of $\mathbf{F} = [x^2, \ 0, \ 3y^2]$ over the portion of the plane $x + y + z = 1$ in the first octant in space. (*AEM Ref.* p. 498)

Pr.9.12 (**Surface integral**) Find the flux integral (as defined in Example 9.5 of this Guide) of $\mathbf{F} = [x^2, y^2, z^2]$ over the **helicoid** $\mathbf{r} = [u \cos v, \ u \sin v, \ 3v]$, where $0 \le u \le 1$, $0 \le v \le 2\pi$. Sketch the surface. Explain its name. (*AEM Ref.* p. 503 (#10))

Pr.9.13 (**Integral over a sphere**) Find the flux integral (as defined in Example 9.5 in this Guide) of $\mathbf{F} = [0, \ x, \ 0]$ over the portion of the sphere $x^2 + y^2 + z^2 = 1$ in the first octant. (*AEM Ref.* pp. 493, 503 (#7))

Pr.9.14 (**Triple integral**) Find the total mass in the box $|x| \le 1$, $|y| \le 3$, $|z| \le 2$ if the mass density is $\sigma = x^2 + y^2 + z^2$. (*AEM Ref.* p. 509 (#1))

Pr.9.15 (**Triple integral. Moment of inertia**) Find the moment of inertia I_x about the x-axis of a mass of density $\sigma = 1$ in the solid circular cylinder of radius a about the x-axis, extending in x-direction from 0 to h. (*AEM Ref.* p. 510 (#11))

Pr.9.16 (**Surface integral. Divergence theorem**) Using the divergence theorem (see Example 9.6 in this Guide), find the integral of the normal component of the vector function $\mathbf{F} = [4x, x^2 y^2, y^2 z^2]$ over the surface of the tetrahedron with vertices $(0, 0, 0)$, $(1, 0, 0)$, $(0, 1, 0)$, $(0, 0, 1)$. (*AEM Ref.* p. 506)

Pr.9.17 (**Surface integral. Divergence theorem**) Using the divergence theorem (see Example 9.6 in this Guide), integrate the normal component of the vector function $\mathbf{F} = [9x, \ y \cosh^2 x, \ -z \sinh^2 x]$ over the ellipsoid $4x^2 + y^2 + 9z^2 = 36$. (*AEM Ref.* p. 515 (#5))

Pr.9.18 (**Laplacian, normal derivative**) If $\mathbf{F} = \operatorname{grad} f$ in the formula of the divergence theorem in Example 9.6 in this Guide, show that $\operatorname{div} \mathbf{F} = \nabla^2 f$ and $\mathbf{F} \bullet \mathbf{n} = \partial f / \partial n$ (the normal derivative). Verify the resulting formula

$$\iiint_T \nabla^2 f \, dV = \iint_S \frac{\partial f}{\partial n} \, dA$$

for $f = (x^2 + y^2 + z^2)^2$ and S a sphere of radius a and center at the origin. *Hint*: Represent the interior of S by $\mathbf{R} = [r \cos u \sin v, \ r \sin u \sin v, \ r \cos v)$ (where r is variable!) and use the corresponding volume element $r^2 \sin v \, dr \, dv \, du$. (*AEM Ref.* pp. 493, 513)

Pr.9.19 (**Surface integral. Divergence theorem**) Using the divergence theorem, find the integral of the normal derivative of $\mathbf{F} = [x^3, \ y^3, \ z^3]$ over the sphere $S : x^2 + y^2 + z^2 = 9$. Represent S as in the previous problem. (*AEM Ref.* p. 510 (#19))

Pr.9.20 (**Stokes's theorem**) Using Stokes's theorem (see Example 9.7 in this Guide), integrate the tangential component of $\mathbf{F} = [e^z, \ e^z \sin y, \ e^z \cos y]$ around the boundary of the surface $S : z = y^2$, where $0 \le x \le 4$, $0 \le y \le 2$. (*AEM Ref.* p. 520 (#3))

PART C. FOURIER ANALYSIS AND PARTIAL DIFFERENTIAL EQUATIONS

Content. Fourier series, integrals, and transforms (Chap. 10)

Partial differential equations (PDE's) (Chap. 11)

This Part concerns initial and boundary value problems for the "big" PDE's of Applied Mathematics, namely, the wave, heat, and Laplace equations, their solution by separating variables, and the use of Fourier series and integrals for obtaining solutions sufficiently general to satisfy all the given physical conditions.

Chapter 10

Fourier Series, Integrals, and Transforms

Content. Fourier series (period 2π) (Exs. 10.1, 10.4, Prs. 10.1-10.5)

Fourier series (any period) (Exs. 10.2, 10.3, Prs. 10.6-10.10, 10.15)

Half-range expansions (Ex. 10.3, Pr. 10.9)

Trigonometric approximation (Ex. 10.5, Prs. 10.11-10.13)

Fourier integral (Ex. 10.6, Pr. 10.14)

The **Fourier series** of a function $f(x)$ of period $p = 2L$ is obtained by typing

```
> f := a[0] + sum(a[n]*cos(n*Pi*x/L) + b[n]*sin(n*Pi*x/L),
  n = 1..infinity);
```

$$f := a_0 + \sum_{n=1}^{\infty} \left(a_n \cos\left(\frac{n\pi x}{L}\right) + b_n \sin\left(\frac{n\pi x}{L}\right) \right)$$

with the **Fourier coefficients** a_n and b_n given by the **Euler formulas**

```
> f := 'f':                                    # Unassign f, make it free.
```

```
> a[0] := 1/(2*L)*int(f(x), x = -L..L);
```

$$a_0 = \frac{1}{2}\frac{\int_{-L}^{L} f(x)\, dx}{L}$$

```
> a[n] := 1/L*int(f(x)*cos(n*Pi*x/L), x = -L..L);
```

$$a_n = \frac{\int_{-L}^{L} f(x) \cos\left(\frac{n\pi x}{L}\right) dx}{L}$$

```
> b[n] := 1/L*int(f(x)*sin(n*Pi*x/L), x = -L..L);
```

$$b_n = \frac{\int_{-L}^{L} f(x) \sin\left(\frac{n\pi x}{L}\right) dx}{L}$$

Functions of period $p = 2\pi$. Replace L by π, obtaining $\cos nx$, $\sin nx$ in all formulas, and $1/(2\pi)$ and $1/\pi$ as factors before the integrals, and integrate from $-\pi$ to π. Write down these formulas for yourself.

Examples for Chapter 10

| **EXAMPLE 10.1** | **FUNCTIONS OF PERIOD** 2π. **EVEN FUNCTIONS. GIBBS PHENOMENON** |

Consider the function $f(x)$ of period 2π which is 0 for $-\pi < x < -\pi/2$, then equals k for $-\pi/2 < x < \pi/2$ and is again 0 for $\pi/2 < x < \pi$ (see the figure, where $k = 1$). You may call this function a **periodic rectangular wave** of period 2π. You obtain the Fourier coefficients from the Euler formulas by integrating from $-\pi/2$ to $\pi/2$ only because $f(x)$ is 0 for $-\pi < x < -\pi/2$ and $\pi/2 < x < \pi$. Hence type

```
> a0 := 1/(2*Pi)*int(k, x = -Pi/2..Pi/2);
```
Resp. $a0 := \frac{1}{2}k$

Note that this is the mean value of $f(x)$ over the interval from $-\pi$ to π (as it is always the case).

```
> an := 1/Pi*int(k*cos(n*x), x = -Pi/2..Pi/2);
```

$$an := 2\frac{\sin(\frac{1}{2}\pi n)k}{\pi n}$$

```
> bn := 1/Pi*int(k*sin(n*x), x = -Pi/2..Pi/2);
```
Resp. $bn := 0$

Hence this Fourier series is a **Fourier cosine series**, that is, it has only cosine terms, no sine terms. The reason is that $f(x)$ is an **even function**, that is, $f(-x) = f(x)$. The command seq gives you the first few coefficients

```
> seq(an, n = 1..8);
```
Resp. $2\frac{k}{\pi}, 0, -\frac{2}{3}\frac{k}{\pi}, 0, \frac{2}{5}\frac{k}{\pi}, 0, -\frac{2}{7}\frac{k}{\pi}, 0$

For plotting you have to choose a definite value of k, say, $k = 1$,

```
> An := subs(k = 1, an);
```
Resp. $An := 2\frac{\sin(\frac{1}{2}\pi n)}{\pi n}$

Find and plot (on common axes) the first few partial sums

```
> S1 := 1/2 + sum(An*cos(n*x), n = 1..1);
```

$$S1 := \frac{1}{2} + \frac{2\cos(x)}{\pi}$$

```
> S3 := 1/2 + sum(An*cos(n*x), n = 1..3);
```

$$S3 := \frac{1}{2} + \frac{2\cos(x)}{\pi} - \frac{2}{3}\frac{\cos(3x)}{\pi}$$

```
> S5 := 1/2 + sum(An*cos(n*x), n = 1..5);
```

$$S5 := \frac{1}{2} + \frac{2\cos(x)}{\pi} - \frac{2}{3}\frac{\cos(3\,x)}{\pi} + \frac{\frac{2}{5}\cos(5\,x)}{\pi}$$

```
> with(plots):                                          # Ignore the warning.
> P1 := plot({S1, S3, S5}, x = -Pi..Pi):
> P2 := plot(1, x = -Pi/2..Pi/2):
> display({P1, P2}, ytickmarks = [0, 0.5, 1]);
```

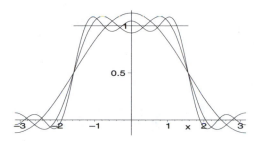

Example 10.1. Given rectangular wave and first three partial sums S1, S3, S5

You see that there are waves near the discontinuities at $-\pi/2$ and $\pi/2$ and you would perhaps expect that these waves become smaller and smaller if you took partial sums with more and more terms – which can easily be done on the computer. However, these waves do not disappear, but they are shifted closer and closer to those points of discontinuity, as is shown in the figure for the partial sum from $n = 1$ to $n = 50$. This is called the **Gibbs phenomenon**. Experiment with $n = 100$ or even larger n.

```
> plot(1/2 + sum(An*cos(n*x), n = 1..50), x = -Pi..Pi);
```

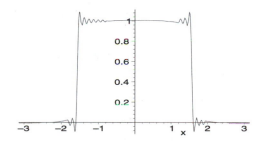

Example 10.1. Gibbs phenomenon

Similar Material in AEM: pp. 529-533

EXAMPLE 10.2	**FUNCTIONS OF ARBITRARY PERIOD.**
	ODD FUNCTIONS

For functions $f(x)$ of arbitrary period $p = 2L$ the determination of the Fourier coefficients is the same in principle as for functions of period 2π; just the formulas look a little more complicated.

For instance, consider the function $f(x)$ of period $p = 2L = 2$ in the figure. (Type `?piecewise` for information.)

```
> with(plots):                                          # Ignore the warning.
> f := piecewise(x < -5, 0, x < -3, x + 4, x < -1, x + 2, x < 1, x,
  x < 3, x - 2, x < 5, x - 4, 0);
```

$$f := \begin{cases} 0 & x < -5 \\ x+4 & x < -3 \\ x+2 & x < -1 \\ x & x < 1 \\ x-2 & x < 3 \\ x-4 & x < 5 \\ 0 & \text{otherwise} \end{cases}$$

```
> plot(f, x = -5..5, scaling = constrained);
```

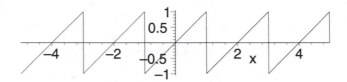

Example 10.2. Given function (**Sawtooth wave**)

Clearly, $f(x) = x$ if $-1 < x < 1$. For this function you obtain from the Euler formulas with $L = 1$ in the chapter opening

```
> a0 := 1/2*int(x, x = -1..1);                          # Resp. a0 := 0
> an := int(x*cos(n*Pi*x), x = -1..1);                  # Resp. an := 0
```

Hence the series has no cosine terms, only sine terms; it is a **Fourier sine series**, because $f(x)$ is an **odd function,** that is, $f(-x) = -f(x)$. The sine terms have the Fourier coefficients

```
> bn := int(x*sin(n*Pi*x), x = -1..1);
```

$$bn := -2 \frac{-\sin(n\,\pi) + n\,\pi\,\cos(n\,\pi)}{n^2\,\pi^2}$$

```
> seq(bn, n = 1..8);
```

$$2\frac{1}{\pi},\ -\frac{1}{\pi},\ \frac{2}{3}\frac{1}{\pi},\ -\frac{1}{2}\frac{1}{\pi},\ \frac{2}{5}\frac{1}{\pi},\ -\frac{1}{3}\frac{1}{\pi},\ \frac{2}{7}\frac{1}{\pi},\ -\frac{1}{4}\frac{1}{\pi}$$

Now type and then plot some partial sums, including a large one (S50), which will again illustrate the **Gibbs phenomenon** at the points of discontinuity -1 and 1, just as in the preceding example.

```
> S1 := sum(bn*sin(n*Pi*x), n = 1..1):
> S2 := sum(bn*sin(n*Pi*x), n = 1..2):
> S3 := sum(bn*sin(n*Pi*x), n = 1..3):
> S50 := sum(bn*sin(n*Pi*x), n = 1..50):
> plot({S1, S2, S3, S50}, x = -1..1, xtickmarks = [-1, -0.5, 0, 0.5, 1],
  ytickmarks = [-1, -0.5, 0, 0.5, 1]);
```

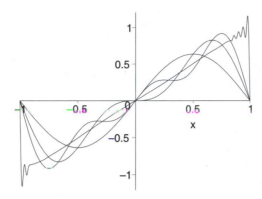

Example 10.2. Approximation of $f(x) = x$ by partial sums of its Fourier series

Similar Material in AEM: pp. 543, 544

| EXAMPLE 10.3 | **HALF-RANGE EXPANSIONS** |

If you develop a function $f(x)$, given for $0 < x < L$, in a Fourier series, then this series represents a periodic function of period L which for $x = 0 \ldots L$ agrees with $f(x)$. In general, this series contains both cosine and sine terms. In most applications it will be much better to first extend $f(x)$ as an **even** function $f_1(x)$ for $-L < x < L$; then $f_1(x)$ has the period $2L$, and its Fourier series contains only cosine terms because $f_1(x)$ is even. This series is called the **cosine half-range expansion** of $f(x)$ because $f(x)$ is given only over one half of the range (that is, over one half of the interval of periodicity).

Similarly, you may extend the given function $f(x)$ to an **odd** function $f_2(x)$ for $-L < x < L$; then $f_2(x)$ has the period $2L$, and its Fourier series contains only sine terms because $f_2(x)$ is odd. Call this series the **sine half-range expansion** of $f(x)$.

For instance, let $f(x) = x^2$ be given for $0 < x < 2$. Figure (A) shows its even and odd extensions to the interval (the "full range") $-2 < x < 2$. Figure (B) shows its even periodic extension $f_1(x)$ of period $p = 2L = 4$. Figure (C) shows its odd periodic extension $f_2(x)$ of period 4. scaling = constrained gives equal scales on the axes.

```
> with(plots):                                      # Ignore the warning.
> P1 := plot(x^2, x = -2..2):
> P2 := plot(-x^2, x = -2..0):
> display(P1, P2, scaling = constrained);           # This is Fig. (A).
> P3 := plot((x + 4)^2, x = -6..-2):
> P4 := plot((x - 4)^2, x = 2..6):
> P5 := plot((x - 8)^2, x = 6..10):
> display(P1, P3, P4, P5, scaling = constrained);   # This is Fig. (B).
> P6 := plot(-(x + 4)^2, x = -6..-4):
> P7 := plot((x + 4)^2, x = -4..-2):
> P8 := plot(x^2, x = 0..2):
> P9 := plot(-(x - 4)^2, x = 2..4):
```

```
> P10 := plot((x - 4)^2, x = 4..6):
> display(P6, P7, P2, P8, P9, P10, scaling = constrained);    # Fig. (C).
```

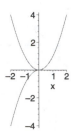

Example 10.3. **(A)** $f(x) = x^2$ (given for $0 < x < 2$)
and its even and odd extensions to $-2 < x < 2$

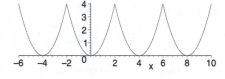

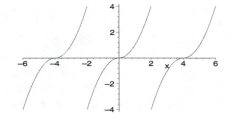

Example 10.3. **(B)** Even periodic
extension $f_1(x)$ of $f(x)$

Example 10.3. **(C)** Odd periodic
extension $f_2(x)$ of $f(x)$

Now determine the Fourier coefficients a_n of the Fourier cosine series of $f_1(x)$ from the Euler formula. Since $f_1(x)$ and the cosine are both even, so is the product (the integrand). Hence the integral from -2 to 2 equals twice the integral from 0 to 2 (the interval on which $f(x)$ is given). Remembering that $L = 2$, type the commands for a_0, a_n, and for a partial sum S. Figure (D) shows that a good approximation can already be obtained with a small number of terms.

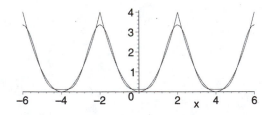

Example 10.3. **(D)** Approximation of $f_1(x)$ by a partial sum of its Fourier series

```
> a0 := 2/(2*2)*int(x^2, x = 0..2);                # Resp. a0 := 4/3
> an := 2/2*int(x^2*cos(n*Pi*x/2), x = 0..2);
```

$$an := 8 \, \frac{-2 \sin(n\,\pi) + n^2\,\pi^2\,\sin(n\,\pi) + 2\,\pi\,n\,\cos(n\,\pi)}{n^3\,\pi^3}$$

```
> S := a0 + sum(an*cos(n*Pi*x/2), n = 1..2);
```

$$S := \frac{4}{3} - \frac{16 \cos(\frac{1}{2}\pi x)}{\pi^2} + \frac{4 \cos(\pi x)}{\pi^2}$$

```
> P11 := plot(S, x = -6..6, scaling = constrained):
> display(P11, P1, P3, P4);                          # This is Fig. (D).
```

For $f_2(x)$ you will need a partial sum consisting of more terms, as Fig. (E) will show, because $f_2(x)$ is discontinuous – and it will be quite interesting to see how a sum of **continuous** terms approximates a **discontinuous** function. Both $f_2(x)$ and the sine in the Euler formula for b_n are odd, hence the integrand is even, so that the integral from -2 to 2 equals twice the integral from 0 to 2. Hence type

```
> bn := 2/2*int(x^2*sin(n*Pi*x/2), x = 0..2);
```

$$bn := -8 \frac{-2 \cos(\pi n) + n^2 \pi^2 \cos(\pi n) - 2\pi n \sin(\pi n) + 2}{n^3 \pi^3}$$

```
> S2 := sum(bn*sin(n*Pi*x/2), n = 1..7);
```

$$S2 := -8 \frac{(4 - \pi^2) \sin(\frac{1}{2}\pi x)}{\pi^3} - 4 \frac{\sin(\pi x)}{\pi} - \frac{\frac{8}{27}(4 - 9\pi^2) \sin(\frac{3}{2}\pi x)}{\pi^3} - 2 \frac{\sin(2\pi x)}{\pi}$$
$$- \frac{\frac{8}{125}(4 - 25\pi^2) \sin(\frac{5}{2}\pi x)}{\pi^3} - \frac{4}{3} \frac{\sin(3\pi x)}{\pi} - \frac{\frac{8}{343}(4 - 49\pi^2) \sin(\frac{7}{2}\pi x)}{\pi^3}$$

```
> P12 := plot(S2, x = -6..6):
> display(P12, P2, P6, P7, P8, P9, P10, scaling = constrained);
```

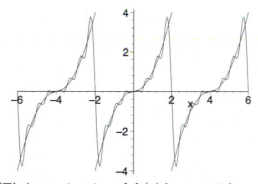

Example 10.3. **(E)** Approximation of $f_2(x)$ by a partial sum of its Fourier series

Similar Material in AEM: pp. 544-546

EXAMPLE 10.4	**RECTIFIER**

A sinusoidal current, for simplicity, $f(t) = \sin t$, where t is time, is passed through a full-wave rectifier which converts the negative half-waves to positive half-waves (and leaves the positive half-waves as they are), resulting in the function $g(t) = |\sin t|$. Find the Fourier series of $g(t)$.

Solution. $g(t)$ is even. Hence you obtain a Fourier cosine series, and the integral for a_n from $t = -\pi$ to π equals twice the integral from 0 to π, an interval on which $g(t) = \sin t$. Using the Euler formulas for a function of period 2π, thus type

```
> a0 := 1/Pi*int(sin(t), t = 0..Pi);
```
$\qquad\qquad\qquad\qquad\qquad\qquad\qquad$ # Resp. $a0 := 2\dfrac{1}{\pi}$

```
> an := 2/Pi*int(sin(t)*cos(n*t), t = 0..Pi);
```

$$an := -2\,\frac{\cos(\pi n) + 1}{\pi\,(1+n)\,(-1+n)}$$

For $n = 1$ the denominator is zero. Hence type a_1 separately and then partial sums.

```
> a1 := 2/Pi*int(sin(t)*cos(t), t = 0..Pi);
```
$\qquad\qquad\qquad\qquad\qquad\qquad\qquad\qquad\quad$ # Resp. $a1 := 0$

```
> S3 := a0 + a1*cos(t) + sum(an*cos(n*t), n = 2..3):
> S7 := a0 + a1*cos(t) + sum(an*cos(n*t), n = 2..7);
```

$$S7 := 2\,\frac{1}{\pi} - \frac{4}{3}\,\frac{\cos(2\,t)}{\pi} - \frac{4}{15}\,\frac{\cos(4\,t)}{\pi} - \frac{4}{35}\,\frac{\cos(6\,t)}{\pi}$$

$S7$ is already fairly accurate and can hardly be distinguished from $g(t)$ in the figure.

```
> with(plots):
```
$\qquad\qquad\qquad\qquad\qquad\qquad\qquad\qquad\qquad$ # Ignore the warning.

```
> P := plot({S3, S7}, t = -3*Pi..3*Pi):
> Pg := plot(abs(sin(t)), t = -3*Pi..3*Pi):
> display({P, Pg});
```

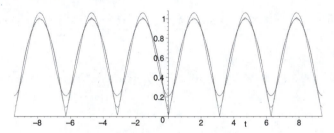

Example 10.4. Full-wave rectification $g(t)$ of $f(t) = \sin t$
and partial sums S_3 and S_7 of its Fourier series

Similar Material in AEM: p. 539

EXAMPLE 10.5	**TRIGONOMETRIC APPROXIMATION.**

MINIMUM SQUARE ERROR

In approximating a given function f by another function F the quality of the approximation can be measured in various ways, depending on the purpose. A quantity that characterizes the *overall goodness of fit* on the interval of interest, say, $-\pi < x < \pi$, is the so-called **total square error** of F (relative to f) defined by

$$E = \int_{-\pi}^{\pi} (f - F)^2\, dx.$$

If

$$F(x) = A_0 + \sum_{n=1}^{N} (A_n \cos nx + B_n \sin nx),$$

then $F(x)$ is called a **trigonometric polynomial** and the approximation is called a **trigonometric approximation**. In this case the total square error of F relative to f (with fixed N) is minimum if and only if you choose the Fourier coefficients of f as the coefficients of F. It can be shown (see AEM, p. 555) that this **minimum square error** is

$$E^* = \int_{-\pi}^{\pi} f^2 \, dx - \pi \left[2 \, a_0{}^2 + \sum_{n=1}^{N} (a_n{}^2 + b_n{}^2) \right].$$

For instance, find the minimum square errors for the approximation of the given function in Example 10.1 in this Guide with $k = 1$ by the partial sums of its Fourier series (illustrated by the figures in that example).

Solution. The b_n are zero since f is even. Type the formula for the a_n, then the formula for the minimum square error, call it Emin, and finally find numerical values of Emin for $N = 10, 20, ..., 250$. (Since f is discontinuous, the convergence of the sequence of the minimum square errors to 0 will be slow.)

```
> an := 1/Pi*int(1*cos(n*x), x = -Pi/2..Pi/2);
```

$$an := 2 \, \frac{\sin \left(\frac{1}{2} \pi n \right)}{\pi n}$$

```
> Emin := int(1, x = -Pi/2..Pi/2) - Pi*(2*(1/2)^2 +sum(an^2,
  n = 1..10*N)):
> seq(evalf(Emin, 5), N = 1..25);
```

.0635, .0318, .0212, .0159, .0128, .0106, .0091, .0080, .0071, .0064, .0058, .0053, .0049, .0046, .0042, .0040, .0038, .0036, .0034, .0032, .0030, .0029, .0028, .0027, .0026

Similar Material in AEM: pp. 554-556

EXAMPLE 10.6	**FOURIER INTEGRAL, FOURIER TRANSFORM**

The **Fourier integral representation** of a given function $f(x)$ defined on the x-axis (and satisfying certain regularity conditions; see AEM, p. 559) is

$$f(x) = \int_0^{\infty} [A(w) \cos wx + B(w) \sin wx] \, dw$$

where

$$A(w) = \frac{1}{\pi} \int_{-\infty}^{\infty} f(v) \cos wv \, dv,$$

$$B(w) = \frac{1}{\pi} \int_{-\infty}^{\infty} f(v) \sin wv \, dv.$$

For instance, find the Fourier integral representation of the function $f(x)$ which equals 1 for $-1 < x < 1$ and zero everywhere else on the x-axis.

Solution. The integration in $A(w)$ and $B(w)$ extends from -1 to 1 only because $f(x)$ is zero elsewhere. Thus type

```
> A := 1/Pi*int(1*cos(w*v), v = -1..1);
```

$$A := 2 \, \frac{\sin(w)}{\pi w}$$

```
> B := 1/Pi*int(1*sin(w*v), v = -1..1);                    # Resp. B := 0
> f := 2/Pi*'int(sin(w)/w*cos(w*x), w = 0..infinity)';
```

$$f := 2 \frac{\displaystyle\int_0^\infty \frac{\sin(w)\,\cos(w\,x)}{w}\,dw}{\pi}$$

Note the ' before and after the integral, because of which Maple leaves the integral unevaluated. Try without to see that then the computer simply reproduces the given function (type ?signum for information).

It can be shown (see AEM, pp. 569, 570) that the Fourier integral may be converted to complex form, and from it one can derive the **Fourier transform** $F(w)$ of $f(x)$ given by

$$F(w) = \frac{1}{\sqrt{2\,\pi}} \int_{-\infty}^\infty f(x)\,e^{-iwx}\,dx$$

and the **inverse Fourier transform** of $F(w)$ given by

$$f(x) = \frac{1}{\sqrt{2\,\pi}} \int_{-\infty}^\infty F(w)\,e^{iwx}\,dw.$$

For instance, find the Fourier transform of $f(x) = e^{-x^2}$.

Solution. Let F denote the Fourier transform of the given $f(x)$. Type

```
> F := 1/(sqrt(2*Pi))*int(exp(-x^2)*exp(-I*w*x),
  x = -infinity..infinity);
```

$$F := \frac{1}{2}\,\sqrt{2}\,e^{\left(-1/4\,w^2\right)}$$

Similar Material in AEM: pp. 570, 571

Problem Set for Chapter 10

Pr.10.1 (**Rectangular wave. Gibbs phenomenon**) Find the Fourier series of the following function of period 2π and make plots that show the Gibbs phenomenon. (*AEM Ref.* p. 532)

$$f(x) = \begin{cases} -3 & \text{if } -\pi < x < 0 \\ 3 & \text{if } 0 < x < \pi \end{cases}$$

Pr.10.2 (**Cosine series**) Find the Fourier series of the function (sketch it) given by

$$f(x) = \begin{cases} 1 & \text{if } -\pi/2 < x < \pi/2 \\ -1 & \text{if } \pi/2 < x < 3\pi/2 \end{cases}$$

and periodic with period 2π. (*AEM Ref.* p. 537 (#13))

Pr.10.3 (**Cosine and sine terms**) Find the Fourier series of $f(x)$ given by $f(x) = (x/(2\pi))^4$ if $0 < x < 2\pi$ and of period 2π. (*AEM Ref.* p. 531)

Pr.10.4 (**Half-wave rectifier**) Pass the current $f(t) = \sin t$ through a half-wave rectifier that clips the negative portion of the wave. Find the Fourier series of the resulting periodic function $g(t)$ and plot some of its partial sums. (*AEM Ref.* p. 439)

Pr.10.5 **(Periodic force, resonance)** If a force $f(t) = t^2$ for $0 < t < 2\pi$ and periodic of period 2π is acting as the driving force on a mechanical system, which term of its Fourier series has the greatest coefficient in absolute value (so that the corresponding frequency should be watched for possible resonance effects)? (*AEM Ref.* p. 536 (#8))

Pr.10.6 **(Behavior near a jump)** Find the Fourier series of the periodic function $f(x) = \pi x^3/2$ $(-1 < x < 1)$ of period $p = 2$. Show by plots that at the jumps the partial sums give the arithmetic mean of the right-hand and left-hand limits of $f(x)$. (*AEM Ref.* p. 540 (#12))

Pr.10.7 **(Continuous function)** Find the Fourier series of the periodic function $f(x) = 3x^2$ $(-1 < x < 1)$ of period $p = 2$. Show by a plot that a partial sum of few terms gives a relatively good approximation (except near the cusps). (*AEM Ref.* p. 540 (#7))

Pr.10.8 **(Triangular wave)** Find the Fourier series of the function (sketch it)

$$f(x) = \begin{cases} 1/2 + x & \text{if} \quad -1/2 < x < 0 \\ 1/2 - x & \text{if} \quad\quad 0 < x < 1/2 \end{cases}$$

of period $p = 1$. Plot $f(x)$ and some partial sums. (*AEM Ref.* p. 540 (#8))

Pr.10.9 **(Half-range expansion)** Find the Fourier cosine and sine series of $f(x) = x$, where $0 < x < L$ and L is arbitrary. (*AEM Ref.* p. 547 (#21))

Pr.10.10 **(Herringbone wave)** Find the Fourier series of the function (sketch it)

$$f(x) = \begin{cases} x & \text{if} \quad 0 < x < 1 \\ 1 - x & \text{if} \quad 1 < x < 2 \end{cases}$$

of period $p = 2$. Plot $f(x)$ and some partial sums. Observe the Gibbs phenomenon. (*AEM Ref.* p. 540 (#10))

Pr.10.11 **(Error distribution)** Find and plot $f(x) - S_5$ for x from $-\pi$ to π, where $f(x) = x^3$ $(-\pi < x < \pi)$ and periodic with period 2π and S_5 is the partial sum of the Fourier series containing terms of $\sin x$ to $\sin 5x$. (*AEM Ref.* pp. 553-556)

Pr.10.12 **(Minimum square error)** Let $f(x) = x^2$ $(-\pi < x < \pi)$ and periodic with period 2π. Find the minimum square error for the trigonometric approximation of $f(x)$ by polynomials of degree $N = 1, ..., 10$ and $N = 100$. (See Example 10.5 in this Guide. *AEM Ref.* p. 556 (#4))

Pr.10.13 **(Minimum square error)** Do the same task as in Pr.10.12 for the periodic function $f(x) = x$ $(-\pi < x < \pi)$ of period 2π.

Pr.10.14 **(Fourier integral, Fourier cosine integral)** Using the Fourier integral representation in Example 10.6 in this Guide, represent

$$f(x) = \begin{cases} \pi e^{-x} & \text{if} \quad x > 0 \\ 0 & \text{otherwise} \end{cases}$$

by a Fourier integral. (*AEM Ref.* p. 563 (#1))

Pr.10.15 **(Experiment on Gibbs phenomenon)** Study and plot the Gibbs phenomenon for functions of your choice. On what does the height of the "spikes" depend? Or is it always the same? Is the speed of the shift of the waves toward the discontinuity different for different functions? On what does it depend? What about the number of "spikes"?

Chapter 11

Partial Differential Equations (PDE's)

Content. Wave equation, animation (Exs. 11.1, 11.2, Prs. 11.1-11.3)
Tricomi, Airy equations, vibrating beam (Prs. 11.6, 11.7)
Heat equation, Laplace equation (Exs. 11.3, 11.4, Prs. 11.8-11.12)
Vibrating membrane (Exs. 11.5, 11.6, Prs. 11.13-11.15)

PDEtools package. Load it by typing `with(PDEtools)`. It may help you in certain tasks (see some of our examples). Type `?PDEtools`.

The PDE's considered are solved by separating variables, which reduces them to ODE's, and by the subsequent use of Fourier series and integrals.

Examples for Chapter 11

EXAMPLE 11.1 **WAVE EQUATION. SEPARATION OF VARIABLES. ANIMATION**

The **one-dimensional wave equation** is $u_{tt} = c^2 u_{xx}$. It governs the vertical vibrations of an **elastic string**, such as a violin string. Then $u(x, t)$ is the displacement from rest along the x-axis at a point x and time t. The string of length L is fixed at its ends $x = 0$ and $x = L$. This gives the boundary conditions $u(0, t) = 0$ and $u(L, t) = 0$ for all t. In separating variables you look for solutions of the form $u(x, t) = F(x)G(t)$ satisfying the boundary conditions. Thus, type this $u(x, t)$ and then the wave equation.

```
> u(x,t) := F(x)*G(t);                              # Resp. u(x, t) := F(x) G(t)
> pde := diff(u(x,t), t, t) = c^2*diff(u(x,t), x, x);
```

$$pde := F(x) \left(\frac{\partial^2}{\partial t^2} G(t) \right) = c^2 \left(\frac{\partial^2}{\partial x^2} F(x) \right) G(t)$$

Now separate the variables by dividing by $c^2 u(x, t) = c^2 F(x) G(t)$ and then set each of the two sides equal to a constant, which must be negative, say $-p^2$, in order to avoid ending up with an identically vanishing solution. First obtain a general solution of the ordinary differential equation for $F(x)$, written as `rhs(eq) = -p^2`. Reduce this solution by the left boundary condition. Then determine p from the right boundary condition. Finally solve the ordinary differential equation for $G(t)$, obtaining a general solution (*solG*, below).

```
> eq := pde/(c^2*u(x,t));
```

$$eq := \frac{\frac{\partial^2}{\partial t^2} G(t)}{c^2 G(t)} = \frac{\frac{\partial^2}{\partial x^2} F(x)}{F(x)}$$

```
> sol1 := dsolve(rhs(eq) = -p^2);
```

$$sol1 := \mathrm{F}(x) = \_C1 \, \cos(p\,x) + \_C2 \, \sin(p\,x)$$

```
> s2 := eval(subs(x = 0, sol1));                    # Resp.
```
$s2 := \mathrm{F}(0) = \_C1$

Hence $\_C1 = 0$ since the string is fixed at its left end $x = 0$. Also, you may take $\_C2 = 1$ since arbitrary constants will be provided by $G(t)$.

```
> s3 := subs(_C1 = 0, _C2 = 1, sol1);
```

$$s3 := \mathrm{F}(x) = \sin(p\,x)$$

```
> s4 := subs(x = L, s3);
```

$$s4 := \mathrm{F}(L) = \sin(p\,L)$$

Hence $F(L) = 0$ implies that $pL = n\pi$ with integer n; thus $p = n\pi/L$, $n = 1, 2, \ldots$. (Since $\sin(-\alpha) = -\sin\alpha$, you need not consider $n = -1, -2, \ldots$.) Type

```
> p := n*Pi/L;
```

$$p := \frac{n\,\pi}{L}$$

```
> solF := s3;
```

$$solF := \mathrm{F}(x) = \sin\left(\frac{n\,\pi\,x}{L}\right)$$

Now obtain the ODE for $G(t)$ by setting the left-hand side of eq equal to $-p^2$ and solve the ODE

```
> solG := dsolve(lhs(eq) = -p^2);
```

$$solG := \mathrm{G}(t) = \_C1 \, \sin\left(\frac{\pi\,c\,n\,t}{L}\right) + \_C2 \, \cos\left(\frac{\pi\,c\,n\,t}{L}\right)$$

From the solutions $F(x)$ and $G(t)$ in solF and solG you obtain $u(x, t) = G(t)\,F(x)$

```
> u(x,t) := rhs(solG)*rhs(solF);
```

$$u(x,\,t) := \left(\_C1 \, \sin\left(\frac{\pi\,c\,n\,t}{L}\right) + \_C2 \, \cos\left(\frac{\pi\,c\,n\,t}{L}\right)\right) \sin\left(\frac{n\,\pi\,x}{L}\right)$$

Taking series of these solutions with $n = 1$ to ∞ and suitable constants, one can determine solutions that satisfy given initial conditions (displacement and velocity), as explained on pp. 590-593 of AEM.

Animation shows the string in motion. The simplest solution $u(x, t)$ is the product $\sin t \sin x$ (when $n = 1$, $c = 1$, $L = \pi$ in the previous formula). Type

```
> with(plots):                                      # Ignore the warning.
> animate(sin(t)*sin(x), x = 0..Pi, t = Pi/2..10*Pi, frames = 100);
```

Press Enter (as usual). The maximally displaced string (for $t = \pi/2$, the beginning) will appear. Click with the mouse anywhere in the figure. A frame will appear, and above (outside the worksheet) you will see a row of symbols (a tool-bar). Click on the black triangle (play). The motion will begin and make 4 1/2 cycles. Click again for a repetition. The fewer frames you choose (e.g. frames = 50), the more rapid the motion will be. (Type ?plots,animate for information.)

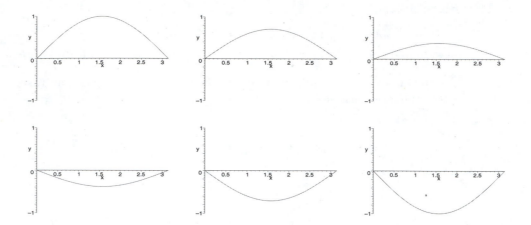

Example 11.1. Fundamental mode of the vibrating string (Animation)

Similar Material in AEM: pp. 587-593

EXAMPLE 11.2 | WAVE EQUATION: D'ALEMBERT'S SOLUTION METHOD. COMMAND pdsolve

D'Alembert's solution method for the one-dimensional wave equation

$$u_{tt} - c^2 u_{xx} = 0$$

consists of transforming the equation by introducing two new independent variables so that one can immediately solve the new equation by two successive integrations. Accordingly, type the wave equation, then the formula tr for the transformed variables and then the transformed equation pde2. (In the response to pde2 the factors c may appear in different positions).

```
> with(PDEtools):
> u := 'u':  v := 'v':  z := 'z':
> pde1 := diff(u(x,t), t, t) - c^2*diff(u(x,t), x, x) = 0;
```

$$pde1 := \left(\frac{\partial^2}{\partial t^2}\, u(x,\, t)\right) - c^2\left(\frac{\partial^2}{\partial x^2}\, u(x,\, t)\right) = 0$$

```
> tr :=   {x = (v + z)/2, t = (v - z)/(2*c)};
```

$$tr := \{x = \frac{1}{2}\, v + \frac{1}{2}\, z,\, t = \frac{1}{2}\, \frac{v-z}{c}\}$$

```
> pde2 := dchange(tr,  pde1, [v, z]);
```

$$pde2 := c\left(c\left(\frac{\partial^2}{\partial v^2}\, u(v,\, z,\, c)\right) - c\left(\frac{\partial^2}{\partial z\, \partial v}\, u(v,\, z,\, c)\right)\right)$$
$$-c\left(c\left(\frac{\partial^2}{\partial z\, \partial v}\, u(v,\, z,\, c)\right) - c\left(\frac{\partial^2}{\partial z^2}\, u(v,\, z,\, c)\right)\right)$$
$$-c^2\left(\left(\frac{\partial^2}{\partial v^2}\, u(v,\, z,\, c)\right) + 2\left(\frac{\partial^2}{\partial z\, \partial v}\, u(v,\, z,\, c)\right) + \left(\frac{\partial^2}{\partial z^2}\, u(v,\, z,\, c)\right)\right) = 0$$

```
> pde2 := simplify(%);
```

$$pde2 := -4\,c^2\left(\frac{\partial^2}{\partial z\,\partial v}\,\mathrm{u}(v,\,z,\,c)\right) = 0$$

Integrate this PDE with respect to z. From the left-hand side you obtain

```
> int(lhs(pde2), z);
```
         `# Resp.` $-4\,c^2\left(\dfrac{\partial}{\partial v}\,\mathrm{u}(v,\,z,\,c)\right)$

Drop $-4\,c^2$, which is not essential. Add an arbitrary integration "constant" $h(v)$ (the result of integrating 0 on the right with respect to z). (Maple puts all integration constants to 0.)

```
> eq := diff(u(v,z), v) - h(v);
```
         `# Resp.` $eq := \left(\dfrac{\partial}{\partial v}\,\mathrm{u}(v,\,z)\right) - \mathrm{h}(v)$

Integrate this with respect to v

```
> int(eq, v);
```
    `# Maple should put parentheses around the whole integrand.`

$$\int\left(\frac{\partial}{\partial v}\,\mathrm{u}(v,\,z)\right) - \mathrm{h}(v)\,dv$$

Add another integration "constant" $-g(z)$. Hence $u - \int h(v)\,dv = g(z)$. Call the last integral $f(v)$. Then in the original variables, since $v = x + ct$, $z = x - ct$, the solution, called **D'Alembert's solution**, is

$$u(x, t) = f(x + ct) + g(x - ct).$$

Practically the same solution can be obtained by the command `pdsolve`, as follows, where _F1 and _F2 are arbitrary functions.

```
> u := 'u':
> sol := pdsolve(pde1, u(x,t));
```

$$sol := \mathrm{u}(x,\,t) = \_F1(c\,t + x) + \_F2(c\,t - x)$$

Similar Material in AEM: pp. 595-597

EXAMPLE 11.3   ONE-DIMENSIONAL HEAT EQUATION

The **one-dimensional heat equation** is $u_t = c^2\,u_{xx}$, where $u(x,t)$ is the temperature in a straight bar or wire at a point x and time t. If the ends of a bar are at $x = 0$ and $x = \pi$ (so that the bar has length π and lies along the x-axis) and are kept at temperature 0, the boundary conditions are $u(0, t) = 0$ and $u(\pi, t) = 0$ for all t. Separation of variables leads to solutions $u = u_n$,

$$u_n(x,\,t) = F_n(x)\,G_n(t) = B_n\,\sin nx\,\exp\left(-(c\,n)^2 t\right)$$

satisfying the boundary conditions. A series of these will satisfy a given initial condition $f(x) = u(x,0)$, for instance, a "triangular" initial temperature (see the figure)

$$f(x) = \begin{cases} x & \text{if} & 0 < x < \pi/2 \\ \pi - x & \text{if} & \pi/2 < x < \pi \end{cases}$$

if the coefficients B_n of the series are the Fourier coefficients of the Fourier sine half-range expansion of $f(x)$ (of period $2\,\pi$). Show this in a plot, which also exhibits the temperatures for various constant values of t. Take $c = 1$ for simplicity.

Solution. Type B_n and then a partial sum of the series of the u_n with these coefficients B_n for various $t = 0, 0.1, 0.2, \ldots$. You will see that even for small t you obtain almost a sine curve because the exponential factors for the second and further terms decrease very rapidly to zero. For information on one of the commands below, type `?piecewise`. Read the command as follows. For $x < 0$ the function f is 0. For $x < \pi/2$ it is x. For $x < \pi$ it is $\pi - x$. Elsewhere it is 0. By the Euler formula,

```
> Bn := simplify(2/Pi*(int(x*sin(n*x), x = 0..Pi/2) +
  int((Pi - x)*sin(n*x), x = Pi/2.:Pi)));
```

$$Bn := 2\,\frac{2\sin\left(\frac{1}{2}\,\pi\,n\right) - \sin(\pi\,n)}{\pi\,n^2}$$

```
> S := sum(Bn*sin(n*x)*exp(-n^2*t), n = 1..13);
```

$$S := 4\,\frac{\sin(x)\,e^{(-t)}}{\pi} - \frac{4}{9}\,\frac{\sin(3\,x)\,e^{(-9\,t)}}{\pi} + \frac{\frac{4}{25}\sin(5\,x)\,e^{(-25\,t)}}{\pi} - \frac{4}{49}\,\frac{\sin(7\,x)\,e^{(-49\,t)}}{\pi}$$

$$+ \frac{\frac{4}{81}\sin(9\,x)\,e^{(-81\,t)}}{\pi} - \frac{4}{121}\,\frac{\sin(11\,x)\,e^{(-121\,t)}}{\pi} + \frac{\frac{4}{169}\sin(13\,x)\,e^{(-169\,t)}}{\pi}$$

```
> S0 := eval(subs(t = 0, S)):
> S1 := eval(subs(t = 0.1, S)):
> S2 := eval(subs(t = 0.2, S)):
> S10 := eval(subs(t = 1.0, S)):
> S20 := eval(subs(t = 2.0, S)):
> S30 := eval(subs(t = 3.0, S)):
> f := piecewise(x < 0,  0, x < Pi/2,  x, x < Pi,  Pi - x, 0):
> plot({S0, S1, S2, S10, S20, S30, f}, x = 0..Pi, color = black);
```

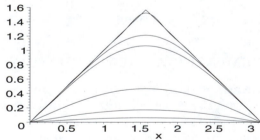

Example 11.3. "Triangular" initial temperature and its decrease with time

Similar Material in AEM: pp. 600-604

| EXAMPLE 11.4 | **HEAT EQUATION, LAPLACE EQUATION** |

Find the steady-state temperature in the rectangular plate $0 \leq x \leq \pi$, $0 \leq y \leq \pi/2$ if the upper edge is kept at temperature $u(x, \pi/2) = f(x) = 1$ and the other three edges are kept at temperature 0. Plot the temperature as a surface over the rectangle.

Solution. The two-dimensional heat equation is $u_t = c^2\,(u_{xx} + u_{yy})$. Since u is assumed to be steady-state (time independent), this equation reduces to **Laplace's**

equation $u_{xx} + u_{yy} = 0$. Separation of variables leads to solutions

$$u_n(x, y) = B_n \sin nx \sinh ny,$$

that satisfy the three zero boundary conditions. It can be shown that a series of these also satisfies $u(x, \pi/2) = f(x) = 1$ if you choose the $B_n \sinh(n\pi/2)$ to be the coefficients of the Fourier sine half-range expansion of $f(x)$. Thus type the following. (Type `?orientation` for information. Actually, you can turn the surface by clicking on any point in the figure and then moving the mouse. Try it.) By the Euler formulas

```
> Bn := 2/(Pi*sinh(n*Pi/2))*int(1*sin(n*x), x = 0..Pi);
```

$$Bn := -2 \frac{\cos(n\pi) - 1}{\pi \sinh\left(\frac{1}{2} n\pi\right) n}$$

```
> u30 := sum(Bn*sin(n*x)*sinh(n*y), n = 1..30):   # Partial sum of 15 terms
> with(plots):                                      # Ignore the warning.

> plot3d(u30, x = 0..Pi, y = 0..Pi/2, axes = NORMAL, orientation =
  [30, 60]);
```

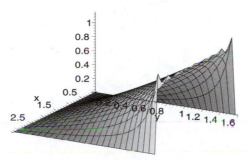

Example 11.4. Temperature given by u30 shown as a surface
over the rectangular plate

To obtain a figure showing the approximation of the temperature along the upper edge, type

```
> w := subs(y = Pi/2, u30):
> plot(w, x = 0..Pi, y = 0..1.2, scaling = constrained);
```

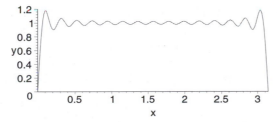

Example 11.4. Temperature given by u30 on the upper edge
Note the beginning Gibbs phenomenon near 0 and π.

Similar Material in AEM: pp. 605-607

EXAMPLE 11.5	RECTANGULAR MEMBRANE.
	DOUBLE FOURIER SERIES

Vibrations of elastic membranes (such as drumheads) are governed by the **two-dimensional wave equation**

$$u_{tt} = c^2 \left(u_{xx} + u_{yy} \right),$$

where $u(x, y, t)$ is the displacement of the membrane at a point (x, y) and time t from its position at rest in the xy-plane. Let the density of the material and the tension be such that $c^2 = 5$. Let the initial velocity be zero and the initial displacement $f(x, y)$ as typed and as shown in the figure. (You can turn the figure by clicking on any point of it and then moving the mouse.)

```
> f := 1/10*(4*x - x^2)*(2*y - y^2);
```

$$f := \frac{1}{10} \left(4x - x^2 \right) \left(2y - y^2 \right)$$

```
> plot3d(f, x = 0..4, y = 0..2, axes = NORMAL);
```

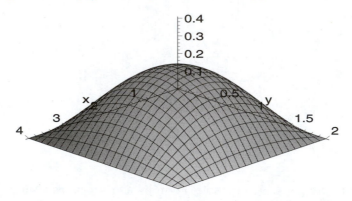

Example 11.5. Initial displacement of the membrane

The solution method is similar to that for the vibrating string. Separation of variables leads to a double sequence of solutions

$$u_{mn} = B_{mn} \cos \left(\lambda_{mn} t \right) \sin \left(\frac{m\pi x}{a} \right) \sin \left(\frac{n\pi y}{b} \right) \qquad (m, \ n = 1, 2, \ldots)$$

satisfying the boundary condition $u = 0$ on the edges of the membrane. A double series of these (a "**double Fourier series**") will satisfy the initial conditions $u(x, y, 0) = f(x, y)$ (given initial displacement) and $u_t(x, y, 0) = 0$ (zero initial velocity) if the B_{mn} are the coefficients of the double Fourier series of $f(x, y)$. Accordingly, type the following. (The resulting terms may come out in a different order.)

```
> a := 4:  b := 2:                      # Lengths of the edges of the membrane
```

The **Euler formulas for a double Fourier series** give

```
> Bmn := 4/(a*b)*int(int(f*sin(m*Pi*x/a)*sin(n*Pi*y/b), x = 0..a),
  y = 0..b);
```

$$Bmn := \frac{128}{5}(-4\cos(n\pi) + 2m\pi\sin(m\pi)\cos(n\pi) - 2n\pi\sin(n\pi)+$$

$$m\pi^2\sin(m\pi)\,n\sin(n\pi) + 4\cos(m\pi)\cos(n\pi) + 2\cos(m\pi)\,n\pi\sin(n\pi)$$

$$-4\cos(m\pi) + 4 - 2m\pi\sin(m\pi))/(m^3\pi^6 n^3)$$

Tell the computer that m and n are integers, so that B_{mn} will simplify. (The tildes after m and n remind you of these assumptions. Type `?assume`.)

```
> assume(m, integer): assume(n, integer):
> Bmn := simplify(Bmn);          # The minus in front may appear in the fraction.
```

$$Bmn := -\frac{512}{5}\frac{(-1)^{n\tilde{}} - 1 - (-1)^{(m\tilde{}+n\tilde{})} + (-1)^{m\tilde{}}}{m\tilde{}^3\pi^6 n\tilde{}^3}$$

For the time function in u_{mn} you need λ_{mn} as follows.

```
> lambda_mn := 5*Pi*sqrt(m^2/a^2 + n^2/b^2);
```

$$lambda\_mn := \frac{5}{4}\pi\sqrt{m\tilde{}^2 + 4n\tilde{}^2}$$

```
> umn := Bmn*cos(lambda_mn*t)*sin(m*Pi*x/a)*sin(n*Pi*y/b);
```

$$umn := -\frac{512}{5}\left(((-1)^{n\tilde{}} - 1 - (-1)^{(m\tilde{}+n\tilde{})} + (-1)^{m\tilde{}})\cos\left(\frac{5}{4}\pi\sqrt{m\tilde{}^2 + 4n\tilde{}^2}\,t\right)\right.$$
$$\left.\sin\left(\frac{1}{4}m\tilde{}\,\pi x\right)\sin\left(\frac{1}{2}n\tilde{}\,\pi y\right)\right)\Big/ m\tilde{}^3\pi^6 n\tilde{}^3$$

Now type a partial sum S of these $u_{mn}(x, y, t)$. Very few terms will do. In fact, let us take just a single term. (Try more terms.) Then type $S0$, which is S for $t = 0$, that is, an approximation of the initial shape of the membrane.

```
> S := sum(sum(umn, m = 1..1), n = 1..1);          # Or use subs(...);.
```

$$S := \frac{2048}{5}\frac{\cos(\frac{5}{4}\pi\sqrt{5}\,t)\,\sin(\frac{1}{4}\pi x)\sin(\frac{1}{2}\pi y)}{\pi^6}$$

```
> S0 := eval(subs(t = 0, S));          # Partial sum for t = 0
```

$$S0 := \frac{2048}{5}\frac{\sin\left(\frac{1}{4}\pi x\right)\sin\left(\frac{1}{2}\pi y\right)}{\pi^6}$$

```
> plot3d(S0, x = 0..4, y = 0..2, axes = NORMAL);
```

Similar Material in AEM: pp. 619-625

| EXAMPLE 11.6 | **LAPLACIAN. CIRCULAR MEMBRANE.**
BESSEL EQUATION |

This concerns the vertical vibrations of a **circular membrane** fixed along its edge $x^2 + y^2 = R^2$ in the xy-plane. So you must first obtain the two-dimensional wave equation $u_{tt} = c^2(u_{xx} + u_{yy})$ in polar coordinates. (Type `?laplacian`, not `?Laplacian`, for information.) Relevant Maple commands are as follows.

```
> with(linalg):          # Ignore the warning.
> lap := laplacian(u(x,y), [x, y]);
```

$$lap := \left(\frac{\partial^2}{\partial x^2} \, u(x, y) \right) + \left(\frac{\partial^2}{\partial y^2} \, u(x, y) \right)$$

```
> with(PDEtools):
> tr := {x = r*cos(theta), y = r*sin(theta)}:
> dchange(tr, lap);                          # Long response not shown.
> combine(%);
```

$$\frac{\left(\frac{\partial^2}{\partial r^2} \, u(r, \theta) \right) r^2 + \left(\frac{\partial}{\partial r} \, u(r, \theta) \right) r + \left(\frac{\partial^2}{\partial \theta^2} \, u(r, \theta) \right)}{r^2}$$

This is the Laplacian in polar coordinates. You may confirm this by typing

```
> laplacian(u(r, theta), [r, theta], coords = polar);
```

$$\frac{\left(\frac{\partial^2}{\partial r^2} \, u(r, \theta) \right) r + \left(\frac{\partial}{\partial r} \, u(r, \theta) \right) + \frac{\frac{\partial^2}{\partial \theta^2} \, u(r, \theta)}{r}}{r}$$

We shall consider vibrations that are rotationally symmetric (i.e., independent of θ). Accordingly. type

```
> u := 'u':
> pde := diff(u(r,t), t, t) = c^2*laplacian(u(r,t), [r, theta], coords
  = polar);
```

$$pde := \frac{\partial^2}{\partial t^2} \, u(r, t) = \frac{c^2 \left(\left(\frac{\partial}{\partial r} \, u(r, t) \right) + r \left(\frac{\partial^2}{\partial r^2} \, u(r, t) \right) \right)}{r}$$

The derivative with respect to θ is gone. Now separate variables. Start from

```
> pde2 := eval(subs(u(r,t) = W(r)*G(t), pde));
```

$$pde2 := W(r) \left(\frac{\partial^2}{\partial t^2} \, G(t) \right) = \frac{c^2 \left(\left(\frac{\partial}{\partial r} \, W(r) \right) G(t) + r \left(\frac{\partial^2}{\partial r^2} \, W(r) \right) G(t) \right)}{r}$$

```
> eq1 := pde2/(c^2*W(r)*G(t));                # Response not shown.
> eq2 := simplify(%);
```

$$eq2 := \frac{\frac{\partial^2}{\partial t^2} \, G(t)}{c^2 \, G(t)} = \frac{\left(\frac{\partial}{\partial r} \, W(r) \right) + r \left(\frac{\partial^2}{\partial r^2} \, W(r) \right)}{W(r) \, r}$$

The variables are separated. Each side must equal a constant, which must be negative, say, $-k^2$, in order to finally obtain solutions that are not identically zero. Set the right-hand side equal to $-k^2$ or write `rhs(eq2) + k^2 = 0`, and then multiply by $W(r) \, r$.

```
> eq3 := rhs(eq2) + k^2 = 0;
```

$$eq3 := \frac{\left(\frac{\partial}{\partial r} \, W(r) \right) + r \left(\frac{\partial^2}{\partial r^2} \, W(r) \right)}{W(r) \, r} + k^2 = 0$$

```
> eq4 := simplify(eq3*W(r)*r);
```

$$eq4 := \left(\frac{\partial}{\partial r}\,\mathrm{W}(r)\right) + r\left(\frac{\partial^2}{\partial r^2}\,\mathrm{W}(r)\right) + k^2\,\mathrm{W}(r)\,r = 0$$

This is the **Bessel equation** with parameter $\nu = 0$ (and independent variable $s = kr$). Assume $k > 0$ (which Maple indicates by a tilde after k) and solve this ordinary differential equation by `dsolve`.

```
> assume(k > 0):
> with(DEtools):                          # Ignore the warning.
> dsolve(eq4, W(r));                       # k~ r means k~ times r
```

$$\mathrm{W}(r) = \_C1\,\mathrm{BesselJ}\,(0,\,k^\sim r) + \_C2\,\mathrm{BesselY}\,(0,\,k^\sim r)$$

Hence a solution is the Bessel function $J_0(kr)$. (The Bessel function Y_0 cannot be used because it becomes infinite as $r \to 0$. Type `?Bessel`, `?BesselJZeros` for information.) k must be determined so that $J_0(kr) = 0$ on the edge $r = R$ of the membrane. For plotting, take $R = 1$ and $c = 1$ (the constant in the wave equation). Then k must equal the first positive zero (this gives the simplest solution), the second positive zero, etc. (See the graph of J_0 in Example 4.6 in this Guide.) To obtain the first three zeros, type

```
> zeros := evalf(BesselJZeros(0, 1..3));
```

$$zeros := 2.404825558,\ 5.520078110,\ 8.653727913$$

```
> z1 := zeros[1];   z2 := zeros[2];   z3 := zeros[3];
```

$$z1 := 2.404825558$$
$$z2 := 5.520078110$$
$$z3 := 8.653727913$$

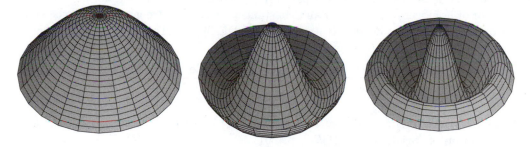

Example 11.6. Circular membrane, simplest forms of vibrations

You can now plot the three simplest solutions by typing

```
> plot3d([r*cos(theta), r*sin(theta), BesselJ(0, z1*r)], r = 0..1,
    theta = 0..2*Pi, scaling = constrained);

> plot3d([r*cos(theta), r*sin(theta), BesselJ(0, z2*r)], r = 0..1,
    theta = 0..2*Pi, scaling = constrained);
```

```
> plot3d([r*cos(theta), r*sin(theta), BesselJ(0, z3*r)], r = 0..1,
   theta = 0..2*Pi, scaling = constrained);
```

Similar Material in AEM: pp. 629-634

Problem Set for Chapter 11

Pr.11.1 (**Normal modes of the vibrating string**) The solutions `F(x)` in `solF` in Example 11.1 in this Guide are called the *normal modes* of the string. Plot them (separately) for $n = 1, 2, 3, 4$ (with $L = \pi$). Multiply them by the corresponding sines of t (with $c = 1$) and apply animation. (See Example 11.1 in this Guide. *AEM Ref.* p. 590)

Pr.11.2 (**Animation**) Show 5 cycles of the motion

$$u(x, t) = \sin x \cos t - (1/9) \sin 3x \cos 3t + (1/25) \sin 5x \cos 5t$$

(approximating the motion when the initial deflection of the string is "triangular"). (See Example 11.1 in this Guide. *AEM Ref.* p. 593)

Pr.11.3 (**Extension of D'Alembert's method**) Solve $u_{xx} + u_{xy} - 2 u_{yy} = 0$ by setting $v = x + y$, $z = 2x - y$. (*AEM Ref.* p. 598 (#14))

Pr.11.4 (**Separation of variables**) Solve $u_{xx} - u_{yy} = 0$, choosing the separation constant positive, zero, and negative. (*AEM Ref.* p. 594 (#18))

Pr.11.5 (**Checking solutions**) Checking is important. The computer may sometimes give you false or insufficient results. Check whether $u = (c_1 e^{kx} + c_2 e^{-kx})(c_3 e^{ky} + c_4 e^{-ky})$ is a solution of $u_{xx} - u_{yy} = 0$. Can you replace k in the functions depending on y by another constant?

Pr.11.6 (**Tricomi equation. Airy equation**) Find solutions $u(x, y) = F(x)G(y)$ of the Tricomi equation $y u_{xx} + u_{yy} = 0$. Show that for $G(y)$ this gives $G'' + kyG = 0$. (With $k = -1$ or 1 this is called *Airy's equation*.) Find a general solution $G(y)$ involving the Airy functions Ai and Bi (see Ref. [1] in Appendix 1 for information). Obtain Ai$(-y)$ from that general solution and plot it. (*AEM Ref.* p. 598)

Pr.11.7 (**Vibrating beam. Command** `pdsolve`) Vertical vibrations of a horizontal elastic beam of homogeneous material and constant cross section are governed by the equation $u_{tt} = c^2 u_{xxxx}$. Solve this fourth-order PDE by `pdsolve`. (Type `?pdsolve` for information. *AEM Ref.* pp. 598, 599)

Pr.11.8 (**Heat equation**) Solve the heat equation in Example 11.3 in this Guide for a bar of length 10 with $c = 1$ and "parabolic" initial temperature $u(x, 0) = x(10 - x)$. (*AEM Ref.* p. 608 (#5))

Pr.11.9 (**Heat equation**) Derive the solutions $u_n(x, t)$ in Example 11.3 in this Guide by separation of variables. (*AEM Ref.* p. 601)

Pr.11.10 (**Heat flow in a long bar. Fourier integral. Error function**) An infinite bar (the x-axis), practically a very long bar, is heated to 100 degrees at some point, practically, between -1 and 1, the rest being kept at 0. At $t = 0$, heating is terminated and heat begins to flow away to both sides. It can be shown that the temperature is given by $100/\sqrt{\pi}$ times the integral of $\exp(-t^2)$ from $-(1+x)/(2c\sqrt{t})$ to $(1-x)/(2c\sqrt{t})$. Study the decrease of the temperature by animation, assuming that $c = 1$. The integral of $\exp(-t^2)$ from 0 to v (times $2/\sqrt{\pi}$) is called the **error function** and is denoted by erf v. (Type `?animate`; see also the instruction on animation in Example 11.1 in this Guide. *AEM Ref.* pp. 613, A56)

Pr.11.11 **(Two-dimensional heat equation)** Solve Example 11.4 in this Guide when the upper edge is kept at the temperature $\sin x$, the other data being as before.
(*AEM Ref.* pp. 605-607)

Pr.11.12 **(Isotherms)** Find the isotherms (curves of constant temperature) in Pr.11.11 and plot some of them. Do these curves look physically reasonable?
(*AEM Ref.* pp. 605-607)

Pr.11.13 **(Animation, rectangular membrane)** Show the motion of the membrane in Example 11.5 in this Guide for u_{22} with $B_{22} = 1$. (*AEM Ref.* p. 625)

Pr.11.14 **(Circular membrane)** Show the motion of the three solutions in the figure of Example 11.6 in this Guide with $c = 1$, $R = 1$, so that $k^2 = z_n{}^2$. Use that $G'' + k^2 G = 0$.
(Type `?BesselJZeros`. *AEM Ref.* p. 632)

Pr.11.15 **(Circular membrane, vibration depending on angle)** Study the motion of the circular membrane given by $u_{11}(r, \theta, t) = \sin kt \, J_1(kr) \cos \theta$, which has the y-axis ($\theta = \pi/2$ and $3\pi/2$) as a nodal line (line along which the membrane does not move). Here, k is the first positive zero of J_1 (type `?BesselJZeros`).
(*AEM Ref.* pp. 635, A85)

PART D. COMPLEX ANALYSIS

Content. Complex numbers and functions, conformal mapping (Chap. 12)
Complex integration (Chap. 13)
Power series, Taylor series (Chap. 14)
Laurent series, singularities, residue integration (Chap. 15)
Potential theory (Chap. 16)

The Maple notation for $i = \sqrt{-1}$ is I. Do not use I otherwise. Symbols $+$, $-$, $*$, $/$, $\hat{\ }$, evalf are the same in complex as in real. Re, Im, abs, argument, and conjugate give the real part, imaginary part, absolute value, argument, and complex conjugate, respectively. evalc (suggesting 'evaluate complex') gives results in the form $a + ib$. This as well as the polar form and the plotting of complex numbers is illustrated in Example 12.1.

Chapter 12

Complex Numbers and Functions.
Conformal Mapping

Content. Complex numbers, polar form, plots (Ex. 12.1, Prs. 12.1-12.5)
Equations, roots, sets (Ex. 12.2, Prs. 12.6-12.11)
Cauchy-Riemann equations, harmonic functions (Ex. 12.3, Pr. 12.12)
Conformal mapping (Ex. 12.4, Prs. 12.13-12.15, 12.17)
Special complex functions (Exs. 12.5, 12.6, Prs. 12.16-12.20)

Examples for Chapter 12

EXAMPLE 12.1   **COMPLEX NUMBERS. POLAR FORM. PLOTTING**

Let us begin with illustrating arithmetic in complex. Let

```
> z1 := 3 + 4*I;
> z2 := 5 - 2*I;
```
\# Resp. $z1 := 3 + 4\,I$
\# Resp. $z2 := 5 - 2\,I$

Then

```
> z1 + z2;   z1 - z2;   z1*z2;   z1/z2;
```

$$8 + 2\,I$$
$$-2 + 6\,I$$
$$23 + 14\,I$$
$$\frac{7}{29} + \frac{26}{29}\,I$$

```
> evalf(z1/z2, 5);
```
\# Resp. $.24138 + .89655\,I$

130

```
> (3.0 + 4*I)/(5 - 2*I);                # Resp. .2413793103 + .8965517241 I
```

If a result is not in the form $a + ib$, but you want it in this form, apply `evalc` (suggesting 'evaluate complex'; type `?evalc`). For instance,

```
> evalc(cos(3 + I));                    # Resp. cos(3) cosh(1) - I sin(3) sinh(1)
> cos(3.0 + I);                         # Resp. -1.527638250 - 1.658444019 I
```

`Re`, `Im`, `abs`, `argument`, and `conjugate` give the real part, imaginary part, absolute value (modulus), argument, and complex conjugate, respectively. For instance,

```
> Re(cos(3 + I));                       # Resp. cos(3) cosh(1)
> evalf(%);                             # Resp. -1.527638250
> Im(cos(3 + I));                       # Resp. - sin(3) sinh(1)
> abs(cos(3 + I));                      # Resp. √(cos(3)² cosh(1)² + sin(3)² sinh(1)²)
> evalc(abs(x + I*y));                  # Resp. √(x² + y²)
> evalc(argument(x + I*y));             # Resp. arctan(y, x)
> conjugate(-4.8 + 0.3*I);              # Resp. -4.8 - .3 I
> evalc(conjugate(x + I*y));            # Resp. x - I y
```

The response equations: `abs(cos(3 + I))` gives $\sqrt{\cos(3)^2\cosh(1)^2 + \sin(3)^2\sinh(1)^2}$; `evalc(abs(x + I*y))` gives $\sqrt{x^2 + y^2}$; `evalc(argument(x + I*y))` gives $\arctan(y, x)$; `evalc(conjugate(x + I*y))` gives $x - I\,y$.

Polar form. Maple gives the **principal value** Arg z of the argument $\theta = \arg z$ of $z = |z| \exp i\theta$, defined by $-\pi < \theta \le \pi$. (Type `?polar` for information.) For instance,

```
> polar(4 + 3*I);
```

$$\mathrm{polar}\left(5,\ \arctan\left(\frac{3}{4}\right)\right)$$

```
> polar(-2*I);
```

$$\mathrm{polar}\left(2,\ -\frac{1}{2}\pi\right)$$

```
> polar(a + I*b);
```

$$\mathrm{polar}\left(|a + I\,b|,\ \mathrm{argument}\,(a + I\,b)\right)$$

To proceed *from polar form to Cartesian form* use `evalc`. For instance,

```
> polar(1 + I);                         # Resp. polar(√2, 1/4 π)
> evalc(polar(sqrt(2), Pi/4));          # Resp. 1 + I
```

Multiplication in polar form can be done as follows.

```
> w := 'w':
> polar(r, t)^2*polar(s, w)^3;          # Resp. polar(r, t)² polar(s, w)³
> simplify(%);                          # Resp. polar(r² s³, 2 t + 3 w)
> simplify(polar(r, t)^5);              # Resp. polar(r⁵, 5 t)
```

The responses: `polar(r, t)^2*polar(s, w)^3` gives $\mathrm{polar}(r, t)^2\,\mathrm{polar}(s, w)^3$; `simplify(%)` gives $\mathrm{polar}\left(r^2 s^3,\ 2t + 3w\right)$; `simplify(polar(r, t)^5)` gives $\mathrm{polar}\left(r^5,\ 5t\right)$.

Plotting complex numbers in the **complex plane.** To obtain $4 + 1.5i$ and $4 - 1.5i$ as points, type

```
> plot([[4, 1.5], [4, -1.5]], x = 0..5, style = point, labels = [x, y]);
```

Here x = 0..5 gives the axis as shown. To obtain these numbers as vectors (line
segments), type

```
> plot({1.5*x/4, -1.5*x/4}, x = 0..4, labels = [x, y]);
```

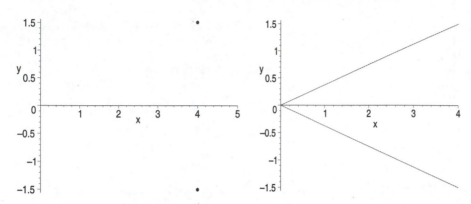

Example 12.1. Complex numbers
plotted as points

Example 12.1. Complex numbers
plotted as vectors (line segments)

To combine both methods of plotting, load the plots package and then use display,
as shown.

```
> with(plots):                              # Ignore the warning.
> P1 := plot([[4, 1.5], [4, -1.5]], x = 0..5, style = point):
> P2 := plot({1.5*x/4, -1.5*x/4}, x = 0..4):
> display({P1, P2}, labels = [x, y]);
```

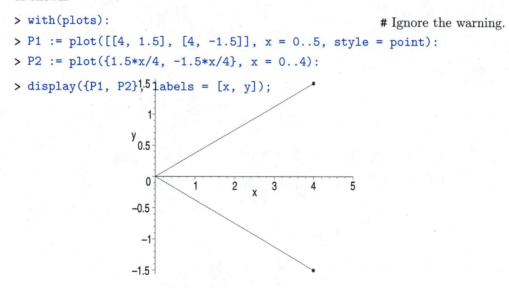

Example 12.1. Combination of the two plots for points

Similar Material in AEM: pp. 652-660

EXAMPLE 12.2 **EQUATIONS. ROOTS. SETS IN THE COMPLEX PLANE**

Quadratic and other equations can be solved by solve or allvalues(RootOf(...)).
(Type ?solve ?fsolve, ?allvalues for information.) For instance, solve the equation $z^2 - (5+i)z + 8 + i = 0$.

Solution.

```
> eq := z^2 - (5 + I)*z + 8 + I = 0;
```

$$eq := z^2 - (5+I)\,z + 8 + I = 0$$

```
> sol := solve(eq, z);                    # Resp. sol := 3 + 2I, 2 - I
> sol[1];                                 # Resp. 3 + 2I
> sol2 := allvalues(RootOf(eq));          # Resp. sol2 := 3 + 2I, 2 - I
```

Roots of unity. By definition, these are the roots of an equation $z^n = 1$, where n is a given integer. For instance, for $n = 4$ you get $1, i, -1, -i$. Find and plot the roots of unity for $n = 9$.

Solution. Type

```
> Sol := solve(z^9 = 1.0, z);
```

$Sol := 1., -.5000000000 + .8660254038\,I, -.5000000000 - .8660254038\,I,$
$\qquad\qquad -.9396926208 + .3420201433\,I, .1736481777 - .9848077530\,I,$
$\qquad\qquad .7660444431 + .6427876097\,I, -.9396926208 - .3420201433\,I,$
$\qquad\qquad .1736481777 + .9848077530\,I, .7660444431 - .6427876097\,I$

Maple cannot plot this *complex* sequence. Convert it to a real sequence of pairs $[\operatorname{Re} z_k, \operatorname{Im} z_k]$, where z_k, $k = 1 \ldots 9$, are the roots.

```
> S := seq([Re(Sol[n]), Im(Sol[n])], n = 1..9);
```

$S := [1., 0], [-.5000000000, .8660254038], [-.5000000000, -.8660254038],$
$\qquad\qquad [-.9396926208, .3420201433], [.1736481777, -.9848077530],$
$\qquad\qquad [.7660444431, .6427876097], [-.9396926208, -.3420201433],$
$\qquad\qquad [.1736481777, .9848077530], [.7660444431, -.6427876097]$

```
> with(plots):          # Try without; see what you get. Ignore the warning.
> P1 := plot([S], style = point):
> P2 := plot([cos(t), sin(t), t = 0..2*Pi], scaling = constrained):
> display(P1, P2);
```

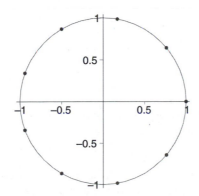

Example 12.2. Roots of unity for $n = 9$
on the unit circle $|z| = 1$

Circles. The circle of radius 1 with center 0 in the figure is called the **unit circle**. Its interior is called the **open unit disk**. From the plotting command you can infer that a circle of radius r and center z_0, for instance, $r = 3$, $z_0 = 2 + i$, can be plotted by the following command. To get a circle instead of an ellipse, add `scaling = constrained`, as before.

```
> plot([2 + 3*cos(t), 1 + 3*sin(t), t = 0..2*Pi], labels = [x, y]);
```

Besides disks, other important regions are those bounded by two concentric circles. Such a region is called an **annulus**. For instance, if the center is $2 + i$ and the radii are 3 (this gives the previous circle) and 1.4, you can type (with P5 for plotting the center)

```
> P3 := plot([2 + 3*cos(t), 1 + 3*sin(t), t = 0..2*Pi]):
> P4 := plot([2 + 1.4*cos(t), 1 + 1.4*sin(t), t = 0..2*Pi]):
> P5 := plot([[2, 1]], style = point):
> display(P3, P4, P5, scaling = constrained, labels = [x, y]);
```

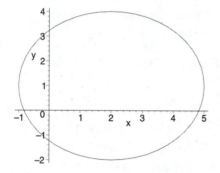   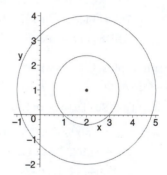

Example 12.2. Arbitrary circle plotted without `scaling = constrained`   *Example 12.2.* Annulus with center $z = 2 + i$ and radii 3 and 1.4

Similar Material in AEM: pp. 660-664

EXAMPLE 12.3 **CAUCHY-RIEMANN EQUATIONS.**
HARMONIC FUNCTIONS

A standard notation for variables and functions is

$$z = x + iy \quad \text{and} \quad f(z) = u(x, y) + iv(x, y)$$

with real u and v. The **Cauchy-Riemann equations** (involving partial derivatives) are

$$u_x = v_y, \qquad u_y = -v_x$$

or, equivalently, $u_x - v_y = 0$, $u_y + v_x = 0$. These are the most important equations in the whole chapter because, roughly speaking, they are necessary and sufficient for $f(z)$ to be an **analytic function**.

For instance, find out whether $e^x(\cos y + i \sin y)$ is analytic.

Solution. Type the function and then the partial derivatives needed.

```
> f := exp(x)*(cos(y) + I*sin(y));
```

$$f := e^x \left(\cos(y) + I \sin(y)\right)$$

```
> ux := diff(evalc(Re(f)), x);          # Resp. ux := e^x cos (y)
> vy := diff(evalc(Im(f)),y);           # Resp. vy := e^x cos (y)
```

Hence the first Cauchy-Riemann equation is satisfied. And so is the second one,

```
> CR2 := diff(evalc(Re(f)), y) + diff(evalc(Im(f)), x);
```

$$CR2 := 0$$

Is $|z|^2 = x^2 + y^2$ analytic?

Solution. $f_2 = u + iv = x^2 + y^2$, hence $u = x^2 + y^2 = f_2$ and $v = 0$. Type

```
> f2 := x^2 + y^2;                       # Resp. f2 := x^2 + y^2
> CR1 := diff(evalc(Re(f2)), x) - diff(evalc(Im(f2)), y);
```

The response is $2x$ (instead of 0). The answer is no. You can stop here.

Harmonic functions. Solutions of Laplace's equation (with continuous second partial derivatives) are called *harmonic functions*. If u and v are harmonic and such that $f = u + iv$ is analytic, then v is called a **conjugate harmonic function** of u.

Show that $u = x^2 - y^2 - y$ is harmonic and find a conjugate harmonic v.

Solution.

```
> u := x^2 - y^2 - y;                    # Resp. u := x^2 - y^2 - y
> with(linalg):                          # Ignore the warning.
> laplacian(u, [x, y]);                  # Resp. 0
```

Hence u is harmonic. Now use the first Cauchy-Riemann equation. Calculate u_x and then integrate $u_x = v_y$ with respect to y, adding an arbitrary "constant" of integration $h(x)$ (not given by Maple).

```
> ux := diff(u, x);                      # Resp. ux := 2x
> v := int(ux, y) + h(x);                # Resp. v := 2xy + h(x)
```

Differentiate this with respect to x, obtaining v_x, and equate it to $-u_y$, thus satisfying the second Cauchy-Riemann equation. Then integrate the result with respect to x.

```
> vx := diff(v, x) = -diff(u, y);
```

$$vx := 2y + \left(\frac{\partial}{\partial x} h(x)\right) = 2y + 1$$

```
> v := int(rhs(vx), x) + c;
```

$$v := (2y + 1)x + c$$

You can now combine this with the given u, obtaining $f(z) = u(x,y) + iv(x,y)$. Using $x = (z + \bar{z})/2$, $y = (z - \bar{z})/2i$, you can express the result in terms of $z = x + iy$ as $f(z) = z^2 + iz + ic$ (c real) because

```
> f := u + I*v;                          # Resp. f := x^2 - y^2 - y + I ((2y + 1) x + c)
```

```
> subs(x = (z + conjugate(z))/2, y = (z - conjugate(z))/(2*I), f);
```

$$\left(\frac{1}{2}z + \frac{1}{2}\,\overline{(z)}\right)^2 + \frac{1}{4}\left(z - \overline{(z)}\right)^2 + \frac{1}{2}I\left(z - \overline{(z)}\right) +$$

$$I\left(\left(-I\left(z - \overline{(z)}\right) + 1\right)\left(\frac{1}{2}z + \frac{1}{2}\,\overline{(z)}\right) + c\right)$$

```
> simplify(%);
```
                              # Resp. $z^2 + Iz + Ic$

Similar Material in AEM: pp. 669-673

EXAMPLE 12.4   **CONFORMAL MAPPING**

The mapping of a region in the z-plane onto a region in the w-plane given by an analytic function $w = f(z) = u(x,y) + iv(x,y)$, $z = x + iy$, is **conformal** (angle-preserving in size and sense), except at points where the derivative $f'(z)$ is zero. You can map rectangles by the command `conformal(f, z = z`$_1$`..z`$_2$`)`. Type `?conformal`. Here z_1 and z_2 are two diagonally opposite vertices of the rectangle. You get the images of 11 lines $x = const$ and 11 lines $y = const$. Or you get m and n lines, respectively, if you add `grid = [m, n]` in your plot command. Add `scaling = constrained` so that the image curves will intersect orthogonally.

```
> with(plots):              # Load the plots package. # Ignore the warning.
```

A 45-degree rotation of the rectangle $0 \leq x \leq 2, 0 \leq y \leq 1$, combined with a dilatation by a factor $\sqrt{2}$, is obtained by typing

```
> conformal((1+I)*z, z = 0..2 + I, grid = [5, 11], scaling =  constrained,
    labels = [u, v]);
```

$w = z^2$ doubles angels at the origin, where $w' = 2z = 0$. For example,

```
> conformal(z^2, z = 0..1 + I, scaling = constrained, labels = [u, v]);
```

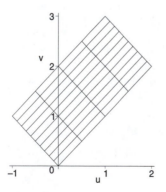 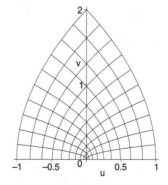

Example 12.4. Conformal mapping   *Example 12.4.* Conformal mapping of the
by $w = (1+i)z$ square $0 \leq x \leq 1, 0 \leq y \leq 1$ by $w = z^2$

An elliptic ring is obtained as the image of the annulus $2 \leq |z| \leq 3$ by using $w = z + 1/z$ and including `coords = polar` in the plot command, so that `z = 2..3 + 2*Pi*I` is interpreted as $2 \leq |z| \leq 3, 0 \leq \theta \leq 2\pi$.

```
> conformal(z + 1/z,  z = 2..3 + 2*Pi*I, grid = [5, 11], labels = [u, v],
    scaling = constrained, coords = polar, numxy = [50, 50]);
```

The **linear fractional transformation** $w = (z-i)/(-iz+1)$ maps a large rectangle in the upper half-plane into the unit disk,

```
> conformal((z - I)/(-I*z + 1), z = -10..10 + 10*I, numxy = [100, 100],
    scaling = constrained, labels = [u, v]);
```

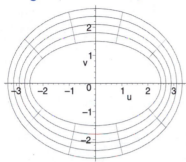

Example 12.4. Mapping of an annulus by $w = z + 1/z$, giving an elliptic ring. Use of polar coordinates

Example 12.4. Mapping of a very large rectangle (theoretically: the upper half-plane) into the unit disk by a linear fractional transformation

An airfoil is obtained by $w = z + 1/z$ as the image of a special circle. Indeed, type

```
> z := 1/10*(1 + I + sqrt(122)*exp(I*t));          # Circle
```

$$z := \frac{1}{10} + \frac{1}{10} I + \frac{1}{10} \sqrt{122}\, e^{(I\,t)}$$

```
> w := z + 1/z;                                     # Airfoil
```

$$w := \frac{1}{10} + \frac{1}{10} I + \frac{1}{10} \sqrt{122}\, e^{(I\,t)} + \frac{1}{\frac{1}{10} + \frac{1}{10} I + \frac{1}{10} \sqrt{122}\, e^{(I\,t)}}$$

```
> plot([Re(w), Im(w), t = 0..2*Pi], scaling = constrained, labels = [u, v]);
```

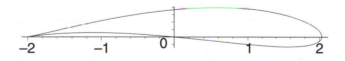

Example 12.4. **Joukowski airfoil** (the image of a special circle under $w = z + 1/z$)

Similar Material in AEM: pp. 675-678, 695

| EXAMPLE 12.5 | **EXPONENTIAL, TRIGONOMETRIC, AND HYPERBOLIC FUNCTIONS** |

Evaluating functions proceeds as in real, except for the possibility of switching back and forth between z and $x + iy$. To return from $z = x + iy$ to z, type the command for unassigning, that is, `z := 'z'` . The following examples will explain the essential commands. These commands produce the given function $f(z)$, its value at $2 + i$, its real part $u(x, y)$, a plot of $u(x, y)$ as a surface, the absolute value $|f(2 + 3i)|$, the unassignment `z:='z':` , and the derivative $f'(z)$.

```
> z := 'z':
> f := -5*z^3 + (12 - 2*I)*z^2 - I*z - 200;
```

$$f := -5\,z^3 + (12 - 2\,I)\,z^2 - I\,z - 200$$

```
> subs(z = 2 + 3*I, f);                          # Resp. -3 + 107 I
> u := evalc(Re(f));                             # Error: z regarded as real
```

$$u := -200 - 5\,z^3 + 12\,z^2$$

```
> z := x + I*y;                                  # Resp. z := x + I y
> u := evalc(Re(f));                             # Try without evalc.
```

$$u := -200 - 5\,x^3 + 15\,x\,y^2 + 12\,x^2 - 12\,y^2 + 4\,x\,y + y$$

```
> plot3d(u, x = 0..5, y = 0..5, axes = NORMAL, labels = [x, y, "f"],
  orientation = [30, 70]);
```

Example 12.5. Real part `u` of `f` plotted as surface over the xy-plane

```
> evalf(subs(z = 2 + 3*I, abs(f)));              # Try without evalf
```

$$107.0420478$$

```
> diff(f, z);
```

`Error, wrong number (or type) of parameters in function diff`

```
> z := 'z':                                      # Return from x + Iy to z.
> diff(f, z);                                    # Resp. -15z^2 + (24 - 4I)z - I
```

Complex exponential function e^z, **also written exp z.** The command for the exponential function with basis $e = 2.71828...$ is `exp(z)` . Accordingly,

```
> exp(1.4 - 0.6*I);              # Resp. 3.346900958 - 2.289738139 I
> exp(2 + I);                                         # Resp. e^(2+I)
> evalc(exp(2 + I));                    # Resp. e^2 cos(1) + I e^2 sin(1)
> Re(exp(2 + I));                             # Resp. e^2 cos(1)
> evalf(exp(2 + I));              # Resp. 3.992324048 + 6.217676312 I
> evalf(exp(1), 50);                 # Base e. Type ?E for information.
```

$$2.7182818284590452353602874713526624977572470937000$$

```
> evalc(exp(I*z));
```
          # Resp. $\cos(z) + I \sin(z)$ **Euler's formula**

Complex trigonometric functions $\cos z$, $\sin z$, $\tan z$, $\cot z$, $\sec z$, $\csc z$. To obtain values of these functions or their real or imaginary parts, etc., use the commands just illustrated. For instance,

```
> z := x + I*y;
```
          # Resp. $z := x + I\,y$
```
> evalc(cos(z));
```
          # Resp. $\cos(x)\cosh(y) - I \sin(x) \sinh(y)$
```
> evalc(sin(z)) ;
```
          # Resp. $\sin(x)\cosh(y) + I \cos(x) \sinh(y)$
```
> evalf(cos(2 + 3*I));
```
          # Resp. $-4.189625691 - 9.109227894\,I$
```
> evalf(tan(1 + 3*I), 4);
```
          # Resp. $.004517 + 1.002\,I$
```
> plot3d(abs(sin(z)), x = 0..2*Pi, y = 0..1.8, axes = NORMAL);
```

The next plot shows the image of the rectangle $-\pi/2+0.5 < x < \pi/2-0.5$, $-1 < y < 1$ under the mapping by $\sin z$.

```
> with(plots):
```
          # Ignore the warning.
```
> conformal(sin(z), z = -Pi/2 + 0.5 - I..Pi/2 - 0.5 + I,
    labels = ["u", "v"]);
```

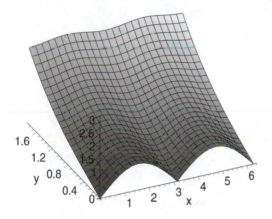 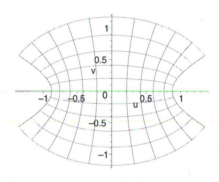

Example 12.5. Surface of the absolute value of $\sin z$ ("**modular surface**")   ***Example 12.5.*** Conformal mapping of a rectangle by $\sin z$

cosine and **sine** are defined in terms of exponential functions. To obtain the defining formulas on the computer, type

```
> z:='z':
> (exp(I*z) + exp(-I*z))/2;
```

$$\frac{1}{2}\,\mathrm{e}^{(I\,z)} + \frac{1}{2}\,\mathrm{e}^{(-I\,z)}$$

```
> convert(%, trig);
```
          # Resp. $\cos(z)$
```
> convert((exp(I*z) - exp(-I*z))/(2*I), trig);
```
          # Resp. $\sin(z)$

Hyperbolic functions $\cosh z$, $\sinh z$, $\tanh z$, $\coth z$. In complex, these functions are closely related to the trigonometric functions. To see this for cosh, type the following. For sinh, tanh, and coth the situation is similar. Try it. Numerical values can be obtained as for the other functions just discussed.

```
> cosh(I*z);                              # Resp. cos(z)
> cos(I*z);                               # Resp. cosh(z)
> simplify(cosh(z)^2 - sinh(z)^2);        # Resp. 1
> tanh(1 - 2*I);                          # Resp. tanh(1 − 2 I)
> evalf(tanh(1 - 2*I), 5);                # Resp. 1.1667 + .24346 I
> z := x + I*y:
> evalc(Re(tanh(z)));
```

$$\text{\# Resp. } \frac{\sinh(x)\cosh(x)}{\sinh(x)^2 + \cos(y)^2}$$

Similar Material in AEM: pp. 679-686

| EXAMPLE 12.6 | **COMPLEX LOGARITHM**

In calculus the natural logarithm $\ln x$ is defined for positive real x only and is single-valued (that is, for each such x it has only one value). In complex, $\ln z$ $(z \neq 0)$ has infinitely many values $\ln z = \ln |z| + i \operatorname{Arg} z \pm 2n\pi i$ $(n = 0, 1, 2, ...)$. These values all have the same real part $\ln |z|$. Their imaginary parts differ by integer multiples of 2π. Maple gives the **principal value** of $\ln z$, by definition corresponding to $n = 0$ and $-\pi < \operatorname{Arg} z \leq \pi$ (called the **principal value** of arg z and denoted by $\operatorname{Arg} z$, as shown; see also Example 12.1 in this Guide). For example,

```
> evalc(ln(x + I*y));
```
$$\text{\# Resp. } \frac{1}{2}\ln\left(x^2 + y^2\right) + I\arctan(y, x)$$

```
> evalc(ln(4 - 3*I));
```
$$\text{\# Resp. } \ln(5) - I\arctan(\frac{3}{4})$$

```
> Re(ln(3 - 4*I));
```
$$\text{\# Resp. } \ln(5)$$

```
> Im(ln(3 - 4*I));
```
$$\text{\# Resp. } -\arctan(\frac{4}{3})$$

```
> evalf(ln(3 - 4*I));                     # Resp. 1.609437912 − .9272952180 I
> ln(-1);                                 # Resp. I π
> evalf(ln(-1), 50);                      # This approximates πi.
```

$$3.1415926535897932384626433832795028841971693993751\, I$$

Derivative. The derivative is

```
> diff(ln(z), z);
```
$$\text{\# Resp. } \frac{1}{z}$$

Plotting. Type `?plot`, `?plot[style]`. Type $\ln(3 - 4i)$ in the form $a + bi$. Type all the values you want to plot; this is the complex sequence `S`.

```
> L := evalc(ln(3 - 4*I) + 2*n*Pi*I);
```

$$L := \ln(5) + I\left(-\arctan(\frac{4}{3}) + 2n\pi\right)$$

```
> S := seq(evalf(L), n = -2..3);
```

$S := 1.609437912 - 13.49366584\, I,\ 1.609437912 - 7.210480526\, I,$
     $1.609437912 - .9272952179\, I,\ 1.609437912 + 5.355890090\, I,$
     $1.609437912 + 11.63907540\, I,\ 1.609437912 + 17.92226070\, I$

```
> plot(S, style = point);
Error, (in plot) invalid arguments
```

So this does not work. You need a real sequence S2 of terms $[\operatorname{Re} L, \operatorname{Im} L]$. The outer brackets in S2 seem to be essential. Try without. n goes from 1 to 6, the numbers of the six terms in S, not from -2 to 3. Try S[-2], S[0], S[1], S[2], etc. separately.

```
> S2 := [seq([Re(S[n]), Im(S[n])], n = 1..6)];
```

$$S2 := [[1.609437912, -13.49366584], [1.609437912, -7.210480526],$$
$$[1.600437012, .9272952179], [1.609437912, 5.355890090],$$
$$[1.609437912, 11.63907540], [1.609437912, 17.92226070]]$$

```
> plot(S2, style = point, labels = [x, y]);
```

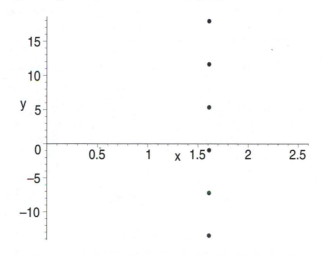

Example 12.6. Some values of $\ln(3 - 4i)$

Similar Material in AEM: pp. 687-690

Problem Set for Chapter 12

Pr.12.1 (Complex arithmetic) Let $z_1 = 8 + 3I$, $z_2 = 9 - 2I$. Find $z_1 + z_2$, $z_1 - z_2$, $z_1 z_2$, z_1/z_2, $|z_1/z_2|$, $|z_1|/|z_2|$, $\operatorname{Re} z_1$, $\operatorname{Im}(z_1{}^2)$, $\operatorname{Arg} z_1$. (*AEM Ref.* pp. 653, 654)

Pr.12.2 (Complex arithmetic) Let $z_1 = 4 + 3i$, $z_2 = 2 - 5i$. Find $z_1 \bar{z}_2$, $\bar{z}_1 z_2$ (why must these two products be conjugate?), $1/|z_1|$, $|z_1| + |z_2| - |z_1 + z_2|$ (why must this be nonnegative?), $\operatorname{Re}(z_1{}^3)$, $(\operatorname{Re} z_1)^3$, $\operatorname{Im}((z_1 - z_2)/(z_1 + z_2))$. (*AEM Ref.* pp. 655-657)

Pr.12.3 (Real and imaginary parts) Obtain the important formulas $x = (z + \bar{z})/2$ and $y = (z - \bar{z})/2i$ on the computer. (*AEM Ref.* p. 656)

Pr.12.4 (Polar form) Using the computer, convert the following complex numbers to polar form and back. $1 + i$, $-2 - 2i$, $(1 + i)(-2 - 2i)$, -10, $(1 - i)/(1 + i)$, $((6 + 8i)/(4 - 3i))^2$. (*AEM Ref.* pp. 657-660)

Pr.12.5 (Plotting complex numbers) Plot $(0.9 + 0.4i)^n$ for integer $n = -20, ..., 20$. Also plot the unit circle on the same axes. Can you find n for each of the 41 points that you see in the plot? (For plotting points see Example 12.6 in this Guide.)

Pr.12.6 (Quadratic equation in z^2) Solve $z^4 - (3 + 6i)z^2 - 8 + 6i = 0$. (*AEM Ref.* p. 662 (#30))

Pr.12.7 (Roots) Find and plot all cube roots of $1 + i$, as well as the circle on which these roots lie. (*AEM Ref.* p. 662 (#21))

Pr.12.8 (Roots of unity) Find and plot the 16th roots of unity. Can you visualize the values before you plot them? (*AEM Ref.* pp. 660-662)

Pr.12.9 (Annulus) Plot the annulus with center $2 + 2i$ whose outer circle passes through the origin and whose inner circle touches the coordinate axes. (*AEM Ref.* p. 664)

Pr.12.10 (Complex plane) Find the curve satisfying $|z + i|/|z - i| = 1$ (a) by a geometric argument without calculation, (b) on the computer. (*AEM Ref.* p. 655)

Pr.12.11 (Domain) Find the domain for whose points the sum of the distances from -1 and 1 is less than $\sqrt{8}$ (a) by a geometrical argument and inspection, (b) on the computer. (*AEM Ref.* pp. 664, 668)

Pr.12.12 (Cauchy-Riemann equations) Is $f(z) = \mathrm{Re}\,(z^2) - i\,\mathrm{Im}\,(z^2)$ analytic? (*AEM Ref.* p. 673 (#12))

Pr.12.13 (Experiment on conformal mapping) Make plots of the images of rectangles under $w = z^n$, $n = 2$, 3, 4, etc. How do these images change as functions of the exponent n? What happens to the image if you change the position of a rectangle in the plane? (*AEM Ref.* pp. 674-677)

Pr.12.14 (Experiment on conformal mapping) Experiment with the image of a square $a \le x \le a+1$, $b \le y \le b+1$ under the mapping $w = z^3$ and characterize the position and form of the image for various $a \ge 0$ and $b \ge 0$. (*AEM Ref.* pp. 674-678)

Pr.12.15 (Inversion in the unit circle) Plot the square $1/2 \le x \le 3/2$, $1/2 \le y \le 3/2$ and its image under $w = 1/z$ as well as the unit circle on common axes. Discuss how the image curves correspond to the straight-line segments in the given square. (*AEM Ref.* p. 692)

Pr.12.16 (Exponential function) Find e^z (in the form $u + iv$) and $|e^z|$ if z equals $2 + 3\pi i$, $1 + i$, $2\pi(1 + i)$, $0.95 - 1.6i$, and $-\pi i/2$. (*AEM Ref.* p. 682 (##1-5))

Pr.12.17 (Conformal mapping by cos z) Find the image of the rectangle $0 \le x \le 2\pi$, $1/2 \le y \le 1$. What are the images of the vertical sides $x = 0$ and $x = 2\pi$? (*Hint*: What will happen if you shorten the x-interval slightly? (*AEM Ref.* p. 687 (#18))

Pr.12.18 (Hyperbolic functions) Obtain the basic formulas $\cosh^2 z - \sinh^2 z = 1$ and $\cosh^2 z + \sinh^2 z = \cosh 2z$ on the computer. (*AEM Ref.* p. 686 (#3))

Pr.12.19 (Natural logarithm ln z) Find the principal value Ln z for $z = -5, -12 - 16i, 1 + i$, $1 - i, -10 + 0.1i, -10 - 0.1i$. What does a comparison of the last two values illustrate? (*AEM Ref.* p. 691 (##5-8))

Pr.12.20 (General powers) Find (in the form $u + iv$) $(2i)^{2i}$, 3^{4-i}, $(1 + 3i)^i$, $i^{1/2}$. (*AEM Ref.* p. 691)

Chapter 13

Complex Integration

Content. Indefinite integration (Ex. 13.1)
  Use of path (Ex. 13.2, Prs. 13.1-13.3, 13.5, 13.7-13.9)
  Cauchy theorem and formula (Ex. 13.3, Pr. 13.6, 13.10)
  Use of partial fractions (Prs. 13.4, 13.5)

Further integration methods follow in Chap. 15.

Examples for Chapter 13

EXAMPLE 13.1	**INDEFINITE INTEGRATION OF ANALYTIC FUNCTIONS**

Let $f(z)$ be analytic in a domain D that is **simply connected** (that is, every closed curve in D encloses only points of D). Then there exists an analytic function $F(z)$ such that $F'(z) = f(z)$ everywhere in D, and for all paths in D from any point z_0 to any point z_1 in D,

$$\int_{z_0}^{z_1} f(z)\, dz = F(z_1) - F(z_0).$$

($F(z)$ is called an **antiderivative** or **indefinite integral** of $f(z)$, and this is the analog of a known formula from calculus, but note well that it applies to *analytic* functions only.) For instance,

```
> int(3*z^2, z = 0..1 + I);                    # Resp. -2 + 2 I
> int(cos(z), z = -Pi*I..Pi*I);                # Resp. 2 I sinh(π)
> evalf(%, 6);                                 # Resp. 23.0974 I
> int(1/z, z = -I..I);                         # Resp. I π
```

In the last integral an antiderivative is the principal value $\text{Ln}\, z$ of $\ln z$, which is not analytic on the negative real axis $x \le 0$, and the paths from $-i$ to i must be restricted accordingly.

      Similar Material in AEM: pp. 707, 708

EXAMPLE 13.2	**INTEGRATION: USE OF PATH. PATH DEPENDENCE**

If the function $f(z)$ you want to integrate is not analytic, the integral will generally depend on path, and you have to use a representation of the path C that is given. Let C be represented by $z = z(t)$, $a \le t \le b$, and piecewise smooth. Let $f(z)$ be continuous on C. Then

$$\int_C f(z)\, dz = \int_a^b f(z(t))\, \dot{z}(t)\, dt \qquad \dot{z} = dz/dt.$$

143

For instance, if you want to integrate $f(z) = 1/z^n$ with any given $n = 1, 2, \ldots$ counterclockwise around the unit circle C, there is no simply connected domain containing C in which $1/z^n$ is analytic. ($1/z^n$ is analytic, for instance, in the annulus $D : 1/2 < |z| < 3/2$, but D is not simply connected.) Represent C, obtain $\dot{z}$, and then integrate from $a = 0$ to $b = 2\pi$.

```
> z := exp(I*t);                              # Resp. z := e^(I t)
> zdot := diff(z, t);                         # Resp. zdot := I e^(I t)
> n:='n':
> J := int(1/z^n*zdot, t = 0..2*Pi);
```

$$J := -\frac{e^{(-2 I n \pi)} - 1}{-1 + n}$$

You see that $n - 1 = 0$ when $n = 1$; hence this formula holds true for any $n \neq 1$. Now tell the computer that n is assumed to be integer.

```
> assume(n, integer);
> evalc(J);                                   # Resp. 0
```

Hence $J = 0$ for $n = 2, 3, \ldots$ (and also for $n = 0, -1, -2, \ldots$, that is, $1, z, z^2, \ldots$). For $n = 1$ you have $f(z) = 1/z$ and obtain the **very important result**

$$\oint_C \frac{1}{z} \, dz = 2\pi i \qquad \text{(counterclockwise around the unit circle)}$$

that is frequently needed throughout complex analysis,

```
> int(1/z*zdot, t = 0..2*Pi);                 # Resp. 2 I π
```

Path dependence of the integral can be illustrated by many examples. For instance, integrate $f(z) = |z|^2$ from 0 to $1 + i$ over two different paths, (A) over the straight segment joining these points, (B) over the parabolic arc $y = x^2$ (see the figure).

Solution. (A)

```
> z1 := t + I*t;          # Resp. z1 := t + I t        Straight segment
> z1dot := diff(z1, t);        # Resp. z1dot := 1 + I
> f1 := abs(z1)^2;             # Function f on the segment
```

$$f1 := |t + I t|^2$$

```
> evalc(%);                                   # Resp. 2 t^2
> int(f1*z1dot, t = 0..1);
```

$$\frac{2}{3} + \frac{2}{3} I$$

(B) The value of the integral taken over the parabolic arc will differ from the value just obtained.

```
> z2 := t + I*t^2;                            # Resp. z2 := t + I t^2
> z2dot := diff(z2, t);                       # Resp. z2dot := 1 + 2 I t
> f2 := abs(z2)^2;
```

$$f2 := \left| t + I\,t^2 \right|^2$$

```
> f2 := evalc(%);                              # Resp. f2 := t² + t⁴
> int(f2*z2dot, t = 0..1);
```

$$\frac{8}{15} + \frac{5}{6}\,I$$

```
> with(plots):                                 # Ignore the warning.
> P1 := plot([t, t, t = 0..1]):
> P2 := plot([t, t^2, t = 0..1]):
> display({P1, P2}, xtickmarks = [0, 0.5, 1], ytickmarks = [0, 0.5, 1]);
```

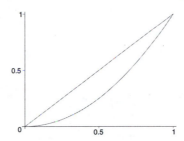

Example 13.2. Paths of the integrals in (A) and (B)

Similar Material in AEM: pp. 708-710

| EXAMPLE 13.3 | **CONTOUR INTEGRATION BY CAUCHY'S INTEGRAL THEOREM AND FORMULA** |

Cauchy's integral theorem. Let $f(z)$ be analytic in a simply connected domain D. Let C be any closed path in D. Then the integral of $f(z)$ around C is zero.

Cauchy's integral formula and derivative formulas. Let $f(z)$ and D be as before. Let C be as before and *simple* (that is, without self-intersections). Let z_0 be any point inside C. Then the integral

$$(1) \qquad \oint_C \frac{f(z)}{(z - z_0)^{n+1}}\,dz$$

(taken counterclockwise) equals $2\pi i f(z_0)$ for $n = 0$ (**"Cauchy's integral formula"**) and ($f^{(n)}$ the nth derivative of f with respect to z)

$$(2) \qquad \frac{2\pi i}{n!} f^{(n)}(z_0) \qquad \text{for } n = 1, 2, \dots.$$

For instance, taking $f(z) = 1$ and $z_0 = 0$, you see without any calculation that for $n = 0$ (that is, for $1/z$) you get $2\pi i$, and for $n = 1, 2, \dots$ the integrals (of $1/z^2$, $1/z^3$, ...) are all zero.

As a second application, find the integral of $\dfrac{e^z}{z\,e^z - 2iz}$ counterclockwise around the circle $|z| = 0.5$.

Solution. From Cauchy's integral formula with $f(z) = \dfrac{e^z}{e^z - 2i}$ and $z_0 = 0$ you obtain

```
> z := 'z':
> f := exp(z)/(exp(z) - 2*I);
```
$$\text{\# Resp. } f := \frac{e^z}{e^z - 2I}$$

```
> 2*Pi*I*subs(z = 0, f);
```
$$\text{\# Resp. } \frac{2I\pi e^0}{e^0 - 2I}$$

```
> eval(%);
```
$$\text{\# Resp. } \left(-\frac{4}{5} + \frac{2}{5}I\right)\pi$$

You still have to make sure that $f(z)$ is analytic everywhere inside and on C. Now the only points at which something could happen are those at which the denominator of $f(z)$ is zero; that is, $e^x = 2i$, $z = \ln 2i = \ln 2 + \pi i/2 \pm 2n\pi i$, but these points lie outside C, as can be seen from

```
> evalc(solve(exp(z) - 2*I = 0, z));
```
$$\text{\# Resp. } \ln(2) + \frac{1}{2}I\pi$$

```
> evalf(abs(%), 4);
```
$$\text{\# Resp. } 1.717$$

and the fact that the other solutions are even greater in absolute value. Your result is now established.

As a third application, find the integral of $(\tan \pi z)/z^6$ counterclockwise around the circle $|z| = 1/4$.

Solution. From (1) and (2) with $f(z) = \tan \pi z$, $z_0 = 0$, and $n + 1 = 6$, hence $n = 5$, you obtain

```
> z := 'z':
> 2*Pi*I/5!*diff(tan(Pi*z), z, z, z, z, z);
```

$$\frac{1}{60}I\pi(88\,(1 + \tan(\pi z)^2)^2\,\pi^5\tan(\pi z)^2 + 16\,(1 + \tan(\pi z)^2)^3\,\pi^5$$

$$+16\tan(\pi z)^4\,(1 + \tan(\pi z)^2)\,\pi^5)$$

```
> eval(subs(z = 0, %));
```
$$\text{\# Resp. } \frac{4}{15}I\pi^6$$

A factor π results from $2\pi i$, and π^5 from the chain rule in the five differentiations.

Note further that the circle of integration is so small that the points $\pm 1/2$, $\pm 3/2$... where $\tan \pi z$ is not analytic lie outside the circle.

Similar Material in AEM: pp. 714, 722, 725 (#14), 726

Problem Set for Chapter 13

Pr.13.1 (Use of path) Integrate $\text{Re}\,z$ over the shortest path from $1 + i$ to $3 + 2i$. (*AEM Ref.* p. 712 (#15))

Pr.13.2 (Contour integral) Using a representation of the path, integrate $3/(z-i) - 6/(z-i)^2$ clockwise around the circle $|z - i| = 5$. Confirm the answer by the method in Example 13.3 in this Guide. (*AEM Ref.* p. 712 (#26))

Pr.13.3 (Use of path) Integrate $\text{Re}\,z$ from $1 + i$ vertically upward to $1 + 2i$ and then horizontally to $3 + 2i$. (*AEM Ref.* p. 712 (#16))

Pr.13.4 (Partial fractions) Integrate $(2z^3 + z^2 + 4)/(z^4 - 4z^2)$ clockwise around the circle of radius 4 and center 2 first as given and then by using partial fractions. (*AEM Ref.* pp. 722, 726)

Pr.13.5 **(Partial fractions)** Integrate $(4z^2 + 17z - 68)/(z^3 - 12z + 16)$ counterclockwise around the circle $|z| = 3$ by using partial fractions. (Type **?parfrac** . *AEM Ref.* pp. 722, 726)

Pr.13.6 **(Derivative formulas)** Integrate $(\tanh z)/z^4$ counterclockwise around the unit circle. (See Example 13.3 in this Guide.)

Pr.13.7 **(Path dependence)** Integrate $\bar{z}$ from 0 to $1 + i$ (A) along the shortest path, (B) along the parabola $y = x^2$. (*AEM Ref.* p. 712 (#18))

Pr.13.8 **(Experiment on path dependence)** Experiment with Pr.13.7 by integrating along $y = x^n$, $n = 2, 3, \ldots$, from 0 to $1 + i$, obtaining the limiting value of the integral as $n \to \infty$, and confirming the result by integration over the "limiting path" from 0 to 1 and then vertically up to $1 + i$.

Pr.13.9 **(Use of contour)** Integrate $\operatorname{Re} z^2$ clockwise around the boundary of the square with vertices 0, i, $1 + i$, 1. (*AEM Ref.* p. 712 (#20))

Pr.13.10 **(Derivative formulas)** Integrate $(z^3 + \sin z)/(z - i)^3$ counterclockwise around the boundary of the square with vertices ± 2 and $\pm 2i$. (*AEM Ref.* p. 729 (#12))

Power Series, Taylor Series

Content. Sequences (Ex. 14.1, Prs. 14.1, 14.2)
Convergence tests (Ex. 14.2, Prs. 14.3-14.5)
Power series (Ex. 14.3, Prs. 14.6, 14.7, 14.11)
Taylor series (Ex 14.4, Prs. 14.8-14.14)
Uniform convergence (Ex. 14.5, Pr. 14.15)

Complex sequences and series are obtained by the same commands as in real, namely, `seq` and `series`; see also the opening to Chap. 4.
Coefficients of polynomials p are accessed as before, by typing `coeffs(p)` or `coeff(p, z^n)`.
Plotting complex sequences (z_n). Convert (z_n) to a sequence of $[\mathrm{Re}\, z_n,\ \mathrm{Im}\, z_n]$. Then plot.

Examples for Chapter 14

EXAMPLE 14.1 | SEQUENCES AND THEIR PLOTS

To obtain a sequence (z_n), you may first type z_n and then apply the command `seq`

```
> zn := ((9 + I)/10)^n;
```

$$ zn := \left(\frac{9}{10} + \frac{1}{10} I \right)^n $$

```
> S1 := seq(zn, n = 1..5);
```

$$ S1 := \frac{9}{10} + \frac{1}{10} I,\ \frac{4}{5} + \frac{9}{50} I,\ \frac{351}{500} + \frac{121}{500} I,\ \frac{1519}{2500} + \frac{36}{125} I,\ \frac{12951}{25000} + \frac{7999}{25000} I $$

or you may do this with one command, inserting the given expression for z_n directly into `seq`. Try it. If you want the terms as decimal fractions with, say, 4 digits, type

```
> evalf(%, 4);
```

$$ .9000 + .1000\, I,\ .8000 + .1800\, I,\ .7020 + .2420\, I,\ .6076 + .2880\, I, .5180 + .3200\, I $$

To plot the sequence, you must convert each term $a + ib$ to the form $[a, b]$. (See also Example 12.6 in this Guide, where the same was done.) Then give the plotting command, which is accepted only after you enclose S2 in brackets, telling the computer that the elements of S2 form a sequence (a "list"; type `?list`). If you omit `style = point` from the plotting command, the points will be joined by a curve consisting of straight segments. Try it.

```
> S2 := seq([Re(zn), Im(zn)], n = 1..100):
> plot(S2, style = point, labels = [Re_z, Im_z]);
```

```
Error, (in plot) invalid arguments
> plot([S2], style = point, labels = [Re_z, Im_z]);
```

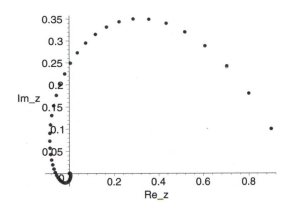

Example 14.1. Plot of the sequence (z_n), where $z_n = (0.9 + 0.1\,i)^n$, 100 terms

Procedure. You can obtain the sequence S2 of real pairs (here denoted by S3) from the complex sequence S1 by a procedure (type ?proc), namely,

```
> S3 := map(proc(zn) [Re(zn), Im(zn)] end, [S1]);
```

$$S3 := \left[\left[\frac{9}{10}, \frac{1}{10}\right], \left[\frac{4}{5}, \frac{9}{50}\right], \left[\frac{351}{500}, \frac{121}{500}\right], \left[\frac{1519}{2500}, \frac{36}{125}\right], \left[\frac{12951}{25000}, \frac{7999}{25000}\right]\right]$$

To understand this, and to see what it does, type first

```
> proc(zn) [Re(zn), Im(zn)] end proc;
```

$$\textbf{proc}\,(zn)\,[\text{Re}\,(zn),\,\text{Im}\,(zn)]\,\textbf{end proc}$$

So the procedure gives from the complex zn its real and imaginary parts. Instead of Re(z) , Im(z) you may sometimes need evalc(Re(z)) , evalc(Im(z)) . Now map (type ?map) has the effect that the conversion from a complex number to a real pair is done for each term of the complex sequence. Again the brackets around S1 are quite important. Try without.
    Similar Material in AEM: pp. 732-735

| EXAMPLE 14.2 | **CONVERGENCE TESTS FOR COMPLEX SERIES**

Ratio test. This is the practically most important test for convergence of a series $z_1 + z_2 + \dots$. It uses the ratios $|z_{n+1}/z_n|$, $n = 1, 2, \dots$. If for all n greater than some N these ratios do not exceed a fixed $q < 1$, the series converges absolutely. If for every $n > N$ the ratios are 1 or greater, the series diverges.

    Very often the sequence of the test ratios will have a limit L. Then if $L < 1$, the series converges absolutely. If $L > 1$, it diverges. If $L = 1$, the test fails, that is, no conclusion is possible.

    Guess whether the following series converges or diverges (consider how fast the terms increase). Then apply the test.

$$\sum_{n=0}^{\infty} \frac{(100 + 75\,i)^n}{n!} = 1 + (100 + 75i) + \frac{1}{2!}(100 + 75\,i)^2 + \dots$$

Solution. The test ratio is

```
> abs((((100 + 75*I)^(n + 1)/(n + 1)!)/((100 + 75*I)^n/n!));
```

$$\left| \frac{(100 + 75\,I)^{(n+1)}\,n!}{(n+1)!\,(100 + 75\,I)^n} \right|$$

```
> ratio := simplify(%);
```
         # Resp. $ratio := 125\,\dfrac{1}{|n+1|}$

The limit is 0. Hence the series converges. In cases in which you cannot see the limit immediately, use `limit` . (Type `?limit`. See also below.)

Root test. This test is usually more difficult to apply. If for every n greater than some N, the nth root of $|z_n|$ does not exceed a fixed $q < 1$, then $z_1 + z_2 + \ldots$ converges absolutely. If for infinitely many n those roots are 1 or greater, the series diverges.

    If the sequence of those roots has a limit $L < 1$, the series converges absolutely. If $L > 1$, it diverges. If $L = 1$, the test fails.

    For instance, for the series just considered,

```
> (abs(100 + 75*I)^n/n!)^(1/n);
```

$$\left(\frac{125^n}{n!} \right)^{\left(\frac{1}{n} \right)}$$

```
> limit(%, n = infinity);
```
         # Resp. 0

This confirms the previous result that the series converges.

    Similar Material in AEM: pp. 737-739

EXAMPLE 14.3   POWER SERIES. RADIUS OF CONVERGENCE

A **power series** is a series of the form

$$(1) \qquad\qquad \sum_{n=0}^{\infty} a_n\,(z - z_0)^n.$$

It may converge for all z (the nicest case) or no $z \neq z_0$ (the useless case) or in a circle

$$(2) \qquad\qquad |z - z_0| = R.$$

If this is the ***smallest*** circle with center z_0 that includes all the points at which the series (1) converges, then its radius is called the **radius of convergence** of (1), and is given by the **Cauchy-Hadamard formula**

$$(3) \qquad\qquad R = \lim_{n \to \infty} \left| \frac{a_n}{a_{n+1}} \right|.$$

    For instance, find the radius of convergence of the series $\displaystyle\sum_{n=0}^{\infty} \frac{(2\,n)!}{(n!)^2}(z - 3\,i)^n$.

Solution. Use (3). You can save work by typing a_n and then getting the quotient by `subs(n = n + 1, ...)` as shown. The command `simplify(%);` would not be necessary. Try without. The radius of convergence will be 1/4.

```
> an := (2*n)!/(n!)^2;
```
          # Resp. $an := \dfrac{(2\,n)!}{(n!)^2}$

```
> an/subs(n = n + 1, an);
```

$$\frac{(2\,n)!\,((n+1)!)^2}{(n!)^2\,(2\,n+2)!}$$

```
> simplify(%);
```

$$\frac{1}{2}\frac{n+1}{2\,n+1}$$

```
> limit(%, n = infinity);
```
          # Resp. $\dfrac{1}{4}$

Similar Material in AEM: pp. 741-745

| EXAMPLE 14.4 | **TAYLOR SERIES** |

A **Taylor series** of a given function $f(z)$ is a power series (1), Example 14.3, with co-efficients $a_n = f^{(n)}(z_0)/n!$, where $f^{(n)}$ is the nth derivative. Maple knows practically all elementary and higher real and complex functions of general practical interest. If such an $f(z)$ has a Taylor series, you can obtain it by the command series (type ?series). For instance, if $f(z) = \cos \pi z$ and $z_0 = 1/2$, type

```
> ser := series(cos(Pi*z), z = 1/2, 7);
```

$$ser := -\pi\left(z - \frac{1}{2}\right) + \frac{1}{6}\pi^3\left(z - \frac{1}{2}\right)^3 - \frac{1}{120}\pi^5\left(z - \frac{1}{2}\right)^5 + O\left(\left(z - \frac{1}{2}\right)^7\right)$$

7 is the order of the error term, whose choice is up to you. Try other values. For computing values or for plotting, drop first the error term by the command convert(f, polynom). Decimal values for the coefficients and the center are obtained by evalf, and a value f0 of f, for instance, at $0.4 + 0.1i$, by the command subs(z = 0.4 + 0.1*I,...).

```
> poly := convert(ser, polynom);
```

$$poly := -\pi\left(z - \frac{1}{2}\right) + \frac{1}{6}\pi^3\left(z - \frac{1}{2}\right)^3 - \frac{1}{120}\pi^5\left(z - \frac{1}{2}\right)^5$$

```
> evalf(%, 4);
```

$$-3.142\,z + 1.571 + 5.171\,(z - .5000)^3 - 2.552\,(z - .5000)^5$$

```
> f0 := evalf(subs(z = 0.4 + 0.1*I, poly));
```

$$f0 := .3243926844 - .3037218332\,I$$

Confirm your result by the coefficient formula mentioned above.

```
> f := cos(Pi*z);
```
          # Resp. $f := \cos(\pi\,z)$

```
> S := seq(eval(subs(z = 1/2, diff(f, z$n)/n!)), n = 1..11);
```

$$S := -\pi,\, 0,\, \frac{1}{6}\pi^3,\, 0,\, -\frac{1}{120}\pi^5,\, 0,\, \frac{1}{5040}\pi^7,\, 0,\, -\frac{1}{362880}\pi^9,\, 0,\, \frac{1}{39916800}\pi^{11}$$

In this command, z$n means z, z, ..., z ($n$ times). Type z$4 to see what $ does: it forms an "expression sequence". (Type ?$.)

Try the command for S with n = 0..5. It will not work because differentiation starts with $n = 1$. Obtain a_0 separately by typing

```
> a0 := eval(subs(z = 1/2, f));                    # Resp. a0 := 0
```

Individual coefficients are obtained from S by typing S[1], S[2], ... Try it. Accordingly, you get a Taylor polynomial, in agreement with the previous result, by typing

```
> s5 := sum(S[n]*(z - 1/2)^n, n = 1..5);
```

$$s5 := -\pi \left(z - \frac{1}{2}\right) + \frac{1}{6}\pi^3 \left(z - \frac{1}{2}\right)^3 - \frac{1}{120}\pi^5 \left(z - \frac{1}{2}\right)^5$$

You can now plot this, together with other Taylor polynomials, for real z, to create a figure as you may have seen it in your calculus book, illustrating how Taylor polynomials approximate given functions.

```
> s7 := sum(S[n]*(z - 1/2)^n, n = 1..7):
> s9 := sum(S[n]*(z - 1/2)^n, n = 1..9):
> plot({s5, s7, s9}, z = 0..2.5, y = -2..2);
```

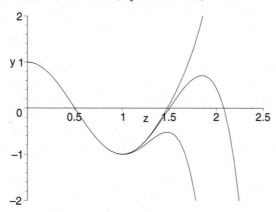

Example 14.4. Taylor polynomials approximating $\cos \pi z$ for real $z = x$
(s7 goes upward.)

Maclaurin series. You can now easily obtain familiar Maclaurin series, which have the same form as those for real functions in calculus. In particular, this holds for the geometric series and the series for e^z, $\cos z$, and $\sin z$. If you want more terms, type Order := 10 or 100 or whatever, and then the command series. (Type ?Order.)

```
> S1 := series(1/(1 - z), z);
```

$$S1 := 1 + z + z^2 + z^3 + z^4 + z^5 + O\left(z^6\right)$$

```
> S2 := series(exp(z), z);
```

$$S2 := 1 + z + \frac{1}{2}z^2 + \frac{1}{6}z^3 + \frac{1}{24}z^4 + \frac{1}{120}z^5 + O\left(z^6\right)$$

```
> S3 := series(cos(z), z);          # Resp. S3 := 1 - (1/2)z^2 + (1/24)z^4 + O(z^6)
```

$$\text{\# Resp. } S3 := 1 - \frac{1}{2}z^2 + \frac{1}{24}z^4 + O\left(z^6\right)$$

```
> Order := 13:
```

```
> S4 := series(sin(z), z);
```

$$S4 := z - \frac{1}{6}z^3 + \frac{1}{120}z^5 - \frac{1}{5040}z^7 + \frac{1}{362880}z^9 - \frac{1}{39916800}z^{11} + \mathrm{O}\left(z^{13}\right)$$

Differentiation and integration of Taylor series is permissible. You can easily do it on the computer. For example, obtain series agreeing with $(\sin z)' = \cos z$. Using series, illustrate other relations of your choice.

```
> diff(S2, z);
```
          # The derivative of e^z is e^z.

$$1 + z + \frac{1}{2}z^2 + \frac{1}{6}z^3 + \frac{1}{24}z^4 + \mathrm{O}\left(z^5\right)$$

```
> int(S3, z);
```
          # Agrees with $\int \cos z \, dz = \sin z + C$, where $C = 0$.

$$z - \frac{1}{6}z^3 + \frac{1}{120}z^5 + \mathrm{O}\left(z^7\right)$$

*Similar Material in **AEM**:* pp. 751-757

| EXAMPLE 14.5 | **UNIFORM CONVERGENCE**

Let the series $f_0(z) + f_1(z) + f_2(z) + \dots$ be convergent for all z in some region G in the z-plane. Let $s(z)$ be its sum and $s_n(z)$ its nth partial sum. Then because of convergence, for a given $\varepsilon > 0$ you can find an N such that for all $n > N$ you have $|s(z) - s_n(z)| < \varepsilon$. Here N will generally depend on both ε and z. If for any given $\varepsilon > 0$ you can find an N, **depending only on epsilon** such that **for all z in G** and all $n > N$ that inequality holds, then the convergence of that series is called **uniform convergence** in G. This concept is important because in the case of uniform convergence in G, if the terms $f_n(z)$ are continuous in G, so is the sum $s(z)$, and termwise integration of the series over any path C in G is permissible, that is, it gives a series whose sum is the integral of $s(z)$ over C. This is not generally true in the case of ordinary convergence.

Show that the following series of continuous terms has a discontinuous sum, and plot the partial sums s_1, s_4, s_{16}, s_{64}, s_{256}, s_{1024}, illustrating the approach to the sum.

$$x^2 + \frac{x^2}{1 + x^2} + \frac{x^2}{(1 + x^2)^2} + \dots$$

Solution. Apart from the factor x^2 this is a geometric series with the quotient $q = 1/(1 + x^2)$ and nth partial sum $(1 - q^{n+1})/(1 - q)$. Accordingly, type

```
> q := 1/(1 + x^2);
```
          # Resp. $q := \dfrac{1}{1 + x^2}$

```
> sn := x^2*(1 - q^(n + 1))/(1 - q);
```

$$sn := \frac{x^2 \left(1 - \left(\frac{1}{1+x^2}\right)^{(n+1)}\right)}{1 - \frac{1}{1+x^2}}$$

```
> sn := simplify(%);
```
          # Resp. $sn := 1 + x^2 - \left(\dfrac{1}{1 + x^2}\right)^n$

For $x = 0$ this is $1 - 1 = 0$ for all n. Hence $s(0) = 0$. For $x \neq 0$ the limit as $n \to \infty$ is $s(x) = 1 + x^2$. Hence the sum of the series is discontinuous at $x = 0$. To plot those partial sums, replace n by 4^n in sn and then plot. Don't omit the brackets [...] shown in the plot command. Try without–it will not work.

```
> s4n := subs(n = 4^n, sn);
```

$$s4n := 1 + x^2 - \left(\frac{1}{1+x^2} \right)^{(4^n)}$$

```
> S := seq(s4n, n = 0..5);
```

$$S := 1 + x^2 - \frac{1}{1+x^2}, \ 1 + x^2 - \frac{1}{(1+x^2)^4}, \ 1 + x^2 - \frac{1}{(1+x^2)^{16}},$$

$$1 + x^2 - \frac{1}{(1+x^2)^{64}}, \ 1 + x^2 - \frac{1}{(1+x^2)^{256}}, \ 1 + x^2 - \frac{1}{(1+x^2)^{1024}}$$

```
> plot([S], x = -1..1);
```

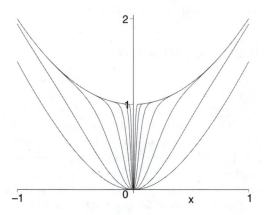

Example 14.5. Partial sums of a convergent series
not converging uniformly in any interval containing 0

Termwise integration giving false results. This is not the fault of the software, but of the mathematics. It can happen if convergence is not uniform. Show that this is the case for the series with the general term $f_m(x) = u_m(x) - u_{m-1}(x)$, where $u_m = mx \exp(-mx^2)$, and integration over x from 0 to 1. (It can be shown that the convergence of this series on the interval of integration is not uniform.)

Solution. Type u_m, u_{m-1}, then f_m, then the partial sum s_n.

```
> um := m*x*exp(-m*x^2);          # Resp. um := m x e^(-m x^2)
```

```
> um1 := subs(m = m - 1, um);     # Resp. um1 := (m - 1) x e^(-(m-1) x^2)
```

```
> fm := um - um1;
```

$$fm := m\,x\,\mathrm{e}^{(-m\,x^2)} - (m-1)\,x\,\mathrm{e}^{(-(m-1)\,x^2)}$$

```
> sn := sum(fm, m = 1..n);
```

$$sn := \frac{\mathrm{e}^{(x^2)}\,x\,n}{\mathrm{e}^{((n+1)\,x^2)}}$$

```
> sn := simplify(%);              # Resp. sn := e^(-x^2 n) x n
```

(Do you see that this is $u_n(x)$? Can you explain why?) Now tell the computer that x is nonnegative real (because you integrate x from 0 to 1) so that **sn** will have a limit.

```
> assume(x >= 0):
> evalf(limit(sn, n = infinity));                    # Resp. 0
```

Hence the sequence of the partial sums has the limit 0 for any fixed x. This is the sum of the series, by definition. Integration of 0 from 0 to 1 gives 0. Now show that by first integrating the partial sums and then taking the limit of the sequence of the integrals you get a different result, namely, 1/2 instead of 0. This shows that termwise integration of the series is not permissible, that is, it leads to a false result.

```
> int(sn, x = 0..1);                    # Resp. $-\dfrac{1}{2}\,e^{(-n)} + \dfrac{1}{2}$
```

```
> limit(%, n = infinity);                    # Resp. $\dfrac{1}{2}$
```

Similar Material in AEM: p. 763

Problem Set for Chapter 14

Pr.14.1 (Complex sequence) Obtain and plot the first terms of the sequence (z_n) with $z_n = 1 - 1/n^2 + i(2 + 4/n)$. (*AEM Ref.* p. 732)

Pr.14.2 (Complex sequence) Plot the first 100 terms of (z_n), where

$$z_n = \left(\frac{20\,i}{21}\right)^{0.1\,n}.$$

(*AEM Ref.* p. 732)

Pr.14.3 (Convergence test) Is the following series

$$\sum_{n=0}^{\infty} \frac{(20 + 30\,i)^n}{n!}$$

convergent or not? Use the ratio test. (*AEM Ref.* p. 737)

Pr.14.4 (Root test) Do Pr.14.3 by the root test. (*AEM Ref.* pp. 739, 740 (#12))

Pr.14.5 (Ratio test) Is the following series

$$\sum_{n=1}^{\infty} \frac{(3\,i)^n\, n!}{n^n}$$

convergent or not? (*AEM Ref.* p. 740 (#17))

Pr.14.6 (Radius of convergence) Find the radius of convergence of the series

$$\sum_{n=0}^{\infty} n\,(n-1)\,2^n\, z^{3\,n}.$$

(See Example 14.3 in this Guide. *AEM Ref.* p. 745 (#8))

Pr.14.7 (Radius of convergence) Find the radius of convergence of the series

$$\sum_{n=0}^{\infty} \frac{(3n)!}{2^n (n!)^3} z^n.$$

(See Example 14.3 in this Guide. *AEM Ref.* p. 745 (#11))

Pr.14.8 (Maclaurin series) The general term $a_n z^n$ of the Maclaurin series of a function $f(z)$ has the coefficient $a_n = f^{(n)}(0)/n!$. Using this formula, obtain the Maclaurin series of e^z (9 terms) and confirm the result by the command `series`.
(*AEM Ref.* p. 755)

Pr.14.9 (Taylor series) Find the Taylor series of Ln z with center $z_0 = 1$ from the coefficient formula $a_n = f^{(n)}(z_0)/n!$ (5 terms) and confirm the result by the command `series`.
(*AEM Ref.* p. 756)

Pr.14.10 (Taylor series) Find the Taylor series of $f(z) = \cos^2 z$ with center $z_0 = \pi/2$ (4 terms). Confirm the result by expressing $f(z)$ in terms of $\cos 2z$ and developing in a series. (*AEM Ref.* p. 759 (# 26))

Pr.14.11 (Integration of power series) The integral of $1/(1 + z^2)$ is arctan z (plus a constant). Using this, obtain the Maclaurin series of arctan z (powers up to z^{19}) by integrating a suitable series. (*AEM Ref.* p. 756)

Pr.14.12 (Experiment on sine integral) The *sine integral* Si (z) is the integral of $(\sin z)/z$ (integration from 0 to z). It is one of the more important integrals occurring in applications that cannot be integrated by the usual methods of calculus. Obtain its Maclaurin series (up to the power z^{15}) and use it to plot Si (z) for real z from 0 to 8. Then plot Si (z) by using the command `Si` (type `?Si` for information). Plot jointly. Finally, find experimentally a value of z at which the Maclaurin series still gives values whose first two decimals are accurate. (*AEM Ref.* pp. 758 (#10), A57)

Pr.14.13 (Bernoulli numbers) The Bernoulli numbers B_n are defined by the Maclaurin series of $z/(e^z - 1)$, written in the form $1 + B_1 z + B_2 z^2/2! + B_3 z^3/3! + \dots$. Find the first seven nonzero Bernoulli numbers. (*AEM Ref.* p. 758 (#13))

Pr.14.14 (Bessel functions) Find the Maclaurin series of $J_0(z)$. (Type `?Bessel` for information.) From the first five partial sums of the series find five approximate values of $J_0(0.5 - 0.5i)$. Note the rapid approach to the 10D value obtained directly by the command `BesselJ(0, 0.5 - 0.5*I)`. (*AEM Ref.* pp. 220, 752)

Pr.14.15 (Lack of uniform convergence) Plot the partial sums S_m, $m = 2^n$, $n = 1, 2, ..., 10$, of the series

$$x^4 \sum_{m=1}^{\infty} (1 + x^4)^{-m}$$

on common axes. (*AEM Ref.* pp. 762, 763)

Chapter 15

Laurent Series. Residue Integration

Content. Laurent series (Ex. 15.1, Prs. 15.1-15.3)
Singularities, zeros (Ex. 15.2, Prs. 15.4-15.7)
Residue integration (Ex. 15.3, Prs. 15.8-15.11)
Real integrals (Exs. 15.4, 15.5, Prs. 15.12-15.15)

Laurent series are infinite series of positive and negative integer powers of $z - z_0$ of the form

$$(1) \quad a_0 + a_1 (z - z_0) + a_2 (z - z_0)^2 + \cdots + b_1 (z - z_0)^{-1} + b_2 (z - z_0)^{-2} + \cdots.$$

They represent functions $f(z)$ analytic in an annulus with center z_0. The series of the negative powers is called the **principal part** of (1). A function $f(z)$ can have several Laurent series with the same center z_0 but valid in different annuli. Particularly important is the series (1) of a given $f(z)$ that converges immediately near z_0 (except at z_0 itself), say, for $0 < |z - z_0| < R$, where $R > 0$. For this series the coefficient b_1 of $1/(z - z_0)$ is called the **residue** of $f(z)$ at z_0. Residues can be used for evaluating contour integrals by the elegant method of **residue integration**. See Example 15.3.

Examples for Chapter 15

EXAMPLE 15.1	**LAURENT SERIES**

Find all Laurent series of $f(z) = 1/(z^3 - z^4)$ with center $z_0 = 0$.

Solution. The first series is S1 (below), which is obtained without difficulty. It converges for $0 < |z| < 1$. (This is a "degenerate" annulus.) Its principal part has three terms. The **residue** of $f(z)$ at 0 is 1 because $1/z$ has the coefficient 1.

```
> f := 1/(z^3 - z^4);
```
$\qquad$ # Resp. $f := \dfrac{1}{z^3 - z^4}$

```
> Order := 12:
> S1 := series(f,z);
```

$$S1 := z^{-3} + z^{-2} + z^{-1} + 1 + z + z^2 + z^3 + z^4 + z^5 + z^6 + z^7 + z^8 + \mathrm{O}(z^9)$$

The second Laurent series S2 is obtained by the **standard trick** of setting $z = 1/w$, developing $f(1/w) = g(w)$ in a series of powers of w, and resubstituting $w = 1/z$ into the series, which converges for $|z| > 1$.

```
> series(f, 1/z);
Error, wrong number (or type) of parameters in function series
```

```
> g := simplify(subs(z = 1/w, f));
```

$$g := \frac{w^4}{w-1}$$

```
> series(g, w);    # Resp. −w^4 − w^5 − w^6 − w^7 − w^8 − w^9 − w^10 − w^11 + O(w^12)
> S2 := subs(w = 1/z, %);
```

$$S2 := -\frac{1}{z^4} - \frac{1}{z^5} - \frac{1}{z^6} - \frac{1}{z^7} - \frac{1}{z^8} - \frac{1}{z^9} - \frac{1}{z^{10}} - \frac{1}{z^{11}} + O\left(\frac{1}{z^{12}}\right)$$

Similar Material in AEM: p. 774

| **EXAMPLE 15.2** | **SINGULARITIES AND ZEROS** |

A function $f(z)$ has a **singularity** at a point z_0 if $f(z)$ is not analytic (perhaps not even defined) at z_0, but every neighborhood of z_0 contains points at which $f(z)$ is analytic. For instance, to find the singularities of $f(z) = e^z/z^3$, type

```
> Order := 6:                    # Needed only if another order has been used before.

> f := exp(z)/z^3;                              # Resp. f := e^z / z^3

> S1 := series(f, z);
```

$$S1 := z^{-3} + z^{-2} + \frac{1}{2}z^{-1} + \frac{1}{6} + \frac{1}{24}z + \frac{1}{120}z^2 + O\left(z^3\right)$$

S1 is the Laurent series of $f(z)$ with center 0. It converges for $|z| > 0$ (thus for $z \neq 0$). It shows that $f(z)$ has a **singularity** (more precisely, a **pole** of third order) at $z = 0$, with residue $1/2$. To see whether $f(z)$ is singular at infinity, develop it in a series in powers of $w = 1/z$.

```
> S2 := g = subs(z = 1/w, S1);
```

$$S2 := g = w^3 + w^2 + \frac{1}{2}w + \frac{1}{6} + \frac{\frac{1}{24}}{w} + \frac{\frac{1}{120}}{w^2} + O(\frac{1}{w^3})$$

This shows that $g(w) = f(1/z)$ has a singularity at $w = 0$ (because of the negative powers of w). Hence $f(z)$ has a **singularity at** ∞, by definition. Confirm these results by typing the following. (Type `?singular`.)

```
> singular(f);
```

$$\{z = 0\}, \{z = \infty\}$$

Zeros. A *zero* of an analytic function $f(z)$ in a domain D is a z_1 in D such that $f(z_1) = 0$. For instance, to find the zeros of $\cosh z$, type

```
> solve(cosh(z) = 0);                          # Resp. \frac{1}{2} I \pi
```

$\cosh z$ has infinitely many zeros. To get them, use the fact that they correspond to singularities of $1/\cosh z$. Type

```
> singular(1/cosh(z));            # Resp. {z = \frac{1}{2} I \pi + I \pi \_Z1~}
```

Here _Z1~ obviously stands for an arbitrary real integer. Be careful! This trick may give false results. Accordingly, verify results by substitution. In the present case,

```
> check := cosh(Pi*I/2 + n*Pi*I);          # Resp. check := − sin (n π)
> assume(n, integer):                        # Type ?assume for information.
> check;                                      # Resp. 0
```

Conversely, if a function $f(z)$ is a quotient of two functions, you may find the **singularities** of $f(z)$ from the zeros of the denominator (type **?denom** for information). For instance, type

```
> f := (-z^2 - 22*z + 8)/(z^3 - 5*z^2 + 4*z);
```

$$f := \frac{-z^2 - 22\,z + 8}{z^3 - 5\,z^2 + 4\,z}$$

```
> den := denom(f);
```

$$den := z\,\left(z^2 - 5\,z + 4\right)$$

```
> sol := solve(den = 0);                     # Resp. sol := 0, 1, 4
```

Make sure that the numerator and the denominator have no common factors (in which case a zero might cancel out and give no singularity). Show that the numerator `num` is not zero at 0, 1, or 4. In the last command below, `sol[n]` gives you 0, 1, 4 for $n = 1, 2, 3$, respectively. Try it separately.

```
> num := numer(f);                           # Resp. num := −z^2 − 22 z + 8
> seq(subs(z = sol[n], num), n = 1..3);     # Resp. 8, −96, −15
```

Similar Material in AEM: pp. 776-780

| **EXAMPLE 15.3** | **RESIDUE INTEGRATION** |

Residue theorem. Let $f(z)$ be analytic inside and on a simple closed path C, except for k points $z_1, z_2, ..., z_k$ inside C. Then the integral of $f(z)$ taken counterclockwise around C equals $2\,\pi\,i$ times the sum of the k residues of $f(z)$ at $z_1, ..., z_k$.

Hence this elegant integration method concerns contour integrals. You can obtain residues from Laurent series of $f(z)$ as in Example 15.2 in this Guide, or by formulas without using a series. There are two such formulas for the residue of $f(z)$ at a **pole of first order** (**"simple pole"**, that is, a singularity at a z_j for which the principal part of the Laurent series converges near z_j has just one term, b_1/z). The first formula is

$$(1) \qquad\qquad \lim_{z \to z_j} (z - z_j)\, f(z).$$

The second formula holds for functions of the form $f(z) = p(z)/q(z)$. Then the residue at $z = z_j$ is

$$(2) \qquad\qquad p(z_j)/q'(z_j) \qquad q' = dq/dz.$$

For instance, integrate $f(z) = (4 - 3z)/(z^2 - z)$ counterclockwise over a simple closed path C for which 0 and 1 lie inside C.

Solution. $z^2 - z = z\,(z - 1) = 0$ for $z = 0$ and 1. At these points, $f(z)$ has simple poles with residues, obtained from (2) by typing

```
> Res := (4 - 3*z)/diff(z^2 - z, z);        # Resp. Res := \frac{4 - 3\,z}{2\,z - 1}
```

```
> answer := 2*Pi*I*(subs(z = 0, Res) + subs(z = 1, Res));
```

$$answer := -6\,I\,\pi$$

You may confirm this by typing

```
> f := (4 - 3*z)/(z^2 - z);                    # Resp. f := \frac{4 - 3z}{z^2 - z}
> residue(f, z = 0);                           # Resp. -4
> residue (f, z = 1);                          # Resp. 1
```

Pole of mth order at z_j. This is a singularity for which the principal part of the Laurent series, convergent near z_j, has $1/(z - z_j)^m$ as the highest negative power. At such a pole the residue of $f(z)$ is

(3)
$$\frac{1}{(m-1)!} \lim_{z \to z_j} \left(\frac{d^{m-1}}{dz^{m-1}} \left[(z - z_j)^m f(z_j) \right] \right).$$

For instance, integrate $f(z) = (\tan 4z)/z^2$ clockwise around the circle $C : |z| = 1/4$.

Solution. $z = 0$ is the only point inside C at which $f(z)$ has a singularity, because $4z = \pi/2$ gives $z = \pi/8 > 1/4$. This is a pole of second order. From (3) with $m = 2$ you obtain the residue 4, so that the answer is $-8\pi\,i$, with the minus sign resulting from *clockwise* integration.

```
> Res := eval(subs(z = 0, diff(tan(4*z), z)));      # Resp. Res := 4

> residue(tan(4*z)/z^2, z = 0);                     # Resp. 4
```

Essential singularity. This is a singularity of $f(z)$ at z_0 such that the Laurent series of $f(z)$, convergent for $0 < |z - z_0| < R$, has infinitely many negative powers. Here we assume that singularity to be **isolated**, that is, it is the only singularity of $f(z)$ in a (possibly very small) neighborhood of z_0. In this case use the Laurent series with center z_0 just mentioned. For instance, integrate $f(z) = \exp(1/(z - 1))$ counterclockwise around the circle $|z| = 2$.

Solution. $f(z)$ has a singularity at $z = 1$, where $z - 1 = 0$. Obtain the Laurent series by setting $1/(z - 1) = w$, thus $z = (w + 1)/w$.

```
> f := exp(1/(z - 1));                         # Resp. f := e^{\left(\frac{1}{z-1}\right)}
> series(f, z - 1);
Error, wrong number (or type) of parameters in function series
> g := simplify(subs(z = (w + 1)/w, f));       # Resp. g := e^w
> S1 := series(g, w);
```

$$S1 := 1 + w + \frac{1}{2}\,w^2 + \frac{1}{6}\,w^3 + \frac{1}{24}\,w^4 + \frac{1}{120}\,w^5 + \mathrm{O}(w^6)$$

```
> S2 := subs(w = 1/(z - 1), S1);
```

$$S2 := 1 + \frac{1}{z-1} + \frac{\frac{1}{2}}{(z-1)^2} + \frac{\frac{1}{6}}{(z-1)^3} + \frac{\frac{1}{24}}{(z-1)^4} + \frac{\frac{1}{120}}{(z-1)^5} + \mathrm{O}\left(\frac{1}{(z-1)^6}\right)$$

You see that the residue of $f(z)$ at $z = 1$ (the coefficient of $1/(z - 1)$) is 1. Hence the answer is $2\,\pi\,i$. Try to obtain the residue by the command residue. It will not work.

The singularity of $f(z)$ at $z = 1$ is an essential singularity. Type `Order := 10` or `Order := 100` or whatever you want, before `S1`. Then `S2` will have as many negative powers as you want.

    Similar Material in AEM: pp. 781-786

EXAMPLE 15.4   **REAL INTEGRALS OF RATIONAL FUNCTIONS OF COS AND SIN**

Certain real integrals can be evaluated by complex contour integration by converting a real interval of integration to a complex contour. For instance, evaluate the integral of the function $1/(\sqrt{2} - \cos t)$ from $t = 0$ to 2π (t real) by complex contour integration.

Solution. The interval of integration is transformed to the unit circle by introducing the new variable of integration $z = e^{it}$. Thus $t = -i \operatorname{Ln} z$ and $dt = dz/iz$, so that the integrand `In` is $(dt/dz)/(\sqrt{2} - \cos t)$. Accordingly, type

```
> t := solve(z = exp(I*t), t);
```
           # Resp. $t := -I \ln(z)$

```
> c := cos(t);
```
           # Resp. $c := \dfrac{1}{2} z + \dfrac{1}{2} \dfrac{1}{z}$

```
> d := diff(t, z);
```
           # Resp. $d := -\dfrac{I}{z}$

```
> In := d/(sqrt(2) - c);
```
           # Resp. $In := \dfrac{-I}{z \left(\sqrt{2} - \frac{1}{2} z - \frac{1}{2} \frac{1}{z} \right)}$

```
> In := simplify(%);
```
           # Resp. $In := \dfrac{2 I}{-2 \sqrt{2} z + z^2 + 1}$

The integrand `In` has singularities at the roots of the quadratic polynomial in the denominator, which you obtain by the command (type `?solve` if necessary)

```
> zeros := solve(denom(In) = 0, z);
```
           # Resp. $zeros := \sqrt{2} + 1, \sqrt{2} - 1$

You need the residue of `In` at the second zero, at which `In` has a pole of first order. The first zero is greater than 1, so that it lies outside the unit circle and is of no interest here. Formula (2) in the previous example gives this residue if you type

```
> quot := numer(In)/diff(denom(In), z);
```
           # Resp. $quot := \dfrac{2 I}{-2 \sqrt{2} + 2 z}$

```
> Res := subs(z = zeros[2], quot);
```
           # Resp. $Res := -I$

Confirm this by typing

```
> residue(In, z = zeros[2]);
```
           # Resp. $-I$

```
> Answer := 2*Pi*I*Res;
```
           # Resp. $Answer := 2 \pi$

Maple can evaluate this kind of integral directly. In the present case, confirm the answer by using the command `int`.

```
> t := 't':
```

```
> eval(int(1/(sqrt(2) - cos(t)), t = 0..2*Pi));
```
           # Resp. 2π

    Similarly,

```
> int((1 + sin(t))/(3 + cos(t)), t = 0..2*Pi);
```
$\qquad$ # Resp. $\dfrac{1}{2}\sqrt{2}\,\pi$

```
> int(cos(t)/(13 - 12*cos(2*t)), t = 0..2*Pi);
```
$\qquad$ # Resp. 0

Similar Material in AEM: pp. 787, 793

$\boxed{\textbf{EXAMPLE 15.5}}$ **IMPROPER REAL INTEGRALS OF RATIONAL FUNCTIONS**

This example concerns the evaluation of **improper integrals** of rational functions $f(x)$ from $-\infty$ to ∞ by complex contour integration under the assumption that the denominator of $f(x)$ is nowhere zero and its degree is at least 2 units higher than that of the numerator. Then the integral of $f(x)$ equals $2\,\pi\,i$ times the sum of the residues at the poles of the ***complex*** function $f(z)$ in the upper half-plane. For instance, show that

$$\int_{-\infty}^{\infty} \frac{1}{1+x^4}\,dx = \frac{\pi}{\sqrt{2}}.$$

Solution. $f(z) = 1/(1 + z^4)$ has poles where $1 + z^4$ has zeros. You obtain them as follows. They may come out in a different order, and you will have to pick those in the upper half-plane (those with a positive imaginary part).

```
> z0 := solve(1 + z^4 = 0, z);
```

$$z0 \ := \ \frac{1}{2}\sqrt{2} + \frac{1}{2}I\sqrt{2},\ \frac{1}{2}I\sqrt{2} - \frac{1}{2}\sqrt{2}, -\frac{1}{2}\sqrt{2} - \frac{1}{2}I\sqrt{2},$$
$$-\frac{1}{2}I\sqrt{2} + \frac{1}{2}\sqrt{2}$$

```
> S := seq([Re(z0[n]), Im(z0[n])], n = 1..4);
```

$$S \ := \ \left[\frac{1}{2}\sqrt{2},\ \frac{1}{2}\sqrt{2}\right],\ \left[-\frac{1}{2}\sqrt{2},\ \frac{1}{2}\sqrt{2}\right],\ \left[-\frac{1}{2}\sqrt{2},\ -\frac{1}{2}\sqrt{2}\right],$$
$$\left[\frac{1}{2}\sqrt{2},\ -\frac{1}{2}\sqrt{2}\right]$$

```
> P1 := plot([S], style = point):
> P2 := plot([cos(t), sin(t), t = 0..2*Pi], scaling = constrained):
> with(plots):
> display(P1, P2);
```

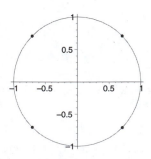

Example 15.5. Poles of $f(z)$ on the unit circle

The four poles are simple. The two poles at S[1] and S[2] lie in the upper half-plane because their imaginary parts are positive. From (2) in Example 15.3 in this Guide you obtain the residues and the answer by typing

```
> Res := 1/diff(1 + z^4, z);
```
$\qquad$ # Resp. $Res := \dfrac{1}{4}\dfrac{1}{z^3}$

```
> 2*Pi*I*(subs(z = z0[1], Res) + subs(z = z0[2], Res));
```

$$2\,I\,\pi\left(\frac{1}{4}\,\frac{1}{\left(\frac{1}{2}\,\sqrt{2}+\frac{1}{2}\,I\,\sqrt{2}\right)^3}+\frac{\frac{1}{4}}{\left(\frac{1}{2}\,I\,\sqrt{2}-\frac{1}{2}\,\sqrt{2}\right)^3}\right)$$

```
> answer := evalc(%);
```
$\qquad$ # Resp. $answer := \dfrac{1}{2}\sqrt{2}\,\pi$

Maple can evaluate the integral directly. Thus, you may confirm your result by typing

```
> int(1/(1 + x^4), x = -infinity..infinity);
```
$\qquad$ # Resp. $\dfrac{1}{2}\,\pi\sqrt{2}$

Integrals as before but with singularities on the real axis. Then under the degree condition the **Cauchy principal value** of the integral equals $2\,\pi\,i$ times the sum of the residues at the poles of $f(z)$ in the upper half-plane (as before) plus $\pi\,i$ times the sum of the residues at the poles of $f(z)$ on the x-axis.

For instance, you will see that

$$f(x) = \frac{1}{(x^2 - 3x + 2)(x^2 + 1)}$$

is of this kind. Type the denominator d with z instead of x and determine its zeros.

```
> d := (z^2 - 3*z + 2)*(z^2 + 1);
```
$\qquad$ # Resp. $d := (z^2 - 3\,z + 2)(z^2 + 1)$
```
> z0 := solve(d = 0, z);
```
$\qquad$ # Resp. $z0 := I,\ -I,\ 2,\ 1$

These are the locations of four simple poles. You need the residues of those at i, 2, and 1. The pole at $-i$ lies in the lower half-plane and is not needed. Use again formula (2) in Example 15.3. Type

```
> Res := simplify(1/diff(d, z));
```

$$Res\ :=\ \frac{1}{4\,z^3 + 6\,z - 9\,z^2\ - 3}$$

```
> answer := 2*Pi*I*subs(z = z0[1], Res) + Pi*I*(subs(z = z0[3], Res)
     + subs(z = z0[4], Res));
```

$$answer\ :=\ \frac{1}{10}\,\pi$$

You may confirm this result by using the command residue, typing

```
> 2*Pi*I*residue(1/d, z = z0[1]) + Pi*I*(residue(1/d, z = z0[3]) +
     residue(1/d, z = z0[4]));
```

$$\frac{1}{10}\,\pi$$

or by applying int(..., x = -infinity..infinity) to $f(x)$.

Similar Material in AEM: pp. 788-793

Problem Set for Chapter 15

Pr.15.1 **(Laurent series)** Find the Laurent series of $(z + \pi i)^{-2} \cosh z$ with center $z_0 = -\pi i$ (4 nonzero terms). (*AEM Ref.* p. 776 (#15))

Pr.15.2 **(Laurent series)** Find all Laurent series of $f(z) = 1/(z^2 + 1)$ with center $z_0 = i$ as well as the residue of $f(z)$ at z_0. (*AEM Ref.* p. 776 (#9))

Pr.15.3 **(Laurent series)** Find the Laurent series of $f(z) = \exp(-1/z^2)/z^2$ with center 0 (eight nonzero terms). What kind of singularity does $f(z)$ have at 0 and what is the corresponding residue? (*AEM Ref.* p. 776 (#7))

Pr.15.4 **(Singularities)** Find the location and kind of the singularities of $f(z) = \tan(\pi z/2)$ and the corresponding residues. (*AEM Ref.* p. 780 (#10))

Pr.15.5 **(Singularities)** Find the location and kind of the singularities and the residues of the function
$$f(z) = \frac{3 z^4 - 18 z^3 + 36 z^2 - 24 z + 1}{z^6 - 10 z^5 + 40 z^4 - 80 z^3 + 80 z^2 - 32 z}.$$
(*AEM Ref.* p. 777)

Pr.15.6 **(Zeros)** Find the zeros of $(z^4 - z^2 - 6)^3$. (*AEM Ref.* p. 780 (#7))

Pr.15.7 **(Zeros)** Find all zeros of $\tan z$. Use the idea in Example 15.2 in this Guide, but verify your result. (*AEM Ref.* p. 780 (#1))

Pr.15.8 **(Residue)** Find the residues of $f(z) = 50z/[(z + 4)(z - 1)^2]$ at all singularities. (*AEM Ref.* p. 784)

Pr.15.9 **(Residue)** Find the residues of $f(z) = 1/(1 - e^z)$ at all singularities. (*AEM Ref.* p. 786 (#5))

Pr.15.10 **(Residue integration)** Integrate $f(z) = \exp(-z^2)/\sin 4z$ counterclockwise around the unit circle. (*AEM Ref.* p. 786 (#20))

Pr.15.11 **(Residue integration)** Integrate $f(z) = (z \cosh \pi z)/(z^4 + 13 z^2 + 36)$ counterclockwise around the circle $|z| = \pi$. (*AEM Ref.* p. 786 (#19))

Pr.15.12 **(Real integral by residue integration)** Integrate $(1 + 4 \cos t)/(17 - 8 \cos t)$ from 0 to 2π by residue integration. Confirm the result by the command `int`. (*AEM Ref.* p. 793 (#8))

Pr.15.13 **(Real integral by residue integration)** Integrate $(\cos t)/(13 - 12 \cos 2t)$ from 0 to 2π by the residue method. Confirm the result by the command `int`. Could you get the result by jointly plotting the numerator and the reciprocal of the denominator and reasoning geometrically? (*AEM Ref.* p. 793 (#7))

Pr.15.14 **(Improper real integral)** Integrate $1/(x^2 - 2x + 5)^2$ from $-\infty$ to ∞ by the residue method. (*AEM Ref.* p. 794 (#14))

Pr.15.15 **(Poles on the real axis)** Integrate $x/(8 - x^3)$ from $-\infty$ to ∞, using the method in Example 15.5 in this Guide. (*AEM Ref.* p. 794 (#23))

Chapter 16

Complex Analysis in Potential Theory

Content. Complex potential (Ex. 16.1, Prs. 16.1-16.3)
Conformal mapping (Ex. 16.2, Prs. 16.4, 16.5)
Fluid flow (Ex. 16.3, Prs. 16.6, 16.7)
Series for potential (Ex. 16.4, Pr. 16.8)
Mean value theorems (Ex. 16.5, Prs. 16.9, 16.10)

Complex potential. $F(z) = \phi(x, y) + i\psi(x, y)$ permits the use of complex analysis for two-dimensional potential problems. Its advantage is that it simultaneously gives equipotential lines $\phi(x, y) = const$ and lines of force (streamlines of flows) $\psi(x, y) = const$.

We use ϕ and ψ instead of Φ and Ψ, used in AEM, because Maple "protects" Ψ, that is, prohibits its general use.

Examples for Chapter 16

EXAMPLE 16.1 COMPLEX POTENTIAL. RELATED PLOTS

Find and plot the equipotential lines and the lines of force of the **complex potential** $F(z) = -i\, z^2$.

Solution. Write $F(z) = \phi(x, y) + i\psi(x, y)$. Type F and ϕ. Then solve the equation $\phi(x, y) = k = const$ for y. Use y for plotting a sequence S of **equipotential lines** (hyperbolas).

```
> F := -I*z^2;                              # Resp. F := -I z^2
> z := x + I*y;                             # Resp. z := x + I y
> phi := evalc(Re(F));                      # Resp. φ := 2 x y
> y := solve(phi = k, y);
```

$$y := \frac{1}{2}\frac{k}{x}$$

```
> S := seq(y, k = 0..10);
```

$$S := 0, \frac{1}{2}\frac{1}{x}, \frac{1}{x}, \frac{3}{2}\frac{1}{x}, 2\frac{1}{x}, \frac{5}{2}\frac{1}{x}, 3\frac{1}{x}, \frac{7}{2}\frac{1}{x}, 4\frac{1}{x}, \frac{9}{2}\frac{1}{x}, 5\frac{1}{x}$$

```
> x := 'x':    y := 'y':
> plot({S}, x = 0..2.5, y = 0..2.5, scaling = constrained);
```

Lines of force can be plotted in a similar fashion. Plotting the sequence S and a sequence S2 of lines of force on the same axes, using `display`, you obtain a rectangular net of curves.

```
> psi := evalc(Im(F));                      # Resp. ψ := -x^2 + y^2
```

```
> y := solve(psi = k2, y);
```
Resp. $y := \sqrt{x^2 + k2},\ -\sqrt{x^2 + k2}$

You see that the equation $\psi = const$ has two solutions. Take the first, `y[1]`, which will give curves in the first quadrant. (`y[2]` gives curves in the fourth; try it.) Write `k2`, avoiding unassigning `k`. For positive (negative) `k2` you get curves above (below) the line $y = x$.

```
> S2 := seq(y[1], k2 = -10..10);
```

$$S2 := \sqrt{x^2 - 10},\ \sqrt{x^2 - 9},\ \sqrt{x^2 - 8},\ \sqrt{x^2 - 7},\ \sqrt{x^2 - 6},\ \sqrt{x^2 - 5},\ \sqrt{x^2 - 4},$$

$$\sqrt{x^2 - 3},\ \sqrt{x^2 - 2},\ \sqrt{x^2 - 1},\ \sqrt{x^2},\ \sqrt{x^2 + 1},\ \sqrt{x^2 + 2},\ \sqrt{x^2 + 3},\ \sqrt{x^2 + 4},$$

$$\sqrt{x^2 + 5},\ \sqrt{x^2 + 6},\ \sqrt{x^2 + 7},\ \sqrt{x^2 + 8},\ \sqrt{x^2 + 9},\ \sqrt{x^2 + 10}$$

```
> x := 'x':    y := 'y':
> P1 := plot({S}, x = 0..2.5, y = 0..2.5):
> P2 := plot({S2}, x = 0..2.5, y = 0..2.5):
> with(plots):
```
Ignore the warning.
```
> display(P1, P2, scaling = constrained);
```

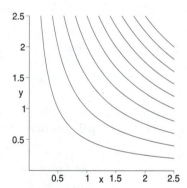

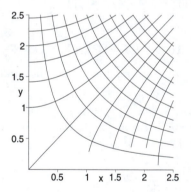

Example 16.1. Equipotential lines $\phi = 2xy = const$ (hyperbolas)

Example 16.1. Equipotential lines and lines of force

Similar Material in AEM: pp. 799-802

EXAMPLE 16.2   USE OF CONFORMAL MAPPING

Conformal mapping provides a second approach to the plotting problem in the previous example. This is based on the command `conformal`. Solving $w = F(z) = -i z^2$ for z, you have $z = \sqrt{i F} = \sqrt{i w}$. Interchanging the notations gives the inverse $w = \sqrt{i z}$. Now `conformal` maps rectangles. Under the mapping by $F(z)$ the vertices 0, 2.5, $2.5 + 2.5i$, $2.5i$ of the square in the figures in the previous example have the images 0, $-6.25i$, 12.5, $6.25i$, respectively. Choose the smallest rectangle (with sides parallel to the axes) that includes these points. Its vertices are $\pm 6.25i$ and $12.5 \pm 6.25i$. The plot (the conformal mapping by the inverse of F) will show a larger region containing the image of this rectangle and equipotential lines and lines of force in this image.

```
> with(plots):                                    # Ignore the warning.
> conformal(sqrt(I*z), z = -6.25*I..12.5 + 6.25*I,
  scaling = constrained, labels = [x, y]);
```

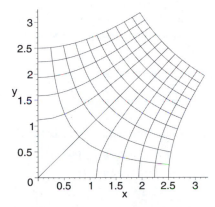

Example 16.2. Plotting equipotential lines and lines of force
by the command `conformal`

Similar Material in AEM: pp. 799-803

| **EXAMPLE 16.3** | **FLUID FLOW** |

Flow around a cylinder. Consider the complex potential $F(z) = z + 1/z$. Find the **stream function** ψ and the **streamlines** $\psi = const$. In particular, show that the streamline $\psi = 0$ consists of the x-axis and the unit circle, so that the flow can be regarded as a flow around a cylinder (intersecting the z-plane in the unit circle), which far enough from the cylinder is practically parallel because $1/z$ in $F(z)$ becomes negligible compared to z when $|z|$ is large. Plot streamlines in the z-plane.

Solution. The physical situation suggests polar coordinates, at least to some extent. Type z, then the complex potential F, and then its imaginary part, which is the stream function $\psi(r, \theta)$.

```
> z := r*exp(I*theta);                            # Resp. z := r e^(I θ)
```

$$\text{\# Resp. } z := r\,e^{(I\,\theta)}$$

```
> F := z + 1/z;
```

$$\text{\# Resp. } F := r\,e^{(I\,\theta)} + \frac{1}{r\,e^{(I\,\theta)}}$$

```
> psi := simplify(evalc(Im(F)));
```

$$\text{\# Resp. } \psi := \frac{\sin(\theta)\,(r^2 - 1)}{r}$$

From this formula you can see that the x-axis ($\theta = 0$ and π) and the unit circle ($r = 1$) give $\psi = 0$, as claimed.

Now get the streamlines by solving $\psi = const$ for $r = r(\theta)$. Denote the constant by k times a numerical factor (0.25) found by trial and error to have a reasonably dense field of streamlines. (Experiment with k and see what happens.)

```
> sol := solve(psi = 0.25*k, r);
```

$$sol := \frac{1}{8}\frac{k + \sqrt{k^2 + 64\sin(\theta)^2}}{\sin(\theta)}, \frac{1}{8}\frac{k - \sqrt{k^2 + 64\sin(\theta)^2}}{\sin(\theta)}$$

In making up a sequence of streamlines for plotting, $\sin\theta$ in the denominator would cause trouble at 0 and π. This is the reason why you stay away from 0 and π, as you see in the next two commands. Try otherwise, and you will see that the figure is no longer usable. Also, the choice of k, positive in S1 and negative in S2 (and 0 in both) has the effect that there are no curves inside the cylinder (which would be physically meaningless). Observe that the streamlines are practically horizontal and parallel in regions far away from the cylinder.

```
> S1 := seq([sol[1], theta, theta = 0.001..Pi - 0.001], k = 0..20):
> S2 := seq([sol[2], theta, theta = 0.001..Pi - 0.001], k = -20..0):
> with(plots):
> P1 := plot({S1}, -9..9, -3..3, coords = polar):
> P2 := plot({S2}, -9..9, -3..3, coords = polar):
> display({P1, P2}, scaling = constrained);
```

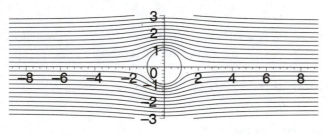

Example 16.3. Flow around a cylinder

Similar Material in AEM: pp. 814, 818

| EXAMPLE 16.4 | **SERIES REPRESENTATION OF POTENTIAL**

This concerns the potential in a disk $|z| \leq R$ for a given boundary potential whose Fourier series is

$$\phi(R, \theta) = a_0 + \sum_{n=1}^{\infty} (a_n \cos n\theta + b_n \sin n\theta).$$

The series for the potential inside the disk is

$$\phi(r, \theta) = a_0 + \sum_{n=1}^{\infty} \left(\frac{r}{R}\right)^n (a_n \cos n\theta + b_n \sin n\theta).$$

For instance, find the series corresponding to the boundary potential $-\theta/\pi$ if $-\pi < \theta < 0$ and θ/π if $0 < \theta < \pi$ on the unit circle $|z| = 1$. (Make a sketch of that potential.)

Solution. Since the boundary potential is an even function, its sine terms are zero, so that you get a Fourier cosine series. The constant term is $a_0 = 1/2$ (the mean value of the boundary potential). The other Fourier cosine coefficients are

```
> an := 2/Pi*int(theta/Pi*cos(n*theta), theta = 0..Pi);
```

$$an := 2\,\frac{\cos(\pi\,n) + n\,\sin(\pi\,n)\,\pi - 1}{\pi^2\,n^2}$$

```
> seq(an, n = 1..10);
```

$$-4\,\frac{1}{\pi^2},\ 0,\ -\frac{4}{9}\,\frac{1}{\pi^2},\ 0,\ -\frac{4}{25}\,\frac{1}{\pi^2},\ 0,\ -\frac{4}{49}\,\frac{1}{\pi^2},\ 0,\ -\frac{4}{81}\,\frac{1}{\pi^2},\ 0$$

Hence the corresponding partial sum of the series representing the potential inside the disk is

```
> phi := 1/2 + sum(an*r^n*cos(n*theta), n = 1..10);
```

$$\phi := \frac{1}{2} - \frac{4\,r\,\cos(\theta)}{\pi^2} - \frac{4}{9}\,\frac{r^3\,\cos(3\,\theta)}{\pi^2} - \frac{4}{25}\,\frac{r^5\,\cos(5\,\theta)}{\pi^2} - \frac{4}{49}\,\frac{r^7\,\cos(7\,\theta)}{\pi^2} - \frac{4}{81}\,\frac{r^9\,\cos(9\,\theta)}{\pi^2}$$

Setting $r = 1$, you can plot a good approximation of the boundary potential by typing

```
> plot(subs(r = 1, phi), theta = -Pi..Pi, scaling = constrained,
  labels = [theta, 'phi']);
```

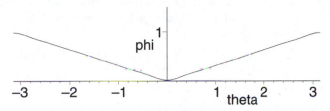

Example 16.4. Boundary potential approximated by the partial sum shown (with $r = 1$)

You may also plot the potential ϕ as a surface over the disk $|z| \le 1$ by the commands

```
> with(plots):                              # Ignore the warning.
> cylinderplot([r, theta, 10*phi], r = 0..1, theta = 0..2*Pi);
```

     Similar Material in AEM: p. 821

| EXAMPLE 16.5 | **MEAN VALUE THEOREM FOR ANALYTIC FUNCTIONS** |

Mean Value Theorem. If $F(z)$ is analytic in a simply connected domain D, then its value at any z_0 in D equals the mean value of $F(z)$ over any circle of radius r in D with center z_0. As a formula,

$$F(z_0) = \frac{1}{2\,\pi}\int_0^{2\,\pi} F(z_0 + r\,e^{i\,t})\,dt.$$

Verify this for $F(z) = (z + 2)^2$ and a circle of radius 1 and center 1.

Solution. Type a representation of the circle, then F on the circle, then the integral. Clearly, it should have the value $F(1) = 9$.

```
> z := 1 + exp(I*t);                        # Resp. z := 1 + e^{(I\,t)}
```

```
> F := (z + 2)^2;                           # Resp. F := (3 + e^{(I\,t)})^2
```

```
> 1/(2*Pi)*(int(F, t = 0..2*Pi));                    # Resp. 9
```

Remark. The present theorem has applications to potential problems (see Pr.16.9).

   Similar Material in AEM: p. 822

Problem Set for Chapter 16

Pr.16.1 **(Complex potential)** Find and plot the equipotential lines of $F(z) = 1/z$. (*AEM Ref.* p. 803 (#16))

Pr.16.2 **(Complex potential)** Find and plot the equipotential lines of the complex potential $F(z) = \arccos z$. Show that this can be interpreted as the potential in a slit on the x-axis from -1 to 1, with the left plate $x \leq -1$ at potential, say, $-a$, and the right plate $x \geq 1$ at potential $+a$. (*AEM Ref.* p. 803 (#13))

Pr.16.3 **(Orthogonal net, maximum principle)** Find and plot equipotential lines and lines of force of $F(z) = \cos z$ jointly in the z-plane. From the plot find the points at which this net is not orthogonal. Plot the real potential ϕ as a surface over a square in the z-plane and convince yourself that ϕ satisfies the *maximum principle*, that is, the maximum and minimum values of ϕ are taken on the boundary of the square. (Type `?plot3d`.) (*AEM Ref.* pp. 799-803, 824)

Pr.16.4 **(Use of conformal mapping)** Obtain a portion of the orthogonal net in Pr.16.3, say, the portion in the rectangle $0 \leq x \leq \pi/2, 0 \leq y \leq 4$, by the command `conformal`. (Type `?conformal`.) See Example 16.2 in this Guide. (*AEM Ref.* pp. 799-803)

Pr.16.5 **(Pair of opposite electrical charges)** The field of two charges, $+1$ at $z = -1$ and -1 at $z = 1$, has the complex potential $F = \text{Ln}\left(-(z+1)/(z-1)\right)$ (the first minus amounting to adding an imaginary constant πi to F). Try to plot a net of equipotential lines and lines of force by using F. The result will be disappointing. Obtain the plot by the command `conformal`. (*AEM Ref.* p. 802)

Pr.16.6 **(Arbitrary cylinder)** Find the complex potential of a flow around a cylinder of radius r. Plot the streamlines for a cylinder of radius 2. (*AEM Ref.* p. 817 (#9))

Pr.16.7 **(Complex potential)** Find and discuss the streamlines of $F(z) = z^3$. You will get these curves in three of six sectors appearing in the plane. Fill the remaining three sectors by applying a suitable rotation. (*AEM Ref.* pp. 799-807)

Pr.16.8 **(Series representation of potential)** Find a series of the potential inside the unit circle if the upper half of the circle is kept at 110 volts and the lower half is grounded (is kept at 0 volt). Plot the potential as a surface. Also plot the boundary potential as a curve, to see how accurately the partial sums used will approximate the actual potential. (*AEM Ref.* p. 822 (#10))

Pr.16.9 **(Mean value of harmonic functions)** By taking the real part on both sides of the formula in Example 16.5 of this Guide you obtain a corresponding mean value theorem for harmonic functions. Verify this theorem for $\phi = (x-1)(y-1)$, the point $(3, -3)$, and a circle of radius 1. (*AEM Ref.* p. 825 (#7))

Pr.16.10 **(Mean value of harmonic functions)** If ϕ is harmonic in a simply connected domain D, its value at a point $P_0 : (x_0, y_0)$ in D equals the mean value of ϕ over any disk D_0 in D with center P_0 (defined as the integral over D_0 divided by the area of D_0). Verify this for the function in Pr.16.9 and a disk of radius 1 with center $(3, -3)$. (*AEM Ref.* p. 825 (#7))

PART E. NUMERICAL METHODS

Content. Basic concepts, solving equations, interpolation, integration (Chap. 17)
            Gauss, Gauss-Seidel methods, LU, matrix inversion, methods for
                eigenvalues (Chap. 18)
            Euler, Runge-Kutta, Adams Moulton methods for ODE's and systems,
                methods for PDE's (Chap. 19)

Chapter 17

Numerical Methods in General

Content. Significant digits, quadratic equations (Ex. 17.1, Pr. 17.1)
            Solution of equations (Newton's method, etc.; Exs. 17.2-17.5,
                Prs. 17.2-17.6)
            Interpolation (Lagrange, Newton, splines; Exs. 17.6, 17.7,
                Prs. 17.7-17.11)
            Integration (Ex. 17.8, Prs. 17.12-17.15)

`evalf(f, k)` gives floating point representations with kS (k **significant digits**).

Examples for Chapter 17

EXAMPLE 17.1 **LOSS OF SIGNIFICANT DIGITS.**
                           QUADRATIC EQUATION

Find the roots of the quadratic equation $x^2 - 40x + 2 = 0$ to 4S (4 significant digits).

Solution. Type a general quadratic equation. Then obtain the usual formula by the command `solve`. Note that also the discriminant could be obtained separately, by `discrim`. (Type `?solve`, `?discrim`.)

```
> eq := a*x^2 + b*x + c = 0;              # Resp. eq := a x^2 + b x + c = 0
```

```
> sol := solve(eq, x);
```

$$sol := \frac{1}{2}\frac{-b + \sqrt{b^2 - 4ac}}{a}, \frac{1}{2}\frac{-b - \sqrt{b^2 - 4ac}}{a}$$

Now substitute the given $a = 1$, $b = -40$, $c = 2$ into the solution formula, obtaining

```
> Sol := evalf(subs({a = 1, b = -40, c = 2}, {sol[1], sol[2]}), 4);
```

$$Sol := \{39.95, .05\}$$

You got a 4S value for the larger root. But the other root shows **loss of significant digits** because it is a small difference of large numbers, $20.00 - 19.95$. An **improved**

171

formula follows by noting that c/a is the product of the two roots. For your special equation it gives 0.050064, correct to 4S. (This is obtained with `evalf(..., 5)`; for `evalf(..., 4)` you would get 0.05008; try it.)

```
> improved := (c/a)/sol[1];
```
$$\text{\# Resp. } improved := 2\,\frac{c}{-b + \sqrt{b^2 - 4\,a\,c}}$$

```
> Imp := evalf(subs({a = 1, b = -40, c = 2}, improved), 5);
```

$$Imp := .050064$$

To avoid misunderstandings, 4S was used for convenience. For more digits the situation is the same in principle. For instance, you get 30S and 27S for the roots, respectively, from the usual formula, but 30S for the smaller root by the improved formula.

```
> S := evalf(solve(x^2 - 40*x + 2 = 0, x), 30);
```

$$S := 39.9499373432600033316538822341,\ .0500626567399966683461177659$$

```
> evalf(2/S[1],30);        # Resp. .0500626567399966683461177658616
```

Similar Material in AEM: p. 836

EXAMPLE 17.2   FIXED-POINT ITERATION

Find a solution of $x^3 + x - 1 = 0$ by **fixed-point iteration.**

Solution. Transform the given $f(x) = x^3 + x - 1 = 0$ algebraically into the form $x = g(x)$ and solve it by iteration, that is, start from a suitable x_0 and calculate

(A) $$x_1 = g(x_0),\quad x_2 = g(x_1),\quad \dots,\quad x_n = g(x_{n-1}),\quad \dots$$

until the desired accuracy is reached. Now $f(x) = 0$ can be cast into the form $x = g(x)$ in several ways. For instance $x = 1 - x^3$ or

(B) $$x = g(x) = 1/(1 + x^2).$$

To find a suitable x_0, sketch or plot $f(x) = x^3 + x - 1$. Type

```
> plot(x^3 + x - 1, x = 0..1.5, xtickmarks = [0, 0.5, 1, 1.5]);
```

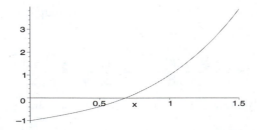

Example 17.2. Given function $f(x)$

Even from a rough sketch you see that there is a zero near 1. Hence you may start the iteration from $x_0 = 1$. You have to do the transformation of $f(x) = 0$ into $x = g(x)$

in such a way that you have convergence. If $|g'(x)| \leq K < 1$ in an interval which contains the solution and the starting value and in which $g'(x)$ is continuous, then you have convergence of the iteration (A). This is true if you choose (B) and, say, $x_0 = 1$. Type

```
> x := 'x':
> x(0) := 1;                                    # Resp. x(0) := 1
> g := 1/(1 + x^2);                             # Resp. g := 1/(1+x^2)
```

$$\text{\# Resp. } g := \frac{1}{1+x^2}$$

Type the iteration as a **do-loop**, which is terminated by od (the letters of do reversed).

```
> for n from 1 to 26 do
>     x(n) := evalf(subs(x = x(n-1), g), 6):
> od:
> seq(x(n), n = 0..26);
```

1, .500000, .800000, .609756, .728969, .652997, .701061, .670470, .689879, .677539, .685373, .680392, .683560, .681547, .682822, .682012, .682529, .682198, .682412, .682273, .682361, .682305, .682342, .682319, .682333, .682324, .682328

The last value is accurate to 6D, as you can see by typing

```
> evalf(solve(x^3 + x = 1, x));
```

     $.6823278040, -.3411639019 + 1.161541401\,I, -.3411639019 - 1.161541401\,I$

To represent the iteration graphically, type

```
> with(plots):                                  # Ignore the warning.
> P1 := plot(x, x = 0..1.2):
> P2 := plot(g(x), x = 0..1.2):
> P3 := plot([[x(0), 0], [x(0), x(1)], [x(1), x(1)], [x(1), x(2)],
    [x(2), x(2)], [x(2), x(3)], [x(3), x(3)], [x(3), x(4)], [x(4), x(4)],
    [x(4), x(5)], [x(5), x(5)]], style = LINE, scaling = constrained):
> display(P1, P2, P3);
```

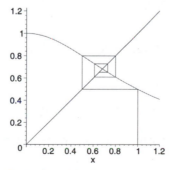

Example 17.2. Graphical representation of the iteration

Similar Material in AEM: pp. 840, 841

EXAMPLE 17.3   **SOLVING EQUATIONS BY NEWTON'S METHOD**

Find the positive solution of $2 \sin x = x$.

Solution. Solve $f(x) = x - 2 \sin x = 0$. To obtain a starting value, sketch $f(x)$. You see that your solution is near 2. As the starting value take $x_0 = 2$. You also need the derivative $f'(x)$. Then set up a do-loop involving the formula $x_{n+1} = x_n - f(x_n)/f'(x_n)$ of Newton's method, whose idea is the approximation of the curve of $f(x)$ by the tangent at x_0, x_1, etc.

```
> f := x - 2*sin(x);                    # Resp. f := x - 2 sin(x)
> plot(f, x = 0..2.5);
```

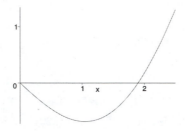

Example 17.3. Curve of the function $f(x)$ whose zero is sought

```
> fprime := diff(f, x);                 # Resp. fprime := 1 - 2 cos(x)
> x:='x':
> x(0) := 2;                            # Resp. x(0) := 2
> for n from 0 to 10 do
>        x(n+1)   := x(n) - evalf(subs(x = x(n), f/fprime));
> od:
> S := seq(x(n), n = 0..5);
```

$$S := 2, 1.900995594, 1.895511645, 1.895494267, 1.895494267, 1.895494267$$

```
> S[4] - 2*sin(S[4]);                   # Resp. 0   # Check
> x:='x':
> fsolve(f = 0, x, 1..2);               # Resp. 1.895494267   # Verification
```

Similar Material in AEM: p. 843

EXAMPLE 17.4   **SOLVING EQUATIONS BY THE SECANT METHOD**

The formula for the secant method is

$$x_{n+1} = x_n - f(x_n) \frac{x_n - x_{n-1}}{f(x_n) - f(x_{n-1})}.$$

Using this method, find the positive solution of $2 \sin x = x$. Compare with the previous example.

Solution. Solve $f(x) = x - 2 \sin x = 0$. You need two starting values x_0 and x_1 at which f has different signs. From a sketch of f (see the figure in the previous example) you see that you may take, say, $x_0 = 2$ and $x_1 = 1.8$. Accordingly, type

```
> x:='x':
> f := x - 2*sin(x);                      # Resp. f := x − 2 sin(x)
> x(0) := 2;                              # Resp. x(0) := 2
> x(1) := 1.8;                            # Resp. x(1) := 1.8
> for n from 1 to 10 do
> x(n + 1) := x(n) - evalf(subs(x = x(n), f)*(x(n) - x(n-1))/(subs(x =
  x(n), f) - subs(x = x(n-1), f)));
> od;
```

$$x(2) := 1.889756961$$
$$x(3) := 1.895834746$$
$$x(4) := 1.895493132$$
$$x(5) := 1.895494267$$
$$x(6) := 1.895494267$$

```
Error, division by zero
```

This is typical. The denominator of the formula of the method is $f(x_n) - f(x_{n-1})$. It becomes 0 as soon as the two x-values are equal within the number of digits used in the calculation.

       Similar Material in AEM: pp. 845, 846

EXAMPLE 17.5   SOLVING EQUATIONS BY THE BISECTION METHOD. PROCEDURE

This explains a procedure typed as `BISECT := proc(f, a, b, acc)`, which will give you the zero if you simply insert the given function `f`, values `a` and `b` ($>$ `a`) at which `f` has different signs, and an error bound `acc` determining the desired accuracy. Since the initial interval has length $b - a$ and is bisected, after n steps you have an interval of length $(b-a)/2^n$, short enough if its length is less than `acc`. Solving this inequality $(b - a)/2^n < acc$ for n gives $n > \ln((b-a)/acc)/\ln 2$. This is used in the procedure, where `trunc(k)` is k truncated to an integer value. (Type `?trunc`.) Bisection of an interval is followed by selecting one of the two resulting subintervals, namely, the one at whose endpoints the function f has different signs. This is the subinterval in which the zero lies. It will be bisected in the next step. Type the procedure as follows. Apply it to specific equations afterward.

```
> BISECT := proc(f, a, b, acc)
> left := a;                                  # Left end of starting interval
> right := b;                                 # Right end of starting interval
> N := evalf(ln((b - a)/acc)/ln(2));     # N determines the number of steps.
> mid := (a + b)/2;                          # First bisection point
> for j from 1 to trunc(N + 1) do
>     if evalf(subs(x = mid, f(x))*subs(x = left, f(x))) > 0
>     then                                    # Right subinterval is next interval.
>         left := mid;          # Left end of new interval is the old midpoint.
>     else                                    # Left subinterval is next interval.
>         right := mid;         # Right end of new interval is the old midpoint.
```

```
>      fi;                                    # End of if-then-else
>      mid := (left + right)/2;               # Bisection point for next step
> od;                                         # End of do loop
> end:                                        # End of procedure
Warning, 'left' is implicitly declared local
Warning, 'right' is implicitly declared local
Warning, 'N' is implicitly declared local
Warning, 'mid' is implicitly declared local
Warning, 'j' is implicitly declared local
```

These warnings tell us that the variables indicated are **local variables**, that is, they have values that are known only within the BISECT procedure. This means that we cannot, for example, use the value of N outside the procedure. To avoid such warnings we could use the command local left, right, N, mid, j; after the line BISECT := proc(f, a, b, acc). This method is used in Example 19.4 of this Guide.

There are many advantages to using a procedure. The first one is that you now merely need to insert your data, that is, the function f, the endpoints a and b of the starting interval, and the required accuracy. For instance, solving $\sin x = 0$ with suitably chosen a and b gives an approximation for π,

```
> BISECT(sin, 0.1, 5, 10^(-10));              # Resp. 3.141592653
```

As another example, let

```
> g := x^2 - 1;                               # Resp. g := x^2 - 1
> BISECT(g, 0.1, 5, 0.0002);
Error, (in BISECT) cannot evaluate boolean
```

Here you need *another kind of representation of a function*, as shown.

```
> g := x -> x^2 - 1;                          # Resp. g := x -> x^2 - 1
> BISECT(g, 0.1, 5, 0.0002);                  # Resp. .9999832158
> BISECT(g, -2, 0.5, 0.0002);                 # Resp. -1.000015259
```

The second advantage is that you can save the procedure for future use. The choice of the name of the file is up to you. The command is

```
> save BISECT, bisect:                        # Or choose any other name for the file.
```

This will **save** the BISECT procedure in the file bisect. (Type ?save for information.) To **use it again** (after having it filed as just shown), type

```
> read bisect:
```

(Type ?read for information.) This will load it, ready for your new use.
     Similar Material in AEM: p. 848

EXAMPLE 17.6   POLYNOMIAL INTERPOLATION

Polynomial interpolation means the following. Given $n + 1$ ordered pairs of numbers $(x_0, y_0), (x_1, y_1), (x_2, y_2), ..., (x_n, y_n)$ (Cartesian coordinates of points in the xy-plane), find the (unique) polynomial $p_n(x)$ of smallest degree that assumes those

y-values at the corresponding x-values. In general, p_n will have degree n, but the degree may sometimes be less, depending on those values (on the location of the given points through which the curve of p_n is supposed to pass). p_n is obtained by the command interp. (Type ?interp.)

For instance, given (9.0, 2.1972), (9.5, 2.2513), (11.0, 2.3979), you obtain a quadratic polynomial p_2 by typing the vector **u** of the x-values, the vector **v** of the y-values, and then a letter for the independent variable of p_2, say, x. Thus,

```
> u := [9.0, 9.5, 11.0];                    # Resp. u := [9.0, 9.5, 11.0]
> v := [2.1972, 2.2513, 2.3979];          # Resp. v := [2.1972, 2.2513, 2.3979]
> p2 := interp(u, v, x);
```

$$p2 := -.005233333333\, x^2 + .2050166667\, x + .775950000$$

If you want a value of p_2 at an x somewhere between the given values, say, at $x = 9.2$, type

```
> evalf(subs(x = 9.2, p2), 7);               # Resp. 2.219155
```

Rounded to 5S, this gives 2.2192. The given values are 5S-values of the natural logarithm, and $\ln 9.2 = 2.219203$. Hence your **quadratic interpolation** has given a correct 5S-value. Greater accuracy would require more given data so that an interpolation polynomial of higher degree could be obtained.

For given data, **Lagrange's** and **Newton's interpolation formulas** give the same polynomial (except, perhaps, for small round-off errors), although both formulas look totally different.

Newton's divided difference formula gives p_2 as follows.

```
> a11 := (v[2] - v[1])/(u[2] - u[1]);         # Resp. a11 := .1082000000
> a12 := (v[3] - v[2])/(u[3] - u[2]);         # Resp. a12 := .09773333333
```

These are the $n = 2$ first divided differences (there would be more if you had more data). Now comes the $n - 1 = 1$ second divided difference (there is only one; again, there would be more if you had more data).

```
> a21 := (a12 - a11)/(u[3] - u[1]);           # Resp. a21 := -.005233333335
> p := v[1] + (x - u[1])*a11 + (x - u[1])*(x - u[2])*a21;
```

$$p := 1.223400000 + .1082000000\, x - .005233333335\,(x - 9.0)(x - 9.5)$$

```
> p := expand(%);   # Resp. p := .7759499999 + .2050166667 x - .005233333335 x²
```

This agrees with the above response to interp. The mysterious 1.2234, in the above, is v[1] - i[1]*a11, a simplification that disguises the actual formula.

Similar Material in AEM: pp. 850, 854

EXAMPLE 17.7 **SPLINE INTERPOLATION**

Higher order polynomials for interpolating data (x_0, y_0), $(x_1, y_1), \ldots, (x_n, y_n)$ tend to oscillate between nodes; see Fig. A, where $n = 8$ and the nodes are at $x = -4$, $-3, \ldots, 4$, and $y(0) = 1$ and the other y's are zero. Obtain this figure by typing (type ?interp)

```
> u := [-4,-3,-2,-1,0,1,2,3,4];     # Resp. u := [-4, -3, -2, -1, 0, 1, 2, 3, 4]
> v := [0,0,0,0,1,0,0,0,0];          # Resp. v := [0, 0, 0, 0, 1, 0, 0, 0, 0]
> p := interp(u, v, x);
```

$$p := \frac{1}{576}x^8 - \frac{5}{96}x^6 + \frac{91}{192}x^4 - \frac{205}{144}x^2 + 1$$

```
> pts := seq([u[j], v[j]], j = 1..9);     # Preparation for plotting
```

$$pts := [-4,\ 0],\ [-3, 0],\ [-2, 0],\ [-1, 0],\ [0, 1],\ [1, 0],\ [2, 0],\ [3, 0],\ [4, 0]$$

```
> P1 := plot({pts}, style = point):
> P2 := plot(p, x = -4..4):
> with(plots):                           # Ignore the warning.
> display(P1, P2, labels = [x, y]);
```

Large oscillations are avoided if instead of a single polynomial of high degree you choose n polynomials p_j of lower degree, say, of degree 3 (cubical polynomials), p_1 between (x_0, y_0) and (x_1, y_1), then p_2 between (x_1, y_1) and (x_2, y_2), and so on, such that the function g consisting of these polynomials interpolates the given data, that is, $g(x_j) = y_j$　$(j = 0, .., n)$, and is twice continuously differentiable. This g is called a (cubic) **spline**. If you impose two further conditions, for instance, $g''(x_0) = 0$, $g''(x_n) = 0$, then g is uniquely determined. A spline satisfying these two additional conditions is called a **natural spline.** (Other additional conditions of practical interest are $g'(x_0) = k_0$, $g'(x_n) = k_n$, where k_0 and k_n are given numbers.) To obtain the natural spline for the above data, type ?spline for information and

```
> spl := spline(u, v, x, 3);             # 3 because the spline is cubic. Try 5.
```

Since $n = 8$, this spline consists of 8 polynomials, as you can see from the response, which we omit for reasons of space. Plot spl by typing

```
> plot(spl, x = -4..4, labels = [x, y]);
```

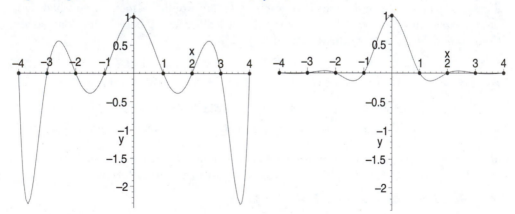

***Example 17.7.* A.** Oscillation of a high-order polynomial

***Example 17.7.* B.** Cubic spline for the data in Fig. A

Similar Material in AEM: pp. 861-866

EXAMPLE 17.8   NUMERICAL INTEGRATION

Integrate $f(x) = \exp(-x^2)$ from 0 to 1, by the usual methods, choosing $n = 10$ steps of size $h = 0.1$.

Solution. Type the integrand

```
> f:='f':
> f(x) := exp(-x^2);                          # Resp. f(x) := e^(-x²)
```

Rectangular rule. This rule is usually not accurate enough. For information type `?leftsum`,`?middlesum`, `?rightsum`. Then type

```
> with(student):
> middlesum(f(x), x = 0..1, 10);
```

$$\frac{1}{10}\left(\sum_{i=0}^{9} e^{\left(-(1/10\,i+1/20)^2\right)}\right)$$

```
> evalf(%, 6);                               # Resp. .747131
```

Trapezoidal rule. Type `?trapezoid` and then

```
> trapezoid(f(x), x = 0..1, 10);
```

$$\frac{1}{20} + \frac{1}{10}\left(\sum_{i=1}^{9} e^{\left(-1/100\,i^2\right)}\right) + \frac{1}{20}\,e^{(-1)}$$

```
> evalf(%, 6);                               # Resp. .746211
```

Simpson's rule. Type `?simpson` and then

```
> simpson(f(x), x = 0..1, 10);
```

$$\frac{1}{30} + \frac{1}{30}\,e^{(-1)} + \frac{2}{15}\left(\sum_{i=1}^{5} e^{\left(-((1/5)\,i-1/10)^2\right)}\right) + \frac{1}{15}\left(\sum_{i=1}^{4} e^{\left(-(1/25)\,i^2\right)}\right)$$

```
> evalf(%);                                  # Resp. .7468249480
```

This is exact to 5S. Indeed,

```
> evalf(int(exp(-x^2), x = 0..1));           # Resp. .7468241330
```

Gauss integration generally requires less work. For nodes and coefficients see [1], pp. 916-919, and AEM, p. 878. (The nodes are not equally spaced.) Setting $x = (t+1)/2$, you integrate over t from -1 to 1; since $dx = dt/2$, you obtain

$$(A) \qquad \int_0^1 \exp(-x^2)\,dx = \frac{1}{2}\int_{-1}^1 \exp\left(-\frac{1}{4}(t+1)^2\right)dt.$$

Choosing $n = 3$, the nodes (t-values) are $-\sqrt{\frac{3}{5}}$, 0, $\sqrt{\frac{3}{5}}$, and the coefficients are $5/9, 8/9, 5/9$. You thus obtain from (a)

```
> 1/2*(5/9*exp(-1/4*(-sqrt(3/5) + 1)^2) + 8/9*exp(-1/4) +
    5/9*exp(-1/4*(sqrt(3/5) + 1)^2));
```

$$\frac{5}{18}\,e^{\left(-1/4\,(-1/5\,\sqrt{15}+1)^2\right)} + \frac{4}{9}\,e^{(-1/4)} + \frac{5}{18}\,e^{\left(-1/4\,(1/5\,\sqrt{15}+1)^2\right)}$$

```
> evalf(%);                                    # Resp. .7468145842
```

The 4D accuracy obtained from this short calculation is astonishing.

Similar Material in AEM: pp. 869-878

Problem Set for Chapter 17

Pr.17.1 (Quadratic equation) Solve $x^2 - 30x + 1 = 0$ by the two methods in Example 17.1 in this Guide. (*AEM Ref.* p. 836 (#6))

Pr.17.2 (Fixed-point iteration) Solve $x \cosh x = 1$ (appearing in connection with vibrations of beams). (*AEM Ref.* p. 847 (#6))

Pr.17.3 (Fixed-point iteration) Find a solution of $x^4 - x - 0.12 = 0$, starting from $x_0 = 1$. (*AEM Ref.* p. 847 (#9))

Pr.17.4 (Newton's iteration method) Design a Newton iteration for cube roots and compute the cube root of 7, starting from $x_0 = 2$. (*AEM Ref.* p. 847 (#11))

Pr.17.5 (Secant method) Find a zero of $f(x) = 1 - x^2/4 + x^4/64 - x^6/2304$ near $x = 2$ by using $x_0 = 2.0$ and $x_1 = 2.5$. ($f(x)$ is a partial sum of the Maclaurin series of the Bessel function $J_0(x)$, whose zeros are important in connection with vibrations of membranes.) (*AEM Ref.* p. 847 (#23))

Pr.17.6 (Bisection method) Solve $x = \cos x$. (See Example 17.5 in this Guide. *AEM Ref.* p. 848 (#25))

Pr.17.7 (Polynomial interpolation) Find the values of the Gamma function at 1.01 and 1.03 by quadratic interpolation of the values 1.0000, 0.9888, 0.9784 at 1.00, 1.02, 1.04, respectively, using the command `interp`. (*AEM Ref.* pp. 860 (#5), A86)

Pr.17.8 (Newton's forward difference interpolation formula) Solve Pr.17.7 by the formula

$$f(x) = f_0 + r\,\Delta\,f_0 + \frac{1}{2!}r\,(r-1)\,\Delta^2\,f_0 + \frac{1}{3!}r\,(r-1)\,(r-2)\,\Delta^3\,f_0 + \dots$$

where $r = (x - x_0)/h$ and h is the distance between the equally spaced adjacent nodes. (*AEM Ref.* pp. 856, 860 (#11))

Pr.17.9 (High-order polynomial) Find the polynomial that interpolates (j, y_j), where $j = -5, -4, \dots 0, \dots 4, 5$ and $y_0 = 1$, $y_j = 0$ otherwise. (*AEM Ref.* p. 862)

Pr.17.10 (Experiment on oscillation of interpolation polynomials) Explore by experimentation and plotting how the oscillations of interpolation polynomials for equally spaced nodes and function values 0 except for a single 1 (as in Example 17.7 in this Guide and in the previous problem) are growing with increasing degree n.

Pr.17.11 (Spline interpolation) Find the natural cubic spline for the data in Pr.17.9. (*AEM Ref.* pp. 861-866)

Pr.17.12 (Rectangular rules) Integrate $\exp(-x^2)$ from 0 to 2 by the rectangular rules with $h = 0.2$ that use the function value (A) at the left endpoint, (B) at the midpoint, (C) at the right endpoint of each of the ten subintervals. Give reasons why (A) yields the largest result and (C) the smallest. Type `?leftsum`, `?middlesum`, `?rightsum` for information. (*AEM Ref.* p. 869)

Pr.17.13 **(Experiment on accuracy of integration rules)** Integrate $\cos(x^2)$ from 0 to $\sqrt{\frac{\pi}{2}}$, first by the command `evalf(int(...))` and then by the rectangular rule ((B) in the previous problem), the trapezoidal rule, and Simpson's rule. In each case find experimentally the number of subintervals that gives 6D accuracy, thereby noting the superiority of Simpson's rule. ($\cos(x^2)$ is the integrand of the **Fresnel integral** $C(x)$. Following [1], p. 300, Maple uses $\cos(\pi t^2/2)$ as the integrand of the Fresnel integral instead of the more common $\cos(x^2)$. Type `?FresnelC` .)
(*AEM Ref.* pp. 860-871, A56)

Pr.17.14 **(Simpson's rule)** Obtain a 10D-value of ln 2 by evaluating a suitable integral by Simpson's rule. (*AEM Ref.* pp. 872-875)

Pr.17.15 **(Gauss integration)** Calculate the sine integral Si(x) (defined as the integral of $(\sin u)/u$ from $u = 0$ to $u = x$) for $x = 1$ by Gauss integration with $n = 3$ nodes and find the accuracy by comparing with the value obtained by the command `int` .
(*AEM Ref.* pp. 877, 878, A57)

Chapter 18

Numerical Linear Algebra

Content. Gauss, Doolittle, Cholesky methods (Exs. 18.1-18.3, Prs. 18.1-18.5)
Gauss-Jordan, Gauss-Seidel methods (Exs. 18.4, 18.5, Prs. 18.6-18.8)
Norms, condition numbers (Ex. 18.6, Prs. 18.9-18.12)
Fit by least squares (Ex. 18.7, Prs. 18.13-18.16)
Approximation of eigenvalues (Exs. 18.8-18.10, Prs. 18.17-18.20)

linalg package. Load it by typing `with(linalg)`. Type `?linalg`. In addition to the commands used in Chaps. 6 and 7 you may use `LUdecomp`, `cholesky`, `gaussjord`, `norm`, `norm(...,1)`, `norm(...,frobenius)`, `leastsqrs`.

Examples for Chapter 18

EXAMPLE 18.1 GAUSS ELIMINATION. PIVOTING

To keep this chapter independent of Chaps. 6 and 7 on matrices and eigenvalue problems, we begin by describing the implementation of the Gauss elimination on the computer.

Solve the linear system

$$
\begin{aligned}
8\,x_2 + 2\,x_3 &= -7 \\
6\,x_1 + 2\,x_2 + 8\,x_3 &= 26\,, \\
3\,x_1 + 5\,x_2 + 2\,x_3 &= 8
\end{aligned}
$$

in matrix form $\mathbf{Ax} = \mathbf{b}$, where

$$
\mathbf{A} = [a_{jk}] = \begin{bmatrix} 0 & 8 & 2 \\ 6 & 2 & 8 \\ 3 & 5 & 2 \end{bmatrix} \quad \text{and} \quad \mathbf{b} = \begin{bmatrix} -7 \\ 26 \\ 8 \end{bmatrix}.
$$

Solution. Load the `linalg` package by typing

```
> with(linalg):                                    # Ignore the warning.
```

Type the matrix and the vector and apply the command `linsolve`. (Type `?linsolve` for information.)

```
> A := matrix([[0, 8, 2],
>              [6, 2, 8],
>              [3, 5, 2]]);
```

$$
A := \begin{bmatrix} 0 & 8 & 2 \\ 6 & 2 & 8 \\ 3 & 5 & 2 \end{bmatrix}
$$

```
> b := [-7, 26, 8];
```
                 # Resp. $b := [-7, 26, 8]$

```
> x := linsolve(A, b);
```
                 # Resp. $x := \left[4, -1, \dfrac{1}{2}\right]$

This gave the solution but no indication on the method by which it was obtained.

The **Gauss elimination** and **back substitution** can be done by the commands `gausselim` (applied to the augmented matrix) and `backsub`. (Type `?gausselim` and `?backsub` for information.) From `B` (below) you see that `gausselim` automatically did **partial pivoting**, which was necessary because $a_{11} = 0$.

```
> A1 := augment(A, b);
```

$$A1 := \begin{bmatrix} 0 & 8 & 2 & -7 \\ 6 & 2 & 8 & 26 \\ 3 & 5 & 2 & 8 \end{bmatrix}$$

```
> B := gausselim(A1);
```

$$B := \begin{bmatrix} 6 & 2 & 8 & 26 \\ 0 & 8 & 2 & -7 \\ 0 & 0 & -3 & \dfrac{-3}{2} \end{bmatrix}$$

```
> x := backsub(B);
```
            # Resp. $x := \left[4, -1, \dfrac{1}{2}\right]$

If you type `gausselim(B, r, d)`, where `r` and `d` are optional, you obtain the rank `r` and the value of the determinant `d` of **A**,

```
> r := 'r':    d := 'd':
> gausselim(B, r, d):
> r;
```
                 # Resp. 3
```
> d;
```
                 # Resp. -144

Details of the Gauss elimination. You must do **partial pivoting**. Use Row 2 as pivot row because it contains the largest entry in the first column. Thus interchange Rows 1 and 2 by typing

```
> A2 := swaprow(A1, 1, 2);
```

$$A2 := \begin{bmatrix} 6 & 2 & 8 & 26 \\ 0 & 8 & 2 & -7 \\ 3 & 5 & 2 & 8 \end{bmatrix}$$

Eliminate x_1 from Row 3 by the command `addrow(A2, 1, 3, -1/2)`, which adds $-3/6 = -1/2$ times Row 1 to Row 3,

```
> A3 := addrow(A2, 1, 3, -3/6);
```

$$A3 := \begin{bmatrix} 6 & 2 & 8 & 26 \\ 0 & 8 & 2 & -7 \\ 0 & 4 & -2 & -5 \end{bmatrix}$$

The pivot row in the next step is Row 2 of `A3`. Add $-4/8 = -1/2$ times Row 2 to Row 3.

> A4 := addrow(A3, 2, 3, -4/8);

$$A4 := \begin{bmatrix} 6 & 2 & 8 & 26 \\ 0 & 8 & 2 & -7 \\ 0 & 0 & -3 & \dfrac{-3}{2} \end{bmatrix}$$

Now **back substitution** begins with x_3 and ends with x_1. Denote the components of the solution vector **x** by x[1] , x[2] , x[3] . Remember that A4[j,k] picks the entry of A4 in Row j and Column k .

> x[3] := 1/A4[3,3]*A4[3,4];   # Resp. $x_3 := \dfrac{1}{2}$

> x[2] := 1/A4[2,2]*(A4[2,4] - A4[2,3]*x[3]);   # Resp. $x_2 := -1$

> x[1] := 1/A4[1,1]*(A4[1,4] - A4[1,3]*x[3] - A4[1,2]*x[2]);

$$x_1 := 4$$

This is the end. The results agree.

Pivot command. The command pivot(A1, j, k) picks equation j as the pivot equation (the pivot row) and the entry in column k as the pivot and eliminates x_k from all the other equations. Then your analysis takes the form (look back at A1 and also note that in using the pivot command the rows are not interchanged)

> B1 := pivot(A1, 2, 1);   # Compare the result with A3

$$B1 := \begin{bmatrix} 0 & 8 & 2 & -7 \\ 6 & 2 & 8 & 26 \\ 0 & 4 & -2 & -5 \end{bmatrix}$$

> B2 := delrows(B1, 2..2);   # Delete Row 2, which was the pivot row

$$B2 := \begin{bmatrix} 0 & 8 & 2 & -7 \\ 0 & 4 & -2 & -5 \end{bmatrix}$$

> B3 := pivot(B2, 1, 2);   # Compare the result with A4

$$B3 := \begin{bmatrix} 0 & 8 & 2 & -7 \\ 0 & 0 & -3 & \dfrac{-3}{2} \end{bmatrix}$$

Do back substitution as before, with the necessary changes of notation.

> x[3] := 1/B3[2, 3]*B3[2, 4];   # Resp. $x_3 := \dfrac{1}{2}$

> x[2] := 1/B2[2, 2]*(B2[2, 4] - B2[2,3]*x[3]);   # Resp. $x_2 := -1$

> x[1] := 1/B1[2,1]*(B1[2,4] - B1[2,3]*x[3] - B1[2, 2]*x[2]);

$$x_1 := 4$$

swaprow moved the first pivot equation up, whereas pivot did not. This causes the difference in notations shown in the present formulas, compared to the previous ones.

Similar Material in AEM: pp. 886-890

EXAMPLE 18.2   DOOLITTLE LU-FACTORIZATION

A linear system $\mathbf{A}\mathbf{x} = \mathbf{b}$ of n equations in n unknowns can be solved by first factorizing $\mathbf{A} = \mathbf{L}\mathbf{U}$, where $\mathbf{L}$ is lower triangular and $\mathbf{U}$ is upper triangular. Then $\mathbf{A}\mathbf{x} = \mathbf{L}(\mathbf{U}\mathbf{x}) = \mathbf{b}$. Set $\mathbf{U}\mathbf{x} = \mathbf{y}$ and solve $\mathbf{L}\mathbf{y} = \mathbf{b}$ for $\mathbf{y}$. Then solve $\mathbf{U}\mathbf{x} = \mathbf{y}$ for $\mathbf{x}$. It can be proved that if $\mathbf{A}$ is nonsingular, its rows can be reordered such that the resulting matrix has an **LU**-factorization. This factorization is obtained by the command LUdecomp , as explained below. Type ?LUdecomp for information.

Solve the linear system in Example 18.1 in this Guide by **LU**-factorization.

Solution. Type the matrix A and the vector b . Load the linalg package. Type the command LUdecomp , in which L = 'l' is optional. The response is **U**. Typing l then gives you **L**.

```
> A := matrix([[0, 8, 2],   [6, 2, 8],   [3, 5, 2]]);
```

$$A := \begin{bmatrix} 0 & 8 & 2 \\ 6 & 2 & 8 \\ 3 & 5 & 2 \end{bmatrix}$$

```
> b := [-7, 26, 8];
```
                                          # Resp. $b := [-7, 26, 8]$

```
> with(linalg):
```
                                          # Ignore the warning.

```
> U := LUdecomp(A, L = 'l');
```

$$U := \begin{bmatrix} 6 & 2 & 8 \\ 0 & 8 & 2 \\ 0 & 0 & -3 \end{bmatrix}$$

```
> L := evalm(l);
```

$$L := \begin{bmatrix} 1 & 0 & 0 \\ 0 & 1 & 0 \\ \frac{1}{2} & \frac{1}{2} & 1 \end{bmatrix}$$

Since $\mathbf{L}$ has the main diagonal 1, 1, 1, this is **Doolittle's factorization**. Note that $\mathbf{U}$ is the triangular matrix as obtained by the Gauss elimination, and $\mathbf{L}$ is the matrix of the multipliers in that method.

Now in the Gauss elimination you had to do partial pivoting. The command LUdecomp did this automatically. To see what was done, type

```
> evalm(L&*U);
```

$$\begin{bmatrix} 6 & 2 & 8 \\ 0 & 8 & 2 \\ 3 & 5 & 2 \end{bmatrix}$$

You now see that Rows 1 and 2 were interchanged. This changes b to

```
> c := [26, -7, 8];
```
                                          # Resp. $c := [26, -7, 8]$

The two triangular systems $\mathbf{L}\mathbf{y} = \mathbf{c}$ and $\mathbf{U}\mathbf{x} = \mathbf{y}$ can now readily be solved, for instance, by

```
> y := linsolve(L, c);
```

$$y := \left[26, -7, \frac{-3}{2}\right]$$

```
> x := linsolve(U, y);
```

$$x := \left[4, -1, \frac{1}{2}\right]$$

Similar Material in AEM: pp. 894-896

EXAMPLE 18.3   **CHOLESKY FACTORIZATION**

Solve the following system by Cholesky's method.

$$\begin{aligned}
4\,x_1 + \ 2\,x_2 + 14\,x_3 &= \ \ \ \ 14 \\
2\,x_1 + 17\,x_2 - \ 5\,x_3 &= -101 \\
14\,x_1 - \ 5\,x_2 + 83\,x_3 &= \ \ \ 155
\end{aligned}$$

Solution. This method solves linear systems $\mathbf{Ax} = \mathbf{b}$ with symmetric and positive definite $\mathbf{A}$. The present matrix is symmetric. You need not check for positive definiteness; if $\mathbf{A}$ were not positive definite, the calculations would lead into complex. The method uses the factorization $\mathbf{A} = \mathbf{LL}^{\mathrm{T}}$ and then solves $\mathbf{Ax} = \mathbf{L}(\mathbf{L}^{\mathrm{T}}\mathbf{x}) = \mathbf{b}$ in two steps as in the previous example, namely, $\mathbf{Ly} = \mathbf{b}$ and $\mathbf{L}^{\mathrm{T}}\mathbf{x} = \mathbf{y}$. Type the coefficient matrix $\mathbf{A}$, then $\mathbf{b}$, and then the factorization command cholesky .

```
> A := matrix([[4, 2, 14],   [2, 17, -5],   [14, -5, 83]]);
```

$$A := \begin{bmatrix} 4 & 2 & 14 \\ 2 & 17 & -5 \\ 14 & -5 & 83 \end{bmatrix}$$

```
> b := [14, -101, 155];                # Resp. b := [14, −101, 155]
> with(linalg):                        # Ignore the warning.
> L := cholesky(A);
```

$$L := \begin{bmatrix} 2 & 0 & 0 \\ 1 & 4 & 0 \\ 7 & -3 & 5 \end{bmatrix}$$

```
> y := linsolve(L, b);                 # Resp. y := [7, −27, 5]
> x : linsolve(transpose(L), y);       # Resp. [3, −6, 1]
```

Similar Material in AEM: pp. 896-898

EXAMPLE 18.4   **GAUSS-JORDAN ELIMINATION. MATRIX INVERSION**

The **Gauss-Jordan method** is not practical for solving linear systems $\mathbf{Ax} = \mathbf{b}$ (because the Jordan reduction needs more calculations than the back substitution does), but the method is useful for obtaining the inverse of a matrix. Gauss's method gives a triangular matrix, and Jordan reduces it further to a diagonal matrix (actually, to the unit matrix). Using the Gauss-Jordan method, solve the following linear system.

As a second, separate task, calculate the inverse of the coefficient matrix $\mathbf{A}$ of the system.

$$
\begin{array}{rcrcrcr}
3\,x_1 &+& 2\,x_2 && &=& 14 \\
12\,x_1 &+& 13\,x_2 &+& 6\,x_3 &=& 40 \\
-3\,x_1 &+& 8\,x_2 &+& 9\,x_3 &=& -28
\end{array}
$$

Solution. The system is of the form $\mathbf{Ax} - \mathbf{b}$. Type $\mathtt{A}$, $\mathtt{b}$ and the augmented matrix $\mathtt{A1}$. Then type $\mathtt{gaussjord\ (A1)}$. Gauss-Jordan reduces $\mathbf{Ax} = \mathbf{b}$ to $\mathbf{A^{-1}Ax} = \mathbf{A^{-1}b}$, that is, to $\mathbf{x} = \mathbf{A^{-1}b}$.

```
> A := matrix([[3, 2, 0], [12, 13, 6], [-3, 8, 9]]);
```

$$
A := \begin{bmatrix} 3 & 2 & 0 \\ 12 & 13 & 6 \\ -3 & 8 & 9 \end{bmatrix}
$$

```
> b := [14, 40, -28];                           # Resp. b := [14, 40, -28]
> with(linalg):                                 # Ignore the warning.
> A1 := augment(A, b);
```

$$
A1 := \begin{bmatrix} 3 & 2 & 0 & 14 \\ 12 & 13 & 6 & 40 \\ -3 & 8 & 9 & -28 \end{bmatrix}
$$

```
> gaussjord(A1);
```

$$
\begin{bmatrix} 1 & 0 & 0 & 2 \\ 0 & 1 & 0 & 4 \\ 0 & 0 & 1 & -6 \end{bmatrix}
$$

From this you can read the answer $x_1 = 2$, $x_2 = 4$, $x_3 = -6$.

Inverse. Gauss-Jordan reduces the 3×6 matrix $\mathbf{B} = [\mathbf{A}, \mathbf{I}]$ ($\mathbf{I}$ the unit matrix) to $\mathbf{A^{-1}B} = [\mathbf{A^{-1}A}, \mathbf{A^{-1}I}] = [\mathbf{I}, \mathbf{A^{-1}}]$, so that you can read the inverse of $\mathbf{A}$ directly from the transformed matrix. Type $\mathtt{?diag}$ for information and then

```
> Unit := diag(1, 1, 1);
```

$$
Unit := \begin{bmatrix} 1 & 0 & 0 \\ 0 & 1 & 0 \\ 0 & 0 & 1 \end{bmatrix}
$$

```
> B1 := augment(A, Unit);
```

$$
B1 := \begin{bmatrix} 3 & 2 & 0 & 1 & 0 & 0 \\ 12 & 13 & 6 & 0 & 1 & 0 \\ -3 & 8 & 9 & 0 & 0 & 1 \end{bmatrix}
$$

```
> B2 := gaussjord(B1);
```

$$
B2 := \begin{bmatrix} 1 & 0 & 0 & \dfrac{-23}{15} & \dfrac{2}{5} & \dfrac{-4}{15} \\[2ex] 0 & 1 & 0 & \dfrac{14}{5} & \dfrac{-3}{5} & \dfrac{2}{5} \\[2ex] 0 & 0 & 1 & -3 & \dfrac{2}{3} & \dfrac{-1}{3} \end{bmatrix}
$$

```
> InvA := submatrix(B2, 1..3, 4..6);
```

$$InvA := \begin{bmatrix} \dfrac{-23}{15} & \dfrac{2}{5} & \dfrac{-4}{15} \\[2ex] \dfrac{14}{5} & \dfrac{-3}{5} & \dfrac{2}{5} \\[2ex] -3 & \dfrac{2}{3} & \dfrac{-1}{3} \end{bmatrix}$$

In the command `submatrix`, `1..3` refers to the rows and `4..6` to the columns of `B2` of which the submatrix of `B2` (the inverse of **A**) consists.

   Similar Material in AEM: p. 898

| EXAMPLE 18.5 | **GAUSS-SEIDEL ITERATION FOR LINEAR SYSTEMS**

Solve the following linear system by **Gauss-Seidel iteration**, starting from the vector $[100 \quad 100 \quad 100 \quad 100]^T$. (Systems of this kind appear in connection with numerical methods for partial differential equations.)

$$\begin{array}{rcrcrcrcl} x_1 & - & 0.25\,x_2 & - & 0.25\,x_3 & & & = & 50 \\ -0.25\,x_1 & + & x_2 & & & - & 0.25\,x_4 & = & 50 \\ -0.25\,x_1 & & & + & x_3 & - & 0.25\,x_4 & = & 25 \\ & - & 0.25\,x_2 & - & 0.25\,x_3 & + & x_4 & = & 25 \end{array}$$

Solution. In $\mathbf{Ax} = \mathbf{b}$ assume that **A** has the main diagonal entries 1, ..., 1 (which can be accomplished by division, provided all the main diagonal entries are originally different from 0). Write $\mathbf{A} = \mathbf{I} + \mathbf{L} + \mathbf{U}$, where **I** is the unit matrix, **L** is lower triangular, and **U** is upper triangular. Then $(\mathbf{I} + \mathbf{L} + \mathbf{U})\mathbf{x} = \mathbf{b}$, hence $\mathbf{x} = \mathbf{b} - \mathbf{Lx} - \mathbf{Ux}$. This gives the iteration $\mathbf{x}^{(m+1)} = \mathbf{b} - \mathbf{L}\mathbf{x}^{(m+1)} - \mathbf{U}\mathbf{x}^{(m)}$ because you always use the latest available approximation, which is the $(m+1)$st below the main diagonal and the mth above it. In terms of components,

$$x_j^{(m+1)} = b_j - \sum_{k=1}^{j-1} a_{jk}\,x_k^{(m+1)} - \sum_{K=j+1}^{n} a_{jK}\,x_K^{(m)}.$$

Hence type $\mathbf{A}, \mathbf{b}, \mathbf{x_0}$ and then the program (which includes divisions to make the main diagonal entries 1). n is the number of unknowns and N is the number of steps. Note that you need no superscripts $(m+1)$ and (m) because if a component has been computed, it is used at once, so that you can overwrite its previous value, which is no longer needed.

```
> with(linalg):                                    # Ignore the warning.
> A := matrix([[1, -0.25, -0.25, 0],   [-0.25, 1, 0, -0.25],
    [-0.25, 0, 1, -0.25],   [0, -0.25, -0.25, 1]]);
```

$$A := \begin{bmatrix} 1 & -.25 & -.25 & 0 \\ -.25 & 1 & 0 & -.25 \\ -.25 & 0 & 1 & -.25 \\ 0 & -.25 & -.25 & 1 \end{bmatrix}$$

```
> b := [50, 50, 25, 25];                    # Resp. b := [50, 50, 25, 25]
> x := [100, 100, 100, 100];              # Resp. x := [100, 100, 100, 100]
> n := 4: N := 7:
> for m from 0 to N-1 do                         # m counts the steps.
>   for j from 1 to n do         # j numbers the components of x.
>     S := sum(A[j,k]*x[k], k=1..j-1)
>        + sum(A[j,K]*x[K], K = j+1..n);
>     x[j] := evalf((b[j]-S)/A[j,j]);
>   od;
> print(x);
> od;
```

$$[100.00, \ 100.0000, \ 75.0000, \ 68.750000]$$
$$[93.750000, \ 90.62500000, \ 65.62500000, \ 64.06250000]$$
$$[89.06250000, \ 88.28125000, \ 63.28125000, \ 62.89062500]$$
$$[87.89062500, \ 87.69531250, \ 62.69531250, \ 62.59765626]$$
$$[87.59765624, \ 87.54882812, \ 62.54882812, \ 62.52441406]$$
$$[87.52441406, \ 87.51220704, \ 62.51220704, \ 62.50610352]$$
$$[87.50610352, \ 87.50305176, \ 62.50305176, \ 62.50152588]$$

The exact solution is $x_1 = 87.5$, $x_2 = 87.5$, $x_3 = 62.5$, $x_4 = 62.5$.

Similar Material in AEM: pp. 900-903

EXAMPLE 18.6   **VECTOR AND MATRIX NORMS.**
                    CONDITION NUMBERS

For information type `?linalg[norm]`, `?linalg[cond]`.

```
> with(linalg):                              # Ignore the warning.
```

Vector norms. There are essentially three practically important vector norms, defined as follows.

```
> x := [x1, x2, x3];                         # Resp. x := [x1, x2, x3]
> norm(x, 1);           # Resp. |x1| + |x2| + |x3|   # l1- norm, notation ||x||1
> norm(x, 2);           # Resp. √(|x1|² + |x2|² + |x3|²)  # Euclidean norm ||x||2
> norm(x, infinity);         # Resp. max(|x1|, |x2|, |x3|)  # l∞- norm ||x||∞
```

Matrix norms. Some matrix norms result from vector norms. The l_1-vector norm leads to the column 'sum' matrix norm. The l_∞-vector norm leads to the row 'sum' matrix norm. The definitions are as follows. The column 'sum' (row 'sum') matrix norm is the largest sum of the n sums of the absolute values of the entries in a column (in a row), respectively.

```
> A := matrix([[a11, a12], [a21, a22]]);         # Resp. A := [ a11  a12 ]
                                                              [ a21  a22 ]
```

```
> norm(A,1);
```
        # Column 'sum' norm $\|\mathbf{A}\|_1$

$$\max\left(|a11|+|a21|,\,|a12|+|a22|\right)$$

```
> norm(A, infinity);
```
        # Row 'sum' norm $\|\mathbf{A}\|_\infty$

$$\max(|a11|+|a12|,\,|a21|+|a22|)$$

A third important matrix norm is the Frobenius norm

```
> norm(A, frobenius);
```
        # Frobenius norm

$$\sqrt{|a11|^2+|a12|^2+|a21|^2+|a22|^2}$$

Similarly for matrices of greater size. For instance, if

```
> A := matrix([[2, 0, -1], [4, 3, -2], [0, 5, 1]]);
```

$$A := \begin{bmatrix} 2 & 0 & -1 \\ 4 & 3 & -2 \\ 0 & 5 & 1 \end{bmatrix}$$

then

```
> norm(A, 1);
```
        # Resp. 8   # $0+3+5=8$
```
> norm(A, infinity);
```
        # Resp. 9   # $4+3+|-2|=9$
```
> norm(A, frobenius);
```
        # Resp. $2\sqrt{15}$

Condition numbers. The definition of the condition number of a matrix $\mathbf{A}$ is $\kappa(A)=\|\mathbf{A}\|\|\mathbf{A}^{-1}\|$ and depends on the choice of a norm, but if $\kappa(A)$ is large in one norm, it tends to be large in others, indicating that $\mathbf{A}$ is **ill-conditioned**. This is illustrated by the following values.

```
> evalf(cond(A, 1),5);
```
        # Resp. 49.333
```
> cond(A, infinity);
```
        # Resp. 54
```
> evalf(cond(A, frobenius), 5);
```
        # Resp. 35.566

        Similar Material in AEM: pp. 909-912

EXAMPLE 18.7	**FITTING DATA BY LEAST SQUARES**

Fit a straight line and a quadratic polynomial through the four points $(-1.3,\ 0.103)$, $(-0.1,\ 1.099)$, $(0.2,\ 0.808)$, $(1.3,\ 1.897)$ by the **principle of least squares**. For information type `?leastsquare`. Also find the interpolation polynomial for these data. Plot the results on common axes.

Solution. Load the `stats` package. Then give the command for obtaining the straight line. Note that in this command you first type all four x-values in brackets [...] and then all four y-values. The equation `y = a*x + b` is optional. If you omit it, you still get the same result. Try it.

```
> a := 'a':  b := 'b':  x := 'x':  y := 'y':
> with(stats):
> data := [[-1.3, -0.1, 0.2, 1.3], [0.103, 1.099, 0.808, 1.897]]:
```

```
> p1 := fit[leastsquare[[x, y], y = a*x + b]](data);
```

$$p1 := y = .6670240700\,x + .9600743982$$

If you want a second or higher order polynomial, you must include the corresponding equation (with arbitrary coefficients) in the command.

```
> c := 'c':
> p2 := fit[leastsquare[[x, y], y = a*x^2 + b*x + c]](data);
```

$$p2 := y = .04532361420\,x^2 + .6680654222\,x + .9211833653$$

If you request a cubic polynomial, since you have four points, you get the unique interpolation polynomial passing precisely through the points.

```
> d:='d':
> p3 := fit[leastsquare[[x, y], y = a*x^3 + b*x^2 + c*x + d]](data);
```

$$p3 := y = 1.\,x^3 + 1. - 1.\,x$$

To compute the data points, type

```
> points := [seq([data[1, j], data[2, j]], j = 1..4)];
```

$$points := \ [[-1.3, .103], [-.1, 1.099], [.2, .808], [1.3, 1.897]]$$

Here, `data[1, 1]` gives the first x-value and `data[2, 1]` gives the first y-value. Since $j = 1, \ldots, 4$, you get 4 pairs of coordinates of the 4 data points. Then plot the three polynomials and the points by the commands following the figure.

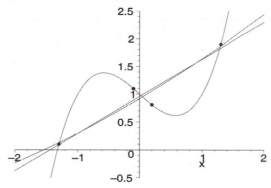

Example 18.7. Least squares straight line and quadratic polynomial.
Cubic interpolation polynomial

```
> P1 := plot(rhs(p1), x = -2..2):
> P2 := plot(rhs(p2), x = -2..2):
> P3 := plot(rhs(p3), x = -2..2, y = -0.5..2.5):
> P4 := plot(points, x = -2..2, style = point):
> with(plots):                                    # Ignore the warning.
> display(P1, P2, P3, P4);
```

Similar Material in AEM: pp. 914-916

| EXAMPLE 18.8 | APPROXIMATION OF EIGENVALUES: COLLATZ METHOD |

Collatz proved the following. Let $\mathbf{A} = [a_{jk}]$ be a real $n \times n$ matrix with all entries a_{jk} positive. Choose any $\mathbf{x} = [x_j]$ with all n components positive. Compute $\mathbf{y} = [y_j] = \mathbf{Ax}$ and the n quotients $p_j = y_j/x_j$. Then the smallest and largest of these quotients are the endpoints of a closed interval J which contains an eigenvalue λ of $\mathbf{A}$. Hence the midpoint of J is an approximation of λ, and half the length of J is a corresponding error bound.

Write a program for this method and, starting from $[1\ 1\ 1]^T$, apply it to the matrix (20 steps)

$$\mathbf{A} = \begin{bmatrix} 0.49 & 0.02 & 0.22 \\ 0.02 & 0.28 & 0.20 \\ 0.22 & 0.20 & 0.40 \end{bmatrix}.$$

Solution. Calculate $\mathbf{y} = \mathbf{Ax}$ and those quotients. Then sort the latter in ascending order, so that p_1 and p_n are the endpoints of that interval. Note that both quotients are positive because all entries of $\mathbf{A}$ and all components of $\mathbf{x}$ are positive. In the last line of the program copy $\mathbf{y}$ to make it the vector $\mathbf{x}$ in the next step. (The response shows steps 1, 2, 10, and 20.)

```
> A := matrix([[0.49, 0.02, 0.22],  [0.02, 0.28, 0.20],
   [0.22, 0.20, 0.40]]);
> x := [1, 1, 1];
> n := 3:                                        # Size of matrix
> N := 20:                                       # Number of steps
> for m from 1 to N do
>     y := evalm(A&*x);
>     for j from 1 to n do
>         p[j] := evalf(y[j]/x[j]);
>     od;
>     p := sort(convert(p, list)):
>     print('Interval' = [p[1], p[n]]);
>     x := copy(y):
> od;
```

$$y := [.73, .50, .82]$$
$$p := [.50, .73, .82]$$
$$Interval = [.50, .82]$$
$$x := [.73, .50, .82]$$
$$y := [.5481, .3186, .5886]$$
$$p := [.6372000000, .7178048780, .7508219178]$$
$$Interval = [.6372000000, .7508219178]$$

$$\vdots$$

$$x := [.05775376208, .02891073442, .05778761526]$$
$$y := [.04159083347, .02080760393, .04160302064]$$
$$p := [.7197189676, .7199297021, .7201406795]$$
$$Interval = [.7197189676, .7201406795]$$
$$\vdots$$
$$x := [.002163090836, .001081546655, .002163092072]$$
$$y := [.001557425698, .0007787132945, .001557426144]$$
$$p := [.7199997253, .7199999317, .7200001372]$$
$$Interval = [.7199997253, .7200001372]$$
$$x := [.001557425698, .0007787132945, .001557426144]$$

```
> with(linalg):                    # Ignore the warning.
> evalf(eigenvalues(A));      # Resp. .09000000000, .3600000000, .7200000000
```

Similar Material in AEM: pp. 923, 924

| EXAMPLE 18.9 | **APPROXIMATION OF EIGENVALUES: POWER METHOD** |

Given a real **symmetric** square matrix $\mathbf{A}$, compute $\mathbf{y} = \mathbf{Ax}$, where $\mathbf{x}$ is any nonzero vector. Compute the dot products $m_0 = \mathbf{x} \bullet \mathbf{x}$, $m_1 = \mathbf{x} \bullet \mathbf{y}$, $m_2 = \mathbf{y} \bullet \mathbf{y}$. Then the **Rayleigh quotient** $q = m_1/m_0$ is an approximation of an eigenvalue of $\mathbf{A}$. A corresponding error bound is $\delta = \sqrt{\frac{m_2}{m_0} - q^2}$.

This was the **first step** of the **power method**. In the **second step**, use $\mathbf{y}$ divided by a component that is largest in absolute value as your new $\mathbf{x}$ and proceed as before. (This **"scaling"** will make one component of the new $\mathbf{x}$ equal to 1 and the others smaller than or equal to 1 in absolute value. It has the effect that the sequence of vectors obtained will converge to an eigenvector of that λ approximated by the Rayleigh quotients.) In the further steps you proceed in the same fashion.

Apply this power method to the matrix in the previous example, starting from $\mathbf{x} = [1 \ 1 \ 1]^T$.

Solution. The matrix $\mathbf{A}$ is symmetric. Hence the error bounds apply. The command before od makes the scaled y the x for the next step. Here sort(y) arranges the components of y in ascending order. sign(sort(y)[1] + sort(y)[3]) equals −1 if the absolutely largest component of y is negative, and equals +1 if that component is positive. Hence in both cases the next x will have a component +1. For reasons of space we show only the responses to steps 1, 2, and 10.

```
> with(linalg):                    # Ignore the warning.
> A := matrix([[0.49, 0.02, 0.22],    [0.02, 0.28, 0.20],
  [0.22, 0.20, 0.40]]);
```

$$A := \begin{bmatrix} .49 & .02 & .22 \\ .02 & .28 & .20 \\ .22 & .20 & .40 \end{bmatrix}$$

```
> x := [1, 1, 1];               # Resp. x := [1, 1, 1]   # Starting vector
```

```
> N := 10:                                              # Number of steps
> for j from 1 to N do
>     y := evalm(A&*x);
>     q := evalf(innerprod(x,y)/innerprod(x,x));        # Rayleigh quotient
>     delta := evalf(sqrt(innerprod(y,y)/innerprod(x,x) - q^2));
>     x := evalm(copy(y)*sign(sort(y)[1] + sort(y)[3])/max(abs(y[1]),
       abs(y[2]), abs(y[3])));
> od;
```

$$y := [.73, .50, .82]$$
$$q := .6833333333$$
$$\delta := .1347425530$$
$$x := [.8902439024, .6097560975, .9999999999]$$
$$y := [.6684146342, .3885365854, .7178048780]$$
$$q := .7160482374$$
$$\delta := .03888718169$$
$$\vdots$$
$$x := [.9994141795, .5002929139, 1.000000000]$$
$$y := [.7197188063, .3600702995, .7199297023]$$
$$q := .7199999451$$
$$\delta := .0001407124728$$
$$x := [.9997070601, .5001464702, .9999999998]$$

Similar Material in AEM: pp. 925-928

| EXAMPLE 18.10 | **APPROXIMATION OF EIGENVALUES: QR-FACTORIZATION** |

This is another iteration method. We explain it for a real symmetric tridiagonal matrix $\mathbf{A} = \mathbf{A}_0$. In the **first step,** factor $\mathbf{A}_0 = \mathbf{Q}_0 \mathbf{R}_0$, where $\mathbf{Q}_0$ is orthogonal and $\mathbf{R}_0$ is upper triangular. Then compute $\mathbf{A}_1 = \mathbf{R}_0 \mathbf{Q}_0$. In the **second step,** factor $\mathbf{A}_1 = \mathbf{Q}_1 \mathbf{R}_1$. Then compute $\mathbf{A}_2 = \mathbf{R}_1 \mathbf{Q}_1$. And so on. These are similarity transformations, so that the spectrum is preserved. If the eigenvalues of $\mathbf{A}$ are different in absolute value, the process converges to a diagonal matrix whose diagonal entries are the eigenvalues of $\mathbf{A}$. (For a proof, see Wilkinson [6] in Appendix 1.)

Obtain approximations of the eigenvalues of the following matrix by the method of QR-factorization.

$$\mathbf{A} = \begin{bmatrix} 6 & -\sqrt{18} & 0 \\ -\sqrt{18} & 7 & \sqrt{2} \\ 0 & \sqrt{2} & 6 \end{bmatrix}$$

Solution. The response shown corresponds to steps 1, 2, and 5 of the iteration.

```
> with(linalg):                                         # Ignore the warning.
```

```
> A := evalf(matrix([[6, -sqrt(18), 0], [-sqrt(18), 7, sqrt(2)],
   [0, sqrt(2), 6]]));
```

$$A := \begin{bmatrix} 6. & -4.242640686 & 0 \\ -4.242640686 & 7. & 1.414213562 \\ 0 & 1.414213562 & 6. \end{bmatrix}$$

```
> for s from 1 to 5 do
>     R := QRdecomp(A, Q = 'q');          # Type ?QRdecomp for information.
>     Q := evalm(q);
>     A:= evalm(R&*Q);
> od;
```

$$R := \begin{bmatrix} 7.34846922770331456 & -7.50555349814506734 & -.816496580568715102 \\ 0 & 3.55902608531176946 & 3.44378412680707502 \\ 0 & 0 & 5.04714614617171086 \end{bmatrix}$$

$$Q := \begin{bmatrix} .8164965810 & .5298129428 & -.2294157337 \\ -.5773502691 & .7492686494 & -.3244428421 \\ 0 & .3973597069 & .9176629356 \end{bmatrix}$$

$$A := \begin{bmatrix} 10.33333333 & -2.054804667 & -.11\,10^{-8} \\ -2.054804668 & 4.035087719 & 2.005532514 \\ 0 & 2.005532513 & 4.631578949 \end{bmatrix}$$

$$R := \begin{bmatrix} 10.53565374965044 & -2.802322412495845 & -.3911458824371833 \\ 0 & 4.083295835066713 & 3.988240277752860 \\ 0 & 0 & 3.068326679330825 \end{bmatrix}$$

$$Q := \begin{bmatrix} .9807965956 & .1698881597 & -.09579170772 \\ -.1950334281 & .8543444593 & -.4817234755 \\ 0 & .4911553299 & .8710720073 \end{bmatrix}$$

$$A := \begin{bmatrix} 10.87987988 & -.7963791839 & -.19\,10^{-8} \\ -.7963791846 & 5.447386641 & 1.507025004 \\ 0 & 1.507025002 & 2.672733479 \end{bmatrix}$$

$$\vdots$$

$$R := \begin{bmatrix} 10.9924358731356406 & -.341792423556139091 & -.00399256570755191724 \\ 0 & 5.99754969861507714 & .264825963159917576 \\ 0 & 0 & 2.00219390570189626 \end{bmatrix}$$

$$Q := \begin{bmatrix} .9997976306 & .02010608811 & -.0006656990975 \\ -.02011710552 & .9992500778 & -.03308450014 \\ 0 & .03309119678 & .9994523364 \end{bmatrix}$$

$$A := \begin{bmatrix} 10.99708721 & -.1206533406 & -.2273\,10^{-8} \\ -.1206533402 & 6.001815411 & .0662549938 \\ 0 & .06625499254 & 2.001097377 \end{bmatrix}$$

```
> eigenvalues(A);              # Resp. 11.00000000, 6.000000000, 2.000000000
```

You see that the accuracy of the diagonal entries of the last matrix $\mathbf{A}$ as approximations of the eigenvalues is greater than you would expect from the size of the off-diagonal entries of $\mathbf{A}$.

    Similar Material in AEM: pp. 933-937

Problem Set for Chapter 18

Pr.18.1 **(Gauss elimination)** Solve the following linear system by each of the three methods in Example 18.1 in this Guide. (*AEM Ref.* p. 893 (#7))

$$\begin{aligned} 6\,x_2 + 13\,x_3 &= 61 \\ 6\,x_1 \qquad\quad - 8\,x_3 &= -38 \\ 13\,x_1 - 8\,x_2 \qquad\quad &= 79 \end{aligned}$$

Pr.18.2 **(Gauss elimination)** Solve the following linear system. (*AEM Ref.* p. 893 (#10))

$$\begin{aligned} 4\,x_1 + 4\,x_2 + 2\,x_3 &= 0 \\ 3\,x_1 - x_2 + 2\,x_3 &= 0 \\ 3\,x_1 + 7\,x_2 + x_3 &= 0 \end{aligned}$$

Pr.18.3 **(Doolittle factorization)** Solve the following linear system by Doolittle's method. (*AEM Ref.* p. 899 (#5))

$$\begin{aligned} 5\,x_1 + 4\,x_2 + x_3 &= 3.4 \\ 10\,x_1 + 9\,x_2 + 4\,x_3 &= 8.8 \\ 10\,x_1 + 13\,x_2 + 15\,x_3 &= 19.2 \end{aligned}$$

Pr.18.4 **(Doolittle factorization)** Solve Pr.18.1 by Doolittle's method. (*AEM Ref.* pp. 894-896)

Pr.18.5 **(Cholesky factorization)** Solve the following linear system by Cholesky's method. (*AEM Ref.* p. 899 (#7))

$$\begin{aligned} 9\,x_1 + 6\,x_2 + 12\,x_3 &= 17.4 \\ 6\,x_1 + 13\,x_2 + 11\,x_3 &= 23.6 \\ 12\,x_1 + 11\,x_2 + 26\,x_3 &= 30.8 \end{aligned}$$

Pr.18.6 **(Gauss-Jordan elimination)** Solve Pr.18.5 by Gauss-Jordan elimination. (*AEM Ref.* p. 898)

Pr.18.7 **(Inverse)** Find the inverse of the coefficient matrix $\mathbf{A}$ in Pr.18.5 by the Gauss-Jordan method. (*AEM Ref.* p. 898)

Pr.18.8 **(Gauss-Seidel iteration)** Solve the following linear system by Gauss-Seidel iteration, starting from $[1\ 1\ 1]^T$. Do 5 steps. (*AEM Ref.* p. 905 (#1))

$$\begin{aligned} 5\,x_1 + x_2 + 2\,x_3 &= 19 \\ x_1 + 4\,x_2 - 2\,x_3 &= -2 \\ 2\,x_1 + 3\,x_2 + 8\,x_3 &= 39 \end{aligned}$$

Pr.18.9 **(Vector norms)** Find the l_1, l_2, l_∞ norms of the vectors $\mathbf{x} = [7\ -12\ 5\ 0]^T$ and $\mathbf{y} = k\mathbf{x}$ on the computer. Here, k is any real number. (*AEM Ref.* p. 912 (#1))

Pr.18.10 (Matrix norms) Find the row "sum", column "sum", and Frobenius norms of the coefficient matrix in Pr.18.8 and of its square. (*AEM Ref.* pp. 903, 909).

Pr.18.11 (Condition number) Find the condition numbers κ of the coefficient matrix in Pr.18.5 for the three matrix norms defined in Example 18.6 in this Guide, by the command cond and check the results by using the definition of κ.
(*AEM Ref.* pp. 910-912)

Pr.18.12 (Experiment on Hilbert matrix) The $n \times n$ Hilbert matrix is $\mathbf{H} = [h_{jk}]$, where $h_{jk} = 1/(j + k - 1)$. Find the condition numbers for the three matrix norms defined in Example 18.6 in this Guide when $n = 3$. Conclude that $\mathbf{H}$ is ill-conditioned. Go on to $n = 4$ and larger n. Study the decrease of the sequence of the values of the determinants of these matrices. (*AEM Ref.* p. 913 (#19))

Pr.18.13 (Least squares, straight line) Fit a straight line to the data $(x, y) = (0, 2)$, $(1, 2)$, $(2, 1)$, $(3, 3)$, $(5, 2)$, $(7, 3)$, $(9, 3)$, $(10, 4)$, $(11, 3)$ by least squares. Plot the points and the straight line. (See Example 18.7 in this Guide.)

Pr.18.14 (Least squares, straight line) Fit a straight line through the points $(x, y) = (400, 580)$, $(500, 1030)$, $(600, 1420)$, $(700, 1880)$, $(750, 2100)$ by least squares. (*AEM Ref.* p. 916 (#6))

Pr.18.15 (Least squares, quadratic parabola) Fit a quadratic parabola to the points $(x, y) = (2, 0)$, $(3, 3)$, $(5, 4)$, $(6, 3)$, $(7, 1)$ by least squares. Plot the points and the parabola. (*AEM Ref.* p. 916 (#11))

Pr.18.16 (Least squares, cubic parabola) Fit a cubic parabola to the points $(x, y) = (-2, -8)$, $(-1, 0)$, $(0, 1)$, $(1, 2)$, $(2, 12)$, $(4, 80)$ by least squares. Plot the points and the parabola. (*AEM Ref.* p. 917 (#17))

Pr.18.17 (Eigenvalues, eigenvectors) Find the eigenvalues and eigenvectors of the matrix in Example 18.8 in this Guide by the commands eigenvalues , eigenvectors . Also find the characteristic polynomial. (Type ?charpoly for information.) Note that Maple's characteristic matrix $\lambda \mathbf{I} - \mathbf{A}$ differs from the more usual $\mathbf{A} - \lambda \mathbf{I}$. Accordingly, *Maple's characteristic polynomial differs from the more usual by a factor* $(-1)^n$, where n is the number of rows and columns of $\mathbf{A}$. (*AEM Ref.* pp. 917-919)

Pr.18.18 (Collatz method) Find an approximation of an eigenvalue of the following matrix by Collatz's method, starting from $[1 \; 1 \; 1]^T$ and doing 10 steps. (See Example 18.8 in this Guide. *AEM Ref.* pp. 923, 924)

$$\mathbf{A} = \begin{bmatrix} 3 & 2 & 3 \\ 2 & 6 & 6 \\ 3 & 6 & 3 \end{bmatrix}$$

Pr.18.19 (Power method for eigenvalues) Find an approximation and error bounds for an eigenvalue of the matrix in Pr.18.18 by the power method, starting from $[1 \; 1 \; 1]^T$ and doing 10 steps. (*AEM Ref.* pp. 925-927)

Pr.18.20 (QR-factorization) Using the QR-method (5 steps), find approximations of the eigenvalues of the following matrix. (*AEM Ref.* p. 938 (#8))

$$\mathbf{A} = \begin{bmatrix} 14.2 & -0.1 & 0 \\ -0.1 & -6.3 & 0.2 \\ 0 & 0.2 & 2.1 \end{bmatrix}$$

Chapter 19

Numerical Methods for Differential Equations

Content. Euler methods (Exs. 19.2, 19.3, Prs. 19.1-19.5)
Runge-Kutta methods (Exs. 19.4, 19.6, 19.7, Prs. 19.6, 19.7, 19.10-19.12)
Multistep method (Ex. 19.5, Prs. 19.8, 19.9)
Laplace equation (Ex. 19.8, Pr. 19.13)
Heat equation (Ex. 19.9, Prs. 19.14, 19.15)

The commands in this chapter are those used in Chaps. 1 and 2 and (for systems of ODE's) in Chaps. 3 and 7.

Examples for Chapter 19

EXAMPLE 19.1 **TWO WAYS OF WRITING A DIFFERENTIAL EQUATION**

In connection with numerical methods for a differential equation $y' = f(x, y)$, writing, for instance,

```
> x := 'x': y := 'y':  g := 'g':
> f := (x, y) -> 5*x^2*y;                    # Resp. f := (x, y) → 5 x² y
```

seems preferable to the more usual

```
> g(x,y) := 5*x^2*y;                         # Resp. g (x, y) := 5 x² y
```

because if, say,

```
> h := 0.1:   x(0) := 1:   y(0) := 3.4:    # This implies y = 3.4 when x = 1
```

then

```
> y(1) := y(0) + h*f(x(0), y(0));     # Resp. y (1) := 5.10 # This is y(1.1).
```

whereas g leaves the expression unevaluated,

```
> y(1) := y(0) + h*g(x(0),y(0));             # Resp. y (1) := 3.4 + .1 g (1, 3.4)
```

Caution! Note well that $y(0)$ means y at $n = 0$ (which is 3.4), not y at $x = 0$.

EXAMPLE 19.2 **EULER METHOD**

The Euler method solves **initial value problems** $y' = f(x, y)$, $y(x_0) = y_0$ numerically, according to the formula

(A) $$y_{n+1} = y_n + h\,f(x_n, y_n) \qquad (n = 0,\ 1,\ 2,\ ...).$$

It is a **step-by-step method** which calculates approximate values of the solution at $x = x_0$, $x_1 = x_0 + h$, $x_2 = x_1 + h = x_0 + 2h$, etc. You choose the **stepsize** h, e.g., $h = 0.1$. Geometrically, the idea of the method is the approximation of the unknown solution curve by a polygon whose first side is tangent to the curve at x_0. The method is too inaccurate in practice, but is a nice illustration of the idea of step-by-step methods.

For instance, solve $y' + 5x^4 y^2 - 0$, $y(0) - 1$. Choose $h = 0.2$. Do $N = 10$ steps. Plot the solution curve and the approximate values obtained.

Solution. Type f in the differential equation $y' = f(x, y) = -5x^4 y^2$ as just explained in the previous example. Type h, N, and the initial condition $y(0) = 1$ in the form $x(0) := 0$, $y(0) := 1$. Then calculate the approximate y-values by a do-loop involving (A). Then use seq to make up a table showing $x(n)$, $y(n)$, and the error. The latter is $1/(1 + x(n)^5) - y(n)$, with the exact solution $y = 1/(1 + x^5)$ obtained by separating variables or by dsolve.

```
> f := (x, y) -> -5*x^4*y^2;              # Resp. f := (x, y) → -5 x^4 y^2
> h := 0.2:     N := 10:
> x(0) := 0:     y(0) := 1:
> for n from 0 to N do
>     y(n+1) := y(n) + h*f(x(n), y(n));
>     x(n+1) := x(n) + h;
> od:
> evalf(seq([x(n), y(n),   1/(1 + x(n)^5) - y(n)], n = 0..N), 5);
```

$[0, 1., 0]$, $[.2, 1., -.00030]$, $[.4, .99840, -.00850]$, $[.6, .97288, -.04506]$,
   $[.8, .85022, -.09704]$, $[1.0, .55413, -.05413]$, $[1.2, .24707, .03960]$,
   $[1.4, .12049, .03629]$, $[1.6, .064718, .022345]$, $[1.8, .037269, .012992]$,
   $[2.0, .022688, .007615]$

If you drop evalf(..., 5), you will get 10S-values. Try it. Now plot the approximate values obtained as well as the solution curve on common axes.

```
> P1 := plot({seq([x(n), y(n)], n = 0..N)}, style = point):
> P2 := plot(1/(1 + x^5), x = 0..2):
> with(plots):                           # Ignore the warning.
> display({P1, P2}, labels = [x, y]);
```

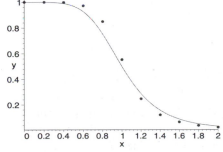

Example 19.2. Solution $y = 1/(1 + x^5)$ and approximation by Euler's method

Similar Material in AEM: pp. 942-944, 951 (#3)

EXAMPLE 19.3 | IMPROVED EULER METHOD

In the **improved Euler method** (or **Euler-Cauchy method**) you compute first
an auxiliary value $y_{n+1}^* = y_n + h\,f(x_n, y_n)$ and then the new value $y_{n+1} = y_n + \frac{1}{2}h[f(x_n, y_n) + f(x_{n+1}, y_{n+1}^*)]$.

For instance, apply this method to the initial value problem $y' = f(x,y) = 1+y^2$,
$y(0) = 0$. Do $N = 10$ steps with $h = 0.1$.

Solution. Type the right-hand side of the differential equation, using `->` as before,

```
> f := (x, y) -> 1 + y^2;                        # Resp. f := (x, y) → 1 + y²
```

Type a do-loop similar to that in the previous example. Solve the initial value problem
exactly by separating variables, obtaining $y = \tan x$, so that you can calculate the
error to find out about the accuracy of the method. You will see that the values
obtained lie almost on the curve of the exact solution. The printout gives the x-
values, y-values, y^*-values, and errors.

```
> h := 0.1:      N := 10:      x(0) := 0:      y(0) := 0:  ystar(0) := 0:
> for n from 0 to N do
>    ystar(n+1) := y(n) + h*f(x(n), y(n));                    # Auxiliary value
>    y(n+1) := y(n) + h/2*(f(x(n), y(n)) + f(x(n) + h, ystar(n+1)));
>    x(n+1) := x(n) + h;                                      # New x-value
> od:
> evalf(seq([x(n), y(n), ystar(n), tan(x(n)) - y(n)], n = 0..N), 5);
```

$[0, 0, 0, 0]$, $[.1, .10050, .1, -.00017]$, $[.2, .20304, .20151, -.00033]$,
$\quad$ $[.3, .30981, .30716, -.00047]$, $[.4, .42341, .41941, -.00062]$,
$\quad$ $[.5, .54702, .54134, -.00072]$, $[.6, .68490, .67695, -.00076]$,
$\quad$ $[.7, .84295, .83181, -.00066]$, $[.8, 1.0299, 1.0140, -.0003]$,
$\quad$ $[.9, 1.2593, 1.2360, .0009]$, $[1.0, 1.5538, 1.5179, .0036]$

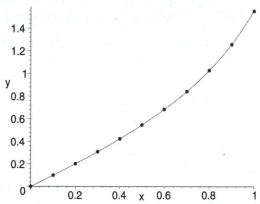

Example 19.3. Solution $y = \tan x$ and values
obtained by the improved Euler method

For plotting the solution curve and the values obtained, type

```
> P1 := plot(tan(x), x = 0..1):
> P2 := plot({seq([x(n), y(n)], n = 0..N)}, style = point):
```

```
> with(plots):                                    # Ignore the warning.
> display({P1, P2}, labels = [x, y]);
```

Similar Material in AEM: pp. 945, 946

EXAMPLE 19.4　　**CLASSICAL RUNGE-KUTTA METHOD (RK). PROCEDURE**

The (classical) Runge-Kutta method is an important and very accurate step-by-step method for solving initial value problems $y' = f(x, y)$, $y(x_0) = y_0$. Using this method, solve $y' = 1 + y^2$, $y(0) = 0$, doing $N = 10$ steps of size $h = 0.1$. Determine the error by using the exact solution $y(x) = \tan x$, obtained by separating variables.

Explanation of the method and solution. In each step of the method you first calculate four auxiliary values k1 , k2 , k3 , k4 by the four formulas in Lines 5-8 of the subsequent program. The new y(n+1) is then obtained from y(n) by adding the linear combination of these k 's in Line 9. In Line 10 you set x to the new value x(n+1) . This is a do-loop which begins in Line 4 and ends with od (do written in reverse order) in Line 11.

Line 3 contains the given initial values, where x(0) and y(0) mean x and y for $n = 0$, as in the previous examples.

Lines 4-11 is the **actual program.** Lines 1-3 and 12 make it into a procedure. The opening Line 1 shows a name for it (whose choice is up to you) and the list of quantities which you must specify later (in Lines 13 and 14).

Line 2 makes the five quantities shown **local variables.** This means that they are used in proc without interference with earlier or later use of these letters.

Line 12 is always needed to mark the end of proc .

How to **"call a procedure"**, that is, how to use it for solving a given initial value problem? This is shown in Lines 14 and 15. In Line 14 you type the name of the procedure and then in parentheses the function f and the values of x0 , y0 , h , and N . In Line 15 you use the command seq to obtain the desired values. And since you know the exact solution (which will *not* be the case in practice), you can obtain the errors of the approximate values computed and see that the Runge-Kutta method is much more accurate than, say, the method in the previous example.

```
> RK := proc(f, x0, y0, h, N)                           # Line 1
> local n, k1, k2, k3, k4;                              # Line 2
> x(0) := x0: y(0) := y0:                               # Line 3
> for n from 0 to N do                                  # Line 4
>     k1 := h*f(x(n), y(n));                             # Line 5
>     k2 := h*f(x(n) + h/2,   y(n) + k1/2);              # Line 6
>     k3 := h*f(x(n) + h/2,   y(n) + k2/2);              # Line 7
>     k4 := h*f(x(n) + h, y(n) + k3);                    # Line 8
>     y(n+1) := y(n) + 1/6*(k1 + 2*k2 + 2*k3 + k4);      # Line 9
>     x(n+1) := x(n) + h;                                # Line 10
> od:                                                   # Line 11
> end:                                                  # Line 12
```

```
> f := (x, y) -> 1 + y^2;                              # Line 13
```

$$f := (x, y) \to 1 + y^2$$

```
> RK(f, 0, 0, 0.1, 10):                                # Line 14
> seq([x(n), y(n), tan(x(n)) - y(n)], n = 0..10);      # Line 15
```

$[0, 0, 0], [.1, .1003345891, .830\,10^{-7}], [.2, .2027098782, .1573\,10^{-6}],$
  $[.3, .3093360394, .2102\,10^{-6}], [.4, .4227929930, .2257\,10^{-6}],$
  $[.5, .5463023078, .1820\,10^{-6}], [.6, .6841367569, .514\,10^{-7}],$
  $[.7, .8422885696, -.1891\,10^{-6}], [.8, 1.029639061, -.504\,10^{-6}],$
  $[.9, 1.260158782, -.564\,10^{-6}], [1.0, 1.557406442, .1283\,10^{-5}]$

What is the **advantage** of a procedure? Well, if you now want to solve another differential equation for other initial values, all you have to do is to change Lines 13−15 accordingly, instead of retyping the whole program.

Future use of a procedure. Save this procedure for future use under some name, say, `rk`, as follows. (Type `?save` for information.)

```
> save RK, rk:
```

where `RK` is the name of the procedure to be saved and `rk` is your choice for the file name. To **use it again** (after having filed it as just shown), type `read rk:` . (Type `?read` for information.) This will load it, ready for your new use. (See Example 19.5 in this Guide.)

Similar Material in AEM: pp. 947-949

| EXAMPLE 19.5 | **ADAMS-MOULTON MULTISTEP METHOD**

The Adams-Moulton method is a step-by-step method for solving initial value problems $y' = f(x, y)$, $y(x_0) = y_0$. It is obtained as follows. Integrate the equation on both sides from x_n to x_{n+1}, obtaining

$$y(x_{n+1}) - y(x_n) = \int_{x_n}^{x_{n+1}} f(x, y(x))\, dx.$$

On the right replace f by an interpolation polynomial p. Then on the left you have an approximation $y_{n+1} - y_n$. For p take the cubic polynomial that at x_n, x_{n-1}, x_{n-2}, x_{n-3} has the values $f(x_n, y_n), ..., f(x_{n-3}, y_{n-3})$, respectively. Then you get the **predictor formula** in Line (A) below. Next take for p the cubic polynomial that at x_{n+1}, x_n, x_{n-1}, x_{n-2} has the values $f(x_{n+1}, y_{n+1}^*)$, $f(x_n, y_n)$, $f(x_{n-1}, y_{n-1})$, $f(x_{n-2}, y_{n-2})$. Then you get the **corrector formula** in Line (B).

The method is called a **multistep method** because in each step it uses the results of several previous steps. Since at the beginning no values are available, the method needs **starting values** at x_0 (there you can take the initial value y_0), x_1, x_2, x_3 which you must obtain by some other method, say, by Runge-Kutta, so that they are as accurate as possible. You can load the RK procedure saved in the previous example, where it was saved by `save RK, rk:` . There it was explained that you can load it for further use by the command `read rk:` Hence you can now design a procedure for Adams-Moulton with built-in Runge-Kutta, as follows. (Type the first command first with ; instead of : , to make sure that the RK procedure is available.)

```
> read rk:
> AdamsMoultonRK := proc(f, x, y, h, N)

> local n;
> RK(f, x(0), y(0), h, 4):                    # IMPORTANT: x(0), y(0), not just x, y
> for n from 3 to N do
> x(n+1) := x(n) + h;
> ystar(n+1) := y(n) + h/24*(55*f(x(n),y(n)) - 59*f(x(n-1),y(n-1)) +
    37*f(x(n-2),y(n-2)) - 9*f(x(n-3),y(n-3)));              # Line (A)
> y(n+1) := y(n) + h/24*(9*f(x(n+1),ystar(n+1)) + 19*f(x(n),y(n)) -
    5*f(x(n-1),y(n-1)) + f(x(n-2),y(n-2)));                # Line (B)
> od:
> end:
```

You can save the procedure for later use by typing

```
> save AdamsMoultonRK, amrk:
```

Solve the initial value problem

$$y' = (y - x - 1)^2 + 2, \qquad y(0) = 1$$

by Adams-Moulton, doing 7 steps of size 0.1.

Solution. Type the right-hand side of the differential equation, then the initial values and $h = 0.1$ and $N = 7$, and then call the Adam-Moulton procedure with built-in Runge Kutta. Finally, use seq to print the values that will be of interest, also showing the accuracy of the method.

```
> f := (x, y) -> (y - x - 1)^2 + 2;
```

$$f := (x, y) \rightarrow (y - x - 1)^2 + 2$$

```
> x(0) := 0:    y(0) := 1:    h := 0.1:    N := 7:

> AdamsMoultonRK(f, x, y, h, N):

> seq([x(n), ystar(n), y(n), tan(x(n))+x(n)+1 - y(n)], n = 0..7);
```

$[0, \text{ystar}(0), 1, 0], [.1, \text{ystar}(1), 1.200334589, .83\,10^{-7}],$
    $[.2, \text{ystar}(2), 1.402709878, .158\,10^{-6}], [.3, \text{ystar}(3), 1.609336039, .211\,10^{-6}],$
    $[.4, 1.822715110, 1.822798081, -.4862\,10^{-5}],$
    $[.5, 2.046197407, 2.046314906, -.000012416],$
    $[.6, 2.283978353, 2.284161051, -.000024243],$
    $[.7, 2.542027365, 2.542331883, -.000043503]$

This table gives x, the predictor y^*, the corrector y, and the error, which you obtain by setting $u = y - x - 1$ and separating variables in the differential equation, or by dsolve, using letters not used above; say, t instead of x and z instead of y,

```
> dsolve({diff(z(t), t) = (z(t) - t - 1)^2 + 2, z(0) = 1});
```

$$z(t) = -\tan(-t) + t + 1$$

```
> simplify(%);                              # Resp. z(t) = tan(t) + t + 1
```

y^* is not needed (is not defined) during the Runge-Kutta starting steps ($n = 1, 2, 3$), so that in the table you see y^* as `ystar(0)`, `ystar(1)`, `ystar(2)`, `ystar(3)`.

   Note that the error is much less in absolute value than the absolute value of the difference between predictor and corrector.

EXAMPLE 19.6   **CLASSICAL RUNGE-KUTTA METHOD FOR SYSTEMS (RKS)**

RKS solves initial value problems for first-order systems $\mathbf{y}' = \mathbf{f}(x, \mathbf{y})$, $\mathbf{y}(x_0) = \mathbf{y}_0$, where $\mathbf{y} = [y_1, y_2, ..., y_m]$ and $\mathbf{f} = [f_1, f_2, ..., f_m]$. (We write m since n will denote the steps of the iteration.)

```
> RKS := proc (f, x, y, h, N)
> local k1, k2, k3, k4, n;
> for n from 0 to N do
>    k1 := h*f(x[n], y[n]);
>    k2 := h*f(x[n] + h/2, y[n] + k1/2);
>    k3 := h*f(x[n] + h/2, y[n] + k2/2);
>    k4 := h*f(x[n] + h, y[n] + k3);
>    y[n+1] := y[n] + (k1 + 2*k2 + 2*k3 + k4)/6;
>    x[n+1] := x[n] + h:
> od:
> end:
```

For later use you can save this procedure by typing

```
> save RKS, rks:
```

Using this procedure, obtain the solution (the **Airy function**) Ai(x) of the **Airy equation** $y'' = xy$. Do 5 steps with $h = 0.2$.

Solution. Setting $y_1 = y$, $y_2 = y_1' = y'$ (the usual conversion of a second-order equation to a system), you obtain $y_1' = y_2$, $y_2' = xy_1$. Hence the vector function $\mathbf{f}$ on the right-hand side of $\mathbf{y}' = \mathbf{f}(x, \mathbf{y})$ is $\mathbf{f} = [y_2, xy_1]$.

```
> f := (x, y) -> [y[2], x*y[1]];          # Resp. f := (x, y) → [y₂, x y₁]
```

Now standard linearly independent solutions of Airy's ODE are Ai(x) and Bi(x) (type `?Ai`: see also Ref. [1], pp. 446 and 475, in Appendix 1). Hence you obtain Ai(x) by choosing the initial conditions $y_1(0) = $ Ai(0), $y_2(0) = $ Ai$'(0)$. Thus type, in vector form,

```
> y[0] := [evalf(AiryAi(0), 8), evalf(subs(x = 0, diff(AiryAi(x), x)), 8)];
```

$$y_0 := [.35502806, -.25881940]$$

```
> x[0] := 0:    h := 0.2:    N := 5:
> RKS(f, x, y, h, N);                                    # Resp. 1.2
> seq([x[n], y[n]], n = 1..N);      # This is x and approximations of y and y'.
```

$[.2, [.3037030415, -.2524046352]], [.4, [.2547421170, -.2358307251]],$

$\quad [.6, [.2097997411, -.2127918437]], [.8, [.1698459708, -.1864117021]],$

$\quad [1.0, [.1352920858, -.1591468635]]$

EXAMPLE 19.7 | CLASSICAL RUNGE-KUTTA-NYSTROEM METHOD (RKN)

The RKN method extends the classical Runge-Kutta method to initial value problems for second-order differential equations $y'' = f(x, y, y')$, $y(x_0) = K_0$, $y'(x_0) = K_1$. The program below shows the following. In each step you have to compute four quantities k_1, k_2, k_3, k_4, in turn involving two quantities K and L (introduced merely for simplifying the notations of the k's). From these you calculate the new y-value y(n+1) and − this is the new feature of RKN over RK − a new value yp(n+1) for the derivative needed in the next step, where p suggests 'prime'. Finally, you type seq to display the values computed.

　　Using RKN (5 steps of size 0.2), solve **Airy's equation** $y'' = xy$ for the initial values $y(0) = 0.35502806$, $y'(0) = -0.25881940$. The exact solution is $\text{Ai}(x)$. Type ?Ai. Obtain the exact values by Maple and calculate the error of the RKN values.

Solution. Type $f = xy$, then the initial values, and then a do-loop that will calculate the auxiliary values as well as the values for y and y'. Finally, use seq to get the results as well as the errors. You may also use dsolve to see a general solution involving the **Airy functions** Ai and Bi.

```
> RKN := proc(f, x, y, yp, h, N)
> local n, k1, k2, k3, k4, K, L;
> for n from 0 to N do
>    k1 := h/2*f(x(n), y(n), yp(n));
>        K := h/2*(yp(n) + k1/2);
>    k2 := h/2*f(x(n)+h/2, y(n)+K, yp(n)+k1);
>    k3 := h/2*f(x(n)+h/2, y(n)+K, yp(n)+k2);
>        L := h*(yp(n) + k3);
>    k4 := h/2*f(x(n)+h, y(n)+L, yp(n)+2*k3);
>    x(n+1) := x(n) + h;
>    y(n+1) := y(n) + h*(yp(n) + 1/3*(k1 + k2 + k3));
>    yp(n+1) := yp(n) + 1/3*(k1 + 2*k2 + 2*k3 + k4);
> od:
> end:

> x := 'x':   y := 'y':   yp := 'yp':
> f := (x, y, yp) -> x*y;                      # Resp. f := (x, y, yp) → x y
> x(0) := 0:    y(0) := 0.35502806:   yp(0) := -0.25881940:
```

```
> h := 0.2:    N := 5:
> RKN(f, x, y, yp, h, N):
> seq([x(n), y(n), evalf(AiryAi(x(n))) - y(n)], n = 0..N);
```

$[0, .35502806, .60\,10^{-8}], [.2, .3037030415, .1128\,10^{-6}], [.4, .2547421170, .2373\,10^{-6}],$
$[.6, .2097997460, .3157\,10^{-6}], [.8, .1698460013, .3161\,10^{-6}],$
$[1.0, .1352921900, .2263\,10^{-6}]$

The exact 10S-value of Ai(1) is 0.1352924163. Your present result is somewhat more accurate than 0.1352920858 in the previous example. For a general solution type

```
> dsolve(diff(z(t), t, t) = t*z(t));
```

$$z(t) = \_C1\ \text{AiryAi}(t) + \_C2\ \text{AiryBi}(t)$$

(_C1 and _C2 may appear interchanged.) You can save the procedure for later use by typing

```
> save RKN, rkn:
```

Similar Material in AEM: p. 960

EXAMPLE 19.8   **LAPLACE EQUATION. BOUNDARY VALUE PROBLEM**

Solve the **Dirichlet problem** for the **Laplace equation** $u_{xx}+u_{yy}=0$ numerically in the square in the figure with the grid shown, when the boundary potential equals 0 on the upper edge and 100 volts on the three other edges (equivalently: the temperature on the upper edge is 0 and on the other edges it is 100 degrees C).

Solution. The Laplace equation is replaced by a difference equation. This gives a linear system of 4 equations in the 4 unknown potentials $u_{jk} = u(x, y) = u(4j, 4k) = u(P_{jk})$, $j = 1, 2$ and $k = 1, 2$, at the 4 inner points of the grid in the figure. At each such point, -4 times the potential plus the sum of the potentials at the 4 closest neighbors equals 0. In these equations you take the 4 inner points in the order $P_{11}, P_{21}, P_{12}, P_{22}$. Thus for P_{11} the equation is $-4u_{11} + u_{10} + u_{21} + u_{12} + u_{01} = -4u_{11} + 100 + u_{21} + u_{12} + 100 = 0$. Hence the coefficient matrix of the linear system $\mathbf{Ax} = \mathbf{b}$ has $[-4\ \ 1\ \ 1\ \ 0]$ as the first row and -200 as the first component of $\mathbf{b}$. Similarly for the other equations. You thus obtain $\mathbf{Ax} = \mathbf{b}$ and its solution (the potential at the 4 inner points) by typing

```
> A := matrix([[-4,1,1,0],  [1,-4,0,1],  [1,0,-4,1],  [0,1,1,-4]]);
```

$$A := \begin{bmatrix} -4 & 1 & 1 & 0 \\ 1 & -4 & 0 & 1 \\ 1 & 0 & -4 & 1 \\ 0 & 1 & 1 & -4 \end{bmatrix}$$

```
> b := [-200, -200, -100, -100];        # Resp. b := [−200, −200, −100, −100]
> with(linalg):                          # Ignore the warning.
> x := evalf(linsolve(A, b), 3);         # Resp. x := [87.5, 87.5, 62.5, 62.5]
```

The figure was obtained by the commands (type `?conformal` and use the function $F(z) = z$)

```
> with(plots):
> conformal(z, z = 0..12 + 12*I, scaling = constrained, grid = [4, 4],
  labels = [x, y]);
```

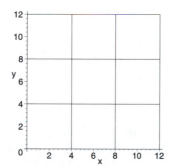

Example 19.8. Grid in the xy-plane

Program. Band matrices. Type `?band`. Type `with(linalg): p := 2`. (A refinement of the grid will follow later by typing `p := 3:`, etc.) Construct `A` as follows (see the commands below, which we are going to explain). Type `c := [1, -4, 1]:` The middle component of `c` gives the diagonal entries of the band matrix `C`. "Double `C` up" to get `E`. Four 1's from `A` are still missing in `E`. To get them, take the zero vector `v` with $2p + 1 = 5$ components. Change its first component and its last component to 1 by the commands `v[1] := 1:` and `v[2*p+1] := 1:`. This gives `v := [1, 0, 0, 0, 1]`. (What looks like a detour will enable you to use it for other $p = 3$, etc. later.) The middle 0 of `v` gives the main diagonal entries of the 4×4 matrix `F`. Hence the first row of `F` is 0, 0, 1 from `v` and another 0. The second row is 0, 0, 0, 1 all from `v`. The next row is 1, 0, 0, 0 all from `v`. The last row is an extra 0 and then 1, 0, 0 from `v`. You now obtain `A := evalm(E + F)`.

```
> A := 'A':    b := 'b':    x := 'x':
> p := 2:
> c := [1, -4, 1]:
> C := band(c, p);                        # −4 gives the diagonal entries of C.
```

$$C := \begin{bmatrix} -4 & 1 \\ 1 & -4 \end{bmatrix}$$

```
> E := diag(seq(C, j = 1..p));                        # Obtain E from C.
```

$$E := \begin{bmatrix} -4 & 1 & 0 & 0 \\ 1 & -4 & 0 & 0 \\ 0 & 0 & -4 & 1 \\ 0 & 0 & 1 & -4 \end{bmatrix}$$

```
> v := vector(2*p + 1, 0);                 # Resp. v := [0, 0, 0, 0, 0]
> v[1] := 1:  v[2*p + 1] := 1:             # Change components.

> evalm(v);      # Resp. [1, 0, 0, 0, 1]    # Gives the new v = [1, 0, 0, 0, 1].
```

```
> F := band(v, p^2);
```

$$F := \begin{bmatrix} 0 & 0 & 1 & 0 \\ 0 & 0 & 0 & 1 \\ 1 & 0 & 0 & 0 \\ 0 & 1 & 0 & 0 \end{bmatrix}$$

```
> A := evalm(E + F);        # The system is Ax = b with b still to be constructed.
```

$$A := \begin{bmatrix} -4 & 1 & 1 & 0 \\ 1 & -4 & 0 & 1 \\ 1 & 0 & -4 & 1 \\ 0 & 1 & 1 & -4 \end{bmatrix}$$

Now turn to the **boundary values**, explaining the commands below . These values were originally on the left-hand sides of the linear equations and were then transferred to the right, hence multiplied by -1. Type them as shown and add them appropriately to the components of the vector b := [b[1], b[2], b[3], b[4]] with $p^2 = 4$ components, which is originally the zero vector b := vector(p^2, 0). Do this by the do-loop shown. To see what is going on, write this program out for $j = 1$ and then for $j = 2$ and recall that b[1] comes from P_{11} and gets bot + left. b[2] comes from P_{21} and gets bot + right. b[3] comes from P_{12} and gets top + left. b[4] comes from P_{22} a nd gets top + right.

```
> top := 0:    left := -100:    bot := -100:    right := -100:
> b := vector(p^2, 0):
> for j from 1 to p do
>    b[j] := b[j] + bot:
>    b[(j-1)*p + 1] := b[(j-1)*p + 1] + left:
>    b[j*p] := b[j*p] + right:
>    b[j + p*(p-1)] := b[j + p*(p-1)] + top:
> od:
> b := evalm(b);                      # Resp. b := [-200, -200, -100, -100]
> x := evalf(linsolve(A, b), 4);      # Resp. x := [87.50, 87.50, 62.50, 62.50]
```

Finally, arrange the result as a $p \times p = 2 \times 2$ matrix so that the location of these values corresponds to the location of the four inner gridpoints. To achieve this, type X and then use swaprow (which is done here in such a way that it can be used for any p). Hence matrix(p, p, x) is a $p \times p$ (= 2×2) matrix with the p^2 (= 4) components of **x** as entries.

```
> X := matrix(p, p, x);
```

$$X := \begin{bmatrix} 87.50 & 87.50 \\ 62.50 & 62.50 \end{bmatrix}$$

```
> for j from 1 to p/2 do
>    swaprow(X, j, p + 1 - j);
> od;
```

$$\begin{bmatrix} 62.50 & 62.50 \\ 87.50 & 87.50 \end{bmatrix}$$

Grid refinement. Consider $p = 3$ (9 inner gridpoints; see the figure on the next page). Type $p := 3:$ and use the program just designed.

```
> p := 3:
> C := band([1, -4, 1], p);
```

$$C := \begin{bmatrix} -4 & 1 & 0 \\ 1 & -4 & 1 \\ 0 & 1 & -4 \end{bmatrix}$$

```
> E := diag(seq(C, j = 1..p));                    # Response deleted
> v := vector(2*p + 1, 0):            # Zero vector with 2p + 1 = 7 components
> v[1] := 1:  v[2*p + 1] := 1# Change of the first component and the last one
> A := evalm(E + band(v, p^2));          # Coefficient matrix of the system
```

$$A := \begin{bmatrix} -4 & 1 & 0 & 1 & 0 & 0 & 0 & 0 & 0 \\ 1 & -4 & 1 & 0 & 1 & 0 & 0 & 0 & 0 \\ 0 & 1 & -4 & 0 & 0 & 1 & 0 & 0 & 0 \\ 1 & 0 & 0 & -4 & 1 & 0 & 1 & 0 & 0 \\ 0 & 1 & 0 & 1 & -4 & 1 & 0 & 1 & 0 \\ 0 & 0 & 1 & 0 & 1 & -4 & 0 & 0 & 1 \\ 0 & 0 & 0 & 1 & 0 & 0 & -4 & 1 & 0 \\ 0 & 0 & 0 & 0 & 1 & 0 & 1 & -4 & 1 \\ 0 & 0 & 0 & 0 & 0 & 1 & 0 & 1 & -4 \end{bmatrix}$$

```
> b := vector(p^2, 0):                    # Zero vector with p^2 = 9 components
> for j from 1 to p do                         # Same do-loop as for p = 2 above
>    b[j]  := b[j] + bot:
>    b[(j-1)*p + 1]  := b[(j-1)*p + 1] + left:
>    b[j*p]  := b[j*p] + right:
>    b[j + p*(p - 1)]  := b[j + p*(p - 1)] + top:
> od:
> b := evalm(b);
```

$$b := [-200, -100, -200, -100, 0, -100, -100, 0, -100]$$

```
> x := evalf(linsolve(A, b), 5);
```

$$x := [92.857, 90.179, 92.857, 81.250, 75., 81.250, 57.143, 47.321, 57.143]$$

```
> X := matrix(p, p, x);
```

$$X := \begin{bmatrix} 92.857 & 90.179 & 92.857 \\ 81.250 & 75. & 81.250 \\ 57.143 & 47.321 & 57.143 \end{bmatrix}$$

```
> for j from 1 to p/2 do                    # Same do-loop as for p = 2 above

>   swaprow(X, j, p + 1 - j);

> od;                          # Potential at the 9 inner points of the grid
```

$$\begin{bmatrix} 57.143 & 47.321 & 57.143 \\ 81.250 & 75. & 81.250 \\ 92.857 & 90.179 & 92.857 \end{bmatrix}$$

The figure of the refined grid is obtained by a slight change of the previous program.

```
> conformal(z, z = 0..12 + 12*I, scaling = constrained, labels = [x, y],
   xtickmarks = [3, 6, 9, 12], ytickmarks = [3, 6, 9, 12], grid = [5, 5]);
```

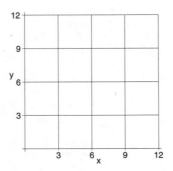

Example 19.8. Refined grid

Similar Material in AEM: pp. 962-966

EXAMPLE 19.9   HEAT EQUATION. CRANK-NICOLSON METHOD

The **one-dimensional heat equation** $u_t = c^2 u_{xx}$ is a **parabolic equation** that governs, for instance, the heat flow in a bar, where $u(x,t)$ is the temperature at a point x and time t. Solve the corresponding difference equation with $c^2 = 1$ on the interval $0 \le x \le 1$ (the bar extending from $x = 0$ to $x = 1$ along the x-axis) subject to the initial temperature $u(x,0) = \sin \pi x$ by the **Crank-Nicolson method** with x-step $h = 0.2$ and time step $k = 0.04$, doing 5 time steps.

Solution. The method proceeds by time steps. For each $t = 0$, 0.04, etc. you have 4 points of the grid at which the 4 values of the temperature are to be determined from a linear system of 4 equations in these unknown values. In this method the discretization of the heat equation is done in such a way that the resulting linear system $\mathbf{Av} = \mathbf{b}$ has the following coefficient matrix (see the previous example for band matrices; type **?band**), where $r = k/h^2$ ($= 1$ in the present case).

```
> with(linalg):                          # Ignore the warning.

> n := 4:   r := 1:    h := 0.2:

> A := band([-r, 2 + 2*r, -r], n);
```

$$A := \begin{bmatrix} 4 & -1 & 0 & 0 \\ -1 & 4 & -1 & 0 \\ 0 & -1 & 4 & -1 \\ 0 & 0 & -1 & 4 \end{bmatrix}$$

Obtain the initial temperature distribution and the preparation for a plot by typing

```
> u := 'u':
> for m from 0 to (n+1) do
>   u[m] := evalf(sin(Pi*m*h));
> od:
> seq(u[m], m = 0..n + 1);
```

$$0, .5877852524, .9510565165, .9510565163, .5877852522, 0$$

```
> P[0] := polygonplot([[0, 0], [0.2, u[1]], [0.4, u[2]], [0.6, u[3]],
    [0.8, u[4]], [1, 0]]):
```

Now obtain the values of the temperature $u(x,t)$ for $t = 0.04, 0.08, ..., 0.20$ by iteration with respect to time, solving $\mathbf{Av} = \mathbf{b}$ in each step (see Line 5). In this program you obtain the mth component $b[m]$ as the sum of two previous values (see Line 3).

```
> b := vector(n, 0):
> for j from 1 to (n + 1) do          # Line 1
>   for m from 1 to n do
>     b[m] := u[m-1] + u[m+1];         # Line 3
>   od;
>   v[j] := evalf(linsolve(A, b));     # Line 5
>   for m from 1 to n do
>     u[m] := v[j][m]:
>   od;
> od;
```

$$v_1 := [.3992737562, .6460385085, .6460385084, .3992737562]$$
$$v_2 := [.2712207082, .4388443243, .4388443244, .2712207082]$$
$$v_3 := [.1842361824, .2981004050, .2981004049, .1842361823]$$
$$v_4 := [.1251488911, .2024951595, .2024951595, .1251488911]$$
$$v_5 := [.08501177538, .1375519420, .1375519420, .08501177538]$$

Obtain polygon plots of these values (and the initial values above) by typing

```
> with(plots):                        #  Ignore the warning.
> PS := plot(sin(Pi*x), x = 0..1):
> for i from 1 to 5 do
>   poly[i] := [[0, 0], [0.2, v[i][1]], [0.4, v[i][2]], [0.6, v[i][3]],
      [0.8, v[i][4]], [1, 0]];
>   P[i] := polygonplot(poly[i], scaling = constrained);
```

```
> od:
```
```
> display({PS, P[0], P[1], P[2], P[3], P[4], P[5]});
```

Each polygon corresponds to one of the times t_j considered, and is obtained by joining the temperatures $v(x, t_j)$ just computed for $x = 0.2, \ 0.4, \ 0.6, \ 0.8$

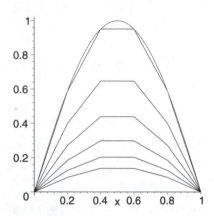

Example 19.9. Temperatures for time $t = 0, \ 0.04, \ 0.08, \ \ldots$

To obtain the grid, type the following, where `[6, 6]` gives the numbers of horizontal and vertical lines of the grid. (Type `?conformal`. The function you use is $F(z) = z$.)

```
> with(plots):                                    # Ignore the warning.
```
```
> conformal(z, z = 0..1+0.2*I, grid = [6, 6], labels = [x, t]);
```

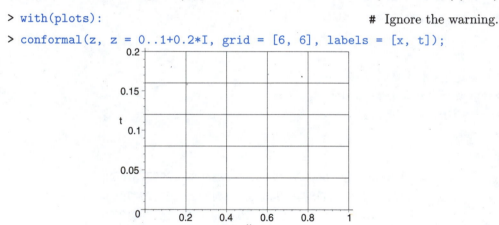

Example 19.9. Grid for a parabolic equation (the heat equation with $c^2 = 1$)

Problem Set for Chapter 19

Pr.19.1 **(Euler's method)** Solve $y' + 0.1y = 0$, $y(0) = 2$, doing 10 steps of size 0.1. Find the error. Plot the solution curve and the approximate values obtained. (*AEM Ref.* p. 951 (#1))

Pr.19.2 **(Euler's method, Riccati equation)** Solve $y' = (x+y)^2$, $y(0) = 0$, 10 steps of size 0.1. Find the exact solution and determine the error. (*AEM Ref.* p. 951 (#4))

Pr.19.3 **(Experiment on stepsize for Euler's method)** In Pr.19.1 find out experimentally for what stepsize the error of $y(1)$ will decrease to 0.0001 (approximately). (*AEM Ref.* p. 951 (#1))

Pr.19.4 **(Improved Euler method. Logistic population)** Solve the **Verhulst equation** $y' = y - y^2$ subject to the initial condition $y(0) = 0.5$. Do 10 steps of size 0.1. Find the exact solution and the error. Plot the exact solution. (*AEM Ref.* p. 951 (#6))

Pr.19.5 **(Experiment on improved Euler method. Unbounded solution)** Try to extend the calculation in Example 19.3 in this Guide to x-values near $\pi/2$ (where the solution becomes infinite). Experiment with $h = 0.1$, 0.05, 0.025 and study the increase of the error with x for constant h and the decrease of the error under halving h for constant x. (*AEM Ref.* pp. 945-947)

Pr.19.6 **(Classical Runge-Kutta method RK)** Solve $y' = (y - x - 1)^2$, $y(0) = 1$ by RK. Do 10 steps of size 0.1. Solve the problem exactly by setting $y - x - 1 = u$ and separating variables, or by dsolve. Calculate the error. (*AEM Ref.* pp. 948-950)

Pr.19.7 **(RK, step halving)** Solve $y' = -0.2xy$, $y(0) = 1$ by RK. Do 50 steps of size 0.2. Do 100 steps of size 0.1. Solve the problem exactly. Compare the errors of the two values for $x = 10$. (*AEM Ref.* p. 952 (#17))

Pr.19.8 **(Adams-Moulton method)** Solve the initial value problem $y' = x/y$, $y(1) = 3$, by the procedure AdamsMoultonRK. Do 10 steps of size 0.2. Find the exact solution. Determine the errors. (*AEM Ref.* p. 955 (#6))

Pr.19.9 **(Multistep method)** Using the procedure in Example 19.5 in this Guide, solve $y' = x + y$, $y(0) = 0$. Do 10 steps of size 0.1. (*AEM Ref.* pp. 954, 955)

Pr.19.10 **(Mass-spring system)** Solve $y'' + 2y' + 0.75y = 0$, $y(0) = 3$, $y'(0) = -2.5$ by RKS (Runge-Kutta for systems). Do 5 steps of size 0.2. Find the exact solution and compute the error. (*AEM Ref.* pp. 957-959)

Pr.19.11 **(RKS Runge-Kutta for systems)** Solve $y_1' = 2y_1 - 4y_2$, $y_2' = y_1 - 3y_2$, $y_1(0) = 3$, $y_2(0) = 0$ by RKS for x from 0 to 5 with $h = 0.1$. Find the exact solution and the errors of y_1 and y_2. Plot y_1 and y_2. (*AEM Ref.* p. 961 (#1))

Pr.19.12 **(Runge-Kutta-Nystroem method)** Solve the initial value problem in Pr.19.10 by RKN (see Example 19.7 in this Guide) and compare the accuracy with that in Pr.19.10. (*AEM Ref.* p. 960)

Pr.19.13 **(Laplace equation. Dirichlet problem)** Solve the Laplace equation numerically for the refined grid in Example 19.8 in this Guide and boundary values 0 on the bottom, 110 V on the right edge, 220 V on the top, 110 V on the left edge. (*AEM Ref.* pp. 964-966)

Pr.19.14 **(Crank-Nicolson. "Triangular" initial temperature)** Solve the heat equation in Example 19.9 in this Guide when the initial temperature is x for $0 \le x \le 1/2$ and $1 - x$ for $1/2 \le x \le 1$, using the same grid and r and h as in that example. (*AEM Ref.* p. 978)

Pr.19.15 **(Crank-Nicolson method, grid refinement)** In Example 19.9 in this Guide choose $n = 24$ (instead of $n = 4$), thus $h = 1/25 = 0.04$, $r = k/h^2 = 1$, thus $k = 1/625 = 0.0016$, and do enough steps so that you obtain values corresponding to those given in Example 19.9 and compare. (*AEM Ref.* pp. 978, 979)

PART F. OPTIMIZATION, GRAPHS

Content. Steepest descent, linear programming (Chap. 20)
Graphs, digraphs, shortest spanning trees, network flows (Chap. 21)

Chapter 20

Unconstrained Optimization. Linear Programming

Content. Steepest descent (Ex. 20.1, Prs. 20.1-20.3)
Linear programming (Ex. 20.2, Prs 20.4, 20.5)

This concerns finding maxima or minima using calculus methods under no constraints (steepest descent) or matrix methods under constraints (restriction of the region in which the maximum or minimum is to be found) (linear programming).

Examples for Chapter 20

EXAMPLE 20.1   METHOD OF STEEPEST DESCENT

This is a method of unconstrained optimization for determining a minimum of a function $f(x, y)$. **"Unconstrained"** means that the variables x, y are not restricted to certain parts (regions) in the xy-plane. The method uses the fact that at each point of the surface S: $(x, y, f(x, y))$ of f over the xy-plane the direction of **steepest descent** is given by $-\text{grad } f$, the **gradient** being $\text{grad } f = [\partial f/\partial x, \ \partial f/\partial y]$. (Exceptional points may occur.) You begin at some point $\mathbf{x}_0 = (X_0, Y_0)$ and follow the straight line in the direction of $-\text{grad } f(x_0)$. This line is given by

(A) $$\mathbf{z}(t) = \mathbf{x}_0 - t \text{ grad } f(\mathbf{x}_0).$$

It gives a curve C : $(z_1, z_2, f(\mathbf{z}))$ on S. Follow the straight line until you reach the lowest point on C, the point at which

(B) $$g(t) = f(\mathbf{z}(t)) \qquad \mathbf{z}(t) = [z_1(t), \ z_2(t)]$$

has a minimum. Determine the latter by finding $t = t_1$ where the derivative is zero,

(C) $$g'(t) = 0.$$

Take $\mathbf{z}(t_1)$ as your new $\mathbf{x} = \mathbf{x_1}$ and begin the next step. The new gradient will be perpendicular to the old. Why? (*Answer.* You reach the level curve of f at $\mathbf{x}_1$ tangentially, whereas the new gradient is perpendicular to that level curve.)

For instance, find a minimum of the function $f(\mathbf{x}) = x_1{}^2 + 3 x_2{}^2$, starting from $\mathbf{x}_0 = (6, 3)$.

Solution. f has a minimum at $\mathbf{x} = \mathbf{0}$, as inspection shows. Hence you can see how the method approaches the solution. The level curves are ellipses. Type `?grad`. A procedure for the method is as follows. Explanations are given afterwards.

```
> with(linalg):                                          # Ignore the warning.
> SD := proc(f, X, Y, N)                                      # Line 1
> local z, g, gprime, j, SOL, T:
> z := evalm([x, y] - t*grad(f(x,y), [x, y]));               # Line 3
> g := f(z[1], z[2]);
> gprime := diff(g, t);                                       # Line 5
> for j from 0 to N do
>    SOL[j] := solve(gprime = 0, t);                          # Line 7
>    T[j] := evalf(subs(x = X[j], y = Y[j], SOL[j]));         # Line 8
>    X[j+1] := subs(x = X[j], y = Y[j], t = T[j], z[1]);
>    Y[j+1] := subs(x = X[j], y = Y[j], t = T[j], z[2]);
> od:                                                         # Line 11
> end:
```

Explanations. You see the data needed, where `N` is the number of steps to be done (Line 1), the local variables (Line 2), the formulas (A), (B), and the derivative in (C) (Lines 3, 4, 5), a do-loop (Lines 6-11) that gives the solution of (C) (Line 7, denoted by `SOL[j]`), the current x and y substituted into `SOL[j]` (Line 8), and the new x and y (Lines 9 and 10). Note that a procedure must end with `end`.

To apply the procedure, type f and the initial data. Then call the procedure and use `seq` to obtain the values $(x,\ y,\ f(x,y))$.

```
> f:= (x, y) -> x^2 + 3*y^2;              # Resp. f := (x, y) → x² + 3y²
> X[0] := 6:  Y[0] := 3:  T[0] := 0;   N := 6:    # Resp. T₀ := 0
> SD(f, X, Y, N):
> S := seq([X[k], Y[k], f(X[k],Y[k])], k = 0..6);
```

$S := [6, 3, 63],\ [3.483870967, -.774193549, 13.93548386],$

$\quad [1.327188942, .663594470, 3.082503346],\ [.7706258366, -.1712501865, .6818440591],$

$\quad [.2935717481, .1467858736, .1508226494],$

$\quad [.1704610149, -.0378802257, .03336169210],$

$\quad [.0649375297, .03246876472, .007379544810]$

Commands for plotting the surface of f and a broken line of segments joining the points $(x_j, y_j, f(x_j, y_j))$ on the surface (exhibiting the change of search direction at those points) are as follows.

```
> with(plots):                                           # Ignore the warning.
> P1 := plot3d(f(x,y), x = -7..7, y  = -6..6, style = WIREFRAME):
> P2 := spacecurve([S], thickness = 5, orientation = [175, 50]):
> display({P1, P2});
```

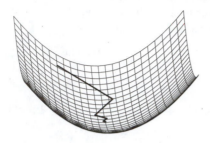

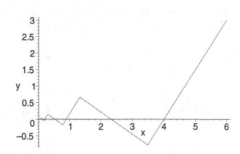

Example 20.1. Surface of $f(x, y)$ and polygonal line marking the descent on it

Example 20.1. Search path of steepest descent in the xy-plane

The second figure shows the search path in the xy-plane. Type

```
> S1 := seq([X[m], Y[m]], m = 0..6);
```

$S1 := [6, 3], [3.483870967, -.774193549], [1.327188942, .663594469],$
  $[.7706258366, -.1712501865], [.2935717481, .1467858736],$
  $[.1704610149, -.0378802257], [.0649375297, .03246876472]$

```
> plot([S1], scaling = constrained, labels = [x, y]);
```

Future use of a procedure

Save the procedure by typing (with sd being your choice for the file name)

```
> save SD, sd:                          # Type ?save for information.
```

To **use it again**, type

```
> read sd:                              # Type ?read for information.
```

Similar Material in AEM: pp. 991-993

EXAMPLE 20.2   SIMPLEX METHOD OF CONSTRAINED OPTIMIZATION

The function $z = f(x_1, x_2)$ to be maximized (profit, e.g.) is linear, say, for instance, $z = 40x_1 + 88x_2$, subject to linear inequalities (and linear equations) that determine a region ("**feasibility region**") in the $x_1 x_2$-plane (bounded by a polygon) to which (x_1, x_2) is restricted and in which the maximum is to be found. For instance, $x_1 \geq 0$, $x_2 \geq 0$ (the first quadrant), and

(A)
$$2x_1 + 8x_2 \leq 60$$
$$5x_1 + 2x_2 \leq 60$$

Find the maximum of z under these **constraints**.

Solution. Convert the inequalities (A) to equations by introducing **slack variables** x_3, x_4. Including the function z to be maximized, this gives the **normal form** of the problem

$$
\begin{aligned}
z \;-\; 40\,x_1 \;-\; 88\,x_2 && &= 0 \\
2\,x_1 \;+\; 8\,x_2 \;+\; x_3 && &= 60 \\
5\,x_1 \;+\; 2\,x_2 && +\; x_4 &= 60
\end{aligned}
$$

with x_1, x_2, x_3, x_4 being nonnegative. This is a linear system of equations. Find the maximal z, that is, find (x_1, x_2) satisfying the constraints and such that $z = f(x_1, x_2)$ is maximum. You can do this by suitably transforming the corresponding **augmented matrix T_0** (also called **simplex table**).

```
> with(linalg):                              # Ignore the warning.
> T[0] := matrix([[1,-40,-88, 0, 0,  0],
>                  [0,  2,  8, 1, 0, 60],
>                  [0,  5,  2, 0, 1, 60]]);
```

$$
T_0 := \begin{bmatrix}
1 & -40 & -88 & 0 & 0 & 0 \\
0 & 2 & 8 & 1 & 0 & 60 \\
0 & 5 & 2 & 0 & 1 & 60
\end{bmatrix}
$$

Proceed by elimination of variables. This is similar to Gauss's method in Chaps. 6 and 18, but the choice of pivots is quite different, as follows. The first *negative* entry in Row 1 is -40 in Column 2, the column of x_1. Eliminate x_1 from 2 of the 3 equations. Choose 5 as the pivot because it gives the smallest of the quotients $60/2 = 30$ (in Row 2) and $60/5 = 12$ (in Row 3). Use `addrow`. (See Example 6.6 in this Guide. Type `?addrow`.) To eliminate x_1 from Row 1 and then from Row 2, add $40/5$ times Row 3 (the pivot row) to Row 1 and then add $-2/5$ of Row 3 to Row 2.

```
> T[1] := addrow(addrow(T[0], 3, 1, 8), 3, 2, -2/5);
```

$$
T_1 := \begin{bmatrix}
1 & 0 & -72 & 0 & 8 & 480 \\
0 & 0 & \dfrac{36}{5} & 1 & \dfrac{-2}{5} & 36 \\
0 & 5 & 2 & 0 & 1 & 60
\end{bmatrix}
$$

The next (and only further) negative entry in Row 1 is -72 in Column 3, the column of x_2. Eliminate x_2 from 2 of the 3 equations. Choose Row 2 as the pivot row because $36/(36/5) = 5$ is the smaller of this quotient and $60/2 = 30$ (in Row 3). To eliminate x_2 from Row 1 and then from Row 3, add $72/(36/5) = 10$ times Row 2 to Row 1 and then add $-2/(36/5) = -10/36$ times Row 2 to Row 3.

```
> T[2] := addrow(addrow(T[1], 2, 1, 10), 2, 3, -10/36);
```

$$
T_2 := \begin{bmatrix}
1 & 0 & 0 & 10 & 4 & 840 \\
0 & 0 & \dfrac{36}{5} & 1 & \dfrac{-2}{5} & 36 \\
0 & 5 & 0 & \dfrac{-5}{18} & \dfrac{10}{9} & 50
\end{bmatrix}
$$

Since Row 1 has no more negative entries, you have reached the maximum. It is 840 (see Row 1) and occurs at $x_1 = 50/5 = 10$ (see Row 3) and $x_2 = 36/(36/5) = 5$ (see Row 2). This is the right upper vertex of the quadrangle in the figure.

```
> with(plots):                               # Ignore the warning.
> polygonplot([[0,0], [12,0], [10,5], [0,7.5]], labels = [x, y],
   scaling = constrained);
```

Example 20.2. Feasibility region

To obtain this result directly, type `?simplex` for information. It shows the functions available. Click on `maximize`, which will show you that you must type

```
> x1:='x1':    x2 := 'x2':
> with(simplex):                                      # Ignore the warning.
> maximize(40*x1 + 88*x2,   {2*x1 + 8*x2 <= 60, 5*x1 + 2*x2 <= 60});
```

$$\{x2 = 5,\ x1 = 10\}$$

So here you have the result, without knowing anything about the method by which it has been obtained.

Similar Material in AEM: pp. 995, 998-1001

Problem Set for Chapter 20

Pr.20.1 **(Steepest descent)** Apply the method of steepest descent to $f(x, y) = x^2 + 10\,y^2$, starting from $(3, -2)$ and plotting the search path in the xy-plane. How many steps do you need to reach a point where $|f(x, y)| < 0.0001$? (*AEM Ref.* pp. 991, 992)

Pr.20.2 **(Steepest descent)** Apply the method of steepest descent to $f(x, y) = xy$, starting from $(1, 2)$ and plotting the search path in the xy-plane. First guess what will happen. Then calculate. (*AEM Ref.* pp. 991, 992)

Pr.20.3 **(Steepest descent)** Apply the method of steepest descent to $f(x, y) = 5\,x^2 - y^2$, starting from $(1/2, 1/2)$ and plotting the search path in the xy-plane. First guess what the path might look like. Then calculate. (*AEM Ref.* pp. 991, 992)

Pr.20.4 **(Linear programming)** Maximize $z = 4\,x_1 + x_2 + 2\,x_3$ subject to $x_1 + x_2 + x_3 \leq 1$, $x_1 + x_2 - x_3 \leq 0$ and x_1, x_2, x_3 nonnegative. Proceed stepwise, showing the matrices. (*AEM Ref.* p. 1007 (#9))

Pr.20.5 **(Linear programming)** Maximize the daily output in producing x_1 glass plates by a process P_1 and x_2 glass plates by a process P_2 subject to the following constraints (labor hours, machine hours, raw material supply). Show the steps as in Example 20.2 in this Guide. (*AEM Ref.* p. 1007 (#3))

$$2\,x_1 + 3\,x_2 \leq 130, \qquad 3\,x_1 + 8\,x_2 \leq 300, \qquad 4\,x_1 + 2\,x_2 \leq 140.$$

Chapter 21

Graphs and Combinatorial Optimization

Content. Graphs, digraphs (Ex. 21.1, Prs. 21.1-21.5)
Shortest spanning trees (Ex. 21.2, Prs. 21.6-21.8)
Network flow (Ex. 21.3, Prs. 21.9, 21.10)

These are optimization problems of a discrete or combinatorial structure, solvable by methods involving graphs or digraphs.

A **graph** G consists of two sets, a set V of **vertices** and a set E of **edges** (pairs of vertices that are endpoints of an edge). A **digraph** (directed graph) is a graph in which each edge has a direction (one of its endpoints is the *initial point* and the other the *terminal point*). Undirected edges are typed with braces, e.g., $\{v_3, v_5\}$. Directed edges are typed with brackets, e.g., $[v_4, v_1]$. In this chapter you will need the **networks package**. Type `?networks` for information. You will see a list of 75 functions (operations) available in this package. Click on any of them for further information. In the following examples you will see which of these functions (`addedge`, `addvertex`, `adjacency`, `connect`, `draw`, `edges`, `new`, `spantree`, ...) you will actually need and how these functions are used.

Examples for Chapter 21

| EXAMPLE 21.1 | **GRAPHS AND DIGRAPHS. THEIR MATRICES** |

Load the networks package. To create a graph G with vertices 1, 2, 3, 4 and edges $e_1 = (1, 2)$, $e_2 = (2, 3)$, $e_3 = (2, 4)$, $e_4 = (3, 4)$, $e_5 = (1, 4)$ (sketch it), type

```
> with(networks):                        # Ignore the warning.
> new(G):
> addvertex({1, 2, 3, 4}, G);                    # Resp. 1, 2, 3, 4
> addedge({{1,2},{2,3},{2,4},{3,4},{1,4}},G); # Resp. e1, e2, e3, e4, e5
```

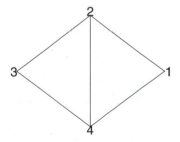

Example 21.1. Graph G

Draw the graph G by the following command, keeping in mind that a graph may be drawn in many ways, depending on how you place and number the vertices.

```
> draw(G);
```

Matrices for graphs. Graphs can be handled on the computer by matrices. Obtain the **adjacency matrix** $\mathbf{A} = [a_{jk}]$ of the graph G by typing the following. Here, $a_{jk} = 1$ if edge (j, k) is in G and 0 if it is not. $\mathbf{A}$ is symmetric (why?). (In the next four responses the Maple output consists merely of the actual matrix.)

```
> A := adjacency(G);
```

$$
A := \begin{array}{c} \quad\quad \text{Vertex} \\ \begin{array}{cccc} 1 & 2 & 3 & 4 \end{array} \\ \begin{bmatrix} 0 & 1 & 0 & 1 \\ 1 & 0 & 1 & 1 \\ 0 & 1 & 0 & 1 \\ 1 & 1 & 1 & 0 \end{bmatrix} \begin{array}{l} \text{Vertex 1} \\ \text{Vertex 2} \\ \text{Vertex 3} \\ \text{Vertex 4} \end{array} \end{array}
$$

The matrix $\mathbf{A}$ shows that G has those five edges.

To obtain the **incidence matrix** $\mathbf{B} = [b_{jk}]$, type the following. Here, $b_{jk} = 1$ if vertex j is an endpoint of edge k and 0 if it is not.

```
> B := incidence(G);
```

$$
B := \begin{array}{c} \quad\quad\quad \text{Edge} \\ \begin{array}{ccccc} e1 & e2 & e3 & e4 & e5 \end{array} \\ \begin{bmatrix} 1 & 0 & 0 & 0 & 1 \\ 1 & 1 & 1 & 0 & 0 \\ 0 & 1 & 0 & 1 & 0 \\ 0 & 0 & 1 & 1 & 1 \end{bmatrix} \begin{array}{l} \text{Vertex 1} \\ \text{Vertex 2} \\ \text{Vertex 3} \\ \text{Vertex 4} \end{array} \end{array}
$$

The columns of $\mathbf{B}$ may come out in a different order. At any rate, you see that vertex 1 is the endpoint of 2 edges, vertex 2 of 3 edges, etc. If you want a detailed description of a graph, type show(G). Try it. We suppress the long response, for reasons of space.

Digraphs (directed graphs) are graphs whose edges have a direction. These directed edges are typed as lists (vectors) [1, 2], etc., in contrast to undirected edges, which are typed {1, 2}, etc. (Type ?list.) For instance, type (make a sketch of this digraph)

Example 21.1. Digraph DiG drawn as graph by the command draw

```
> new(DiG):
> addvertex({1, 2, 3, 4}, DiG);                          # Resp. 1, 2, 3, 4
> e1:='e1':   e2:='e2':   e3:='e3':   e4:='e4':    e5:='e5':
> addedge([[1,2], [2,1], [2,4], [3,2], [4,3], [3,4]], DiG);
```

e1, e2, e3, e4, e5, e6

```
> draw(DiG);
```

Matrices for digraphs. Obtain the **adjacency matrix** $\mathbf{A} = [a_{jk}]$ as before, with $a_{jk} = 1$ if DiG has a directed edge $[j,\ k]$ from vertex j to vertex k. $\mathbf{A}$ is generally not symmetric (why?).

```
> A := adjacency(DiG);
```

$$A := \begin{array}{c} \\ \\ \end{array} \overset{\begin{array}{cccc} & \text{To Vertex} & & \\ 1 & 2 & 3 & 4 \end{array}}{\begin{bmatrix} 0 & 1 & 0 & 0 \\ 1 & 0 & 0 & 1 \\ 0 & 1 & 0 & 1 \\ 0 & 0 & 1 & 0 \end{bmatrix}} \begin{array}{l} \text{From Vertex 1} \\ \text{Vertex 2} \\ \text{Vertex 3} \\ \text{Vertex 4} \end{array}$$

$a_{12} = 1$ shows that DiG has an edge $[1,2]$, etc.

To obtain the **incidence matrix** $\mathbf{B} = [b_{jk}]$ of DiG, type the following. Here, b_{jk} equals -1 if edge e_k leaves vertex j, equals $+1$ if edge k enters vertex j, and is 0 otherwise.

```
> B := incidence(DiG);
```

$$B := \overset{\begin{array}{cccccc} & & \text{Edge} & & & \\ e1 & e2 & e3 & e4 & e5 & e6 \end{array}}{\begin{bmatrix} -1 & 1 & 0 & 0 & 0 & 0 \\ 1 & -1 & -1 & 1 & 0 & 0 \\ 0 & 0 & 0 & -1 & 1 & -1 \\ 0 & 0 & 1 & 0 & -1 & 1 \end{bmatrix}} \begin{array}{l} \text{Vertex 1} \\ \text{Vertex 2} \\ \text{Vertex 3} \\ \text{Vertex 4} \end{array}$$

From Column 1 of $\mathbf{B}$ you can read that e_1 goes from vertex 1 to vertex 2, from Column 2 that e_2 goes from vertex 2 to vertex 1, from Column 3 that e_3 goes from vertex 2 to vertex 4, from Column 4 that e_4 goes from vertex 3 to vertex 2, from Column 5 that e_5 goes from vertex 4 to vertex 3, and from Column 6 that e_6 goes from vertex 3 to vertex 4.

Similar Material in AEM: pp. 1012-1015

EXAMPLE 21.2   **SHORTEST SPANNING TREES**

This concerns graphs G each of whose edges has a given **length** (a positive number, often called "*cost*" or "*weight*"). This may be an actual length or travel cost or time, but it may also be something entirely different. The problem is to find the **shortest spanning tree** T in G. A **tree** is a graph without cycles. It **spans** G means that it includes all the vertices of G. Its **length** L is the sum of the lengths of its edges. And it is **shortest** if L is minimum compared to the length of any other spanning tree in G.

For instance, find a shortest spanning tree in the graph G consisting of vertices 1, 2, 3, 4, and edges (1,2), (1,3), (1,4), (2,3), (2,4) with length 8, 5, 7, 1, 2, respectively.

Solution. Load the networks package and then type the graph. Note that the lengths (weights) are typed as a list after the list of the edges in corresponding order in the command addedge .

```
> with(networks):
> new(G):
> addvertex({1, 2, 3, 4}, G);                    # Resp. 1, 2, 3, 4
> addedge([ {1,2}, {1,3}, {1,4}, {2,3}, {2,4}],
    weights=[8, 5, 7, 1, 2], G);
```

$$e1, e2, e3, e4, e5$$

Now come the two commands that solve the problem. (Type ?spantree, ?ancestor, ?daughter for information.) Vertex 1 is the root of the tree. From the second subsequent command you see that vertex 1 has no ancestor, vertex 2 has the ancestor 3, vertex 3 has the ancestor 1, and vertex 4 has the ancestor 2. That is, the tree consists of the edges $(1,3)$, $(2,3)$, and $(2,4)$.

```
> T := spantree(G, 1, L):
> ancestor(T);          # Resp. table([1 = { }, 2 = {3}, 3 = {1}, 4 = {2}])
> L;                                # Resp. 8   # Length of T
```

You may confirm the result by typing

```
> daughter(T);          # Resp. table([1 = {3}, 2 = {4}, 3 = {2}, 4 = { }])
```

1 has the daughter 3, vertex 3 the daughter 2, vertex 2 the daughter 4, and 4 has no daughters.

```
> draw(G);
> draw(T);
```

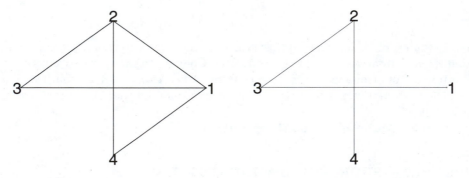

Example 21.2. Given graph G ***Example 21.2.*** Tree T of shortest paths in G

Similar Material in AEM: pp. 1022, 1023

| EXAMPLE 21.3 | **FLOW IN NETWORKS**

This concerns digraphs; call them G for simplicity. Each edge has a direction and a **capacity** (= maximum flow that can go through that edge). One vertex in G is the **source** s, where the flow starts. Another vertex is the **target** (or sink) t of the flow, where the flow disappears. Type ?flow for information. The command flow(G, s, t); gives the maximum flow in G. If you type flow(G, s, t, se);, and then se; , you also get the set of **saturated edges** (= edges used fully to capacity).

For instance, let G have 6 vertices 1, 2, 3, 4, 5, 6, where $s = 1$ is the source and $t = 6$ the target. Let the (directed!) edges be [1,2], [1,4], [2,3], [5,2], [3,5], [3,6], [4,5], [5,6], with capacities 20, 10, 11, 4, 5, 13, 7, 3, respectively (see the figure, but make a sketch that also shows directions and capacities). Find the maximum flow.

Solution. Load the networks package and create the digraph G.

```
> with(networks):
> new(G):
> addvertex({1, 2, 3, 4, 5, 6}, G);                    # Resp. 1, 2, 3, 4, 5, 6
> addedge([[1,2], [1,4], [2,3], [5,2], [3,5], [3,6], [4,5], [5,6]],
    weights = [20, 10, 11, 4, 5, 13, 7, 3], G);
```

$$e1, \ e2, \ e3, \ e4, \ e5, \ e6, \ e7, \ e8$$

In a larger graph it may be good to know that you can obtain vertices of edges by typing ends; . For instance,

```
> ends(e3, G);                                         # Resp. [2, 3]
```

Now type the command for the maximum flow as explained before. This is a procedure based on the idea of **flow augmenting paths,** by which a flow (e.g. the zero flow) is increased stepwise until no more flow augmenting paths can be found and the maximum flow is reached.

```
> flow(G, 1, 6, se);                                   # Resp. 14
> se;                              # Resp. {{5, 6}, {2, 3}}  #  Saturated edges
```

The **Max-flow min-cut theorem** states that the maximum flow in G equals the capacity of a minimum cut set in G. A *cut set* is a set of edges whose deletion disconnects all paths from s to t. A **minimum cut set** is a cut set of the smallest number of edges. To get it, type

```
> mincut(G, 1, 6, f);                                  # Resp. {e3, e8}
> f;                                                   # Resp. 14
> draw(G);
```

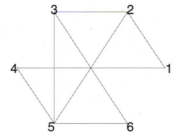

Example 21.3. Given network ($s = 1$ source, $t = 6$ target)

Here, $f = 14$ is the **capacity** of the minimum cut set (= the maximum possible flow through that set). This illustrates the theorem and motivates the present concepts.

From above you see that the third edge is e3 = [2,3] and the eighth is e8 = [5,6] . The figure looks as if you could still reach 6 from 1 by going 1, 2, 5, 3, 6, but this is not true because there is no edge [2,5] but only [5,2] (and the figure shows no directions of edges, so it is not very informative − and you should make a sketch of your own, showing directions and capacities).

  Similar Material in AEM: pp. 1031-1041

Problem Set for Chapter 21

Pr.21.1 **(Graph)** Use Maple to draw the graph with 8 vertices 1, 2, ..., 8 and 12 edges (1,2), (2,3), (3,4), (4,1), (5,6), (6,7), (7,8), (8,5), (1,5), (2,6), (3,7), (4,8). (*AEM Ref.* p. 1012)

Pr.21.2 **(Adjacency matrix)** Find the adjacency matrix of the graph in Pr.21.1. Find its eigenvalues and eigenvectors. Why is 3 an eigenvalue? (*AEM Ref.* p. 1013)

Pr.21.3 **(Incidence matrix)** Find the incidence matrix of the graph in Pr.21.1. Using Row 1 of it, sketch the subgraph of the 3 edges incident with vertex 1. Similarly for vertex 2, then for vertex 3, etc.

Pr.21.4 **(Adjacency matrix of a digraph)** Find the adjacency matrix of the digraph DiG with vertices 1, 2, 3, 4, 5 and edges [1, 2], [2, 3], [2, 4], [4, 2], [2, 5], [5, 2], [4, 5], [5, 4]. Let Maple draw DiG as a graph and complete the figure by indicating directions. (*AEM Ref.* p. 1014 (#10))

Pr.21.5 **(Cycle)** Cycles are closed paths. Load the networks package and type the command cn := cycle(n); and ends(cn);, for $n = 3, 4, ..., 10$. This gives cycles with n vertices and n edges. Draw these cycles. (*AEM Ref.* p. 1015)

Pr.21.6 **(Shortest spanning tree)** Let the edges (1,2), (1,3), (1,4), (2,3), (2,4), (3,4) of the complete graph G with 4 vertices have the lengths 6, 20, 18, 14, 8, 4, respectively. Find a shortest spanning tree T with origin 1 (a) by inspection (sketch G), (b) as in Example 21.2 in this Guide. (*AEM Ref.* p. 1023 (#1))

Pr.21.7 **(Shortest spanning tree)** Find a shortest spanning tree for the graph in Pr.21.1, assuming that all the edges have length 1. (*AEM Ref.* pp. 1020-1030)

Pr.21.8 **(Experiment on shortest spanning tree)** What shortest spanning tree do you obtain in Pr.21.7 if the edges, in the order given in that problem, have the lengths 12, 5, 12, 3, 3, 7, 8, 2, 9, 11, 13, 10? (*AEM Ref.* pp. 1020-1027)

Pr.21.9 **(Network flow)** Find the maximum flow and the set of saturated edges in the network consisting of the vertices $s = 1$ (the source), $2, 3, 4, 5, 6 = t$ (the target) and the edges (the third number being the capacity) (1,2; 5), (1,3; 8), (1,4; 6), (2,4; 4), (3,5; 11), (4,3; 2), (4,5; 5), (4,6; 4), (5,6; 13). Sketch the network. Find a minimum cut set and the maximum flow through its edges. (*AEM Ref.* p. 1037 (#13))

Pr.21.10 **(Network flow)** Consider the network consisting of the vertices $s = 1$ (the source), $2, 3, 4, 5 = t$ (the target) and the edges (the third number being the capacity) (1,2; 8), (1,3; 9), (2,3; 7), (2,4; 5), (2,5; 4), (3,5; 16), (4,3; 10), (4,5; 3). Find the maximum flow, the saturated edges, and a minimum cut set and its capacity. (*AEM Ref.* p. 1040 (#4))

PART G. PROBABILITY AND STATISTICS

Content. Data analysis, binomial, Poisson, hypergeometric, normal distributions
(Chap. 22)
Random numbers, confidence intervals, tests, regression (Chap. 23)

stats package. Load it by typing `with(stats):` Type `?describe` and click on terms of the long list shown as you need them, e.g. `mean`, `variance`, `median`, etc. Type `?distributions` to see the binomial, normal, Student's t-, Fisher's, chi-square, and other distributions. Type `?statevalf` for information on probability functions, densities, and distribution functions and their numerical values.

Chapter 22

Data Analysis. Probability Theory

Content. Data analysis of samples (Exs. 22.1, 22.2, Prs. 22.1-22.5)
Discrete distributions (Ex. 22.3, Prs. 22.7-22.12)
Normal distribution (Ex. 22.4, Prs. 22.13-22.15)

Probability theory provides models (probability distributions) of random experiments in this chapter and the mathematical justification of the statistical methods in the next chapter.

Examples for Chapter 22

EXAMPLE 22.1   DATA ANALYSIS: MEAN, VARIANCE, STANDARD DEVIATION

Given a sample S of size n, that is, consisting of n values,

$$S = [x_1, x_2, ..., x_n],$$

you can arrange these values in ascending or descending order by the command `sort`. For instance,

```
> S := [89, 84, 87, 81, 89, 86, 91, 90, 78, 89, 87, 99, 83, 89];
```

$$S := [89, 84, 87, 81, 89, 86, 91, 90, 78, 89, 87, 99, 83, 89]$$

```
> sort(%);          # Resp. [78, 81, 83, 84, 86, 87, 87, 89, 89, 89, 89, 90, 91, 99]
> -sort(-S);        # Resp. [99, 91, 90, 89, 89, 89, 89, 87, 87, 86, 84, 83, 81, 78]
```

The **sample mean** $\bar{x}$ is

$$(1) \qquad\qquad \bar{x} = \frac{1}{n} \sum_{j=1}^{n} x_j.$$

This is the arithmetic mean of the sample values. It measures the average size of these values. You obtain it (for the above S, which has size $n = 14$) by loading the **statistics package**, typing,

```
> with(stats):                          # Ignore the warning.
```

and then typing (type `?mean` for information)

```
> xbar := evalf(describe[mean](S), 5);          # Resp. xbar := 87.286
```

The **sample variance** s^2 is

$$(2) \qquad s^2 = \frac{1}{n-1} \sum_{j=1}^{n} (x_j - x)^2.$$

It measures the spread of the sample values. You obtain it (for the above S) by loading the statistics package and typing (type `?variance` for information)

```
> var := evalf(14/13*describe[variance](S), 5);
```

$$var := 25.143$$

The **sample standard deviation** is the nonnegative square root of the sample variance. Thus,

```
> sdev := sqrt(var);                   # Resp. sdev := 5.014279609
```

or directly (note that restricting `var` to 5D has resulted in a difference due to round-off),

```
> evalf(sqrt(14/13)*describe[standarddeviation](S));
```

$$5.014265361$$

Maple definition of s^2. Maple has the factor $1/n$ in (2) instead of the more common $1/(n-1)$. This means that the factor $n/(n-1)$ (in this example 14/13) must be used in obtaining (2). We shall use this factor $n/(n-1)$ throughout this Guide. Similarly the factor $\sqrt{n/(n-1)}$ appears in the command for the standard deviation.
  Similar Material in AEM: pp. 1053, 1054

EXAMPLE 22.2   **DATA ANALYSIS: HISTOGRAMS, BOXPLOTS**

These are graphical representations of data. A **histogram** of a sample shows the **absolute frequencies** $a(x) = $ *Number of times the value x occurs in that sample*. Actually, to obtain a better general impression of the essential features of the sample, *group it into classes*. For instance, take the sample in Example 22.1 and order it,

```
> S := [89, 84, 87, 81, 89, 86, 91, 90, 78, 89, 87, 99, 83, 89]:
> S := sort(%); # Resp. S := [78, 81, 83, 84, 86, 87, 87, 89, 89, 89, 89, 90, 91, 99]
```

Look at the data and decide that good **class marks** (midpoints of class intervals) would be 80, 85, 90, 95, 100. That is, take the class intervals 77.5-82.5, 82.5-87.5, 87.5-92.5, 92.5-97.5, 97.5-102.5. The corresponding **absolute class frequencies** are 2 (because 78 and 81 are in the first interval), 5, 6, 0, 1, respectively. Now plot a histogram by typing (type `?histogram` for information)

```
> with(stats):                                              # Ignore the warning.
> data := [Weight(77.5..82.5, 2), Weight(82.5..87.5, 5),
>           Weight(87.5..92.5, 6), Weight(92.5..97.5, 0),
>           Weight(97.5..102.5, 1)]:

> statplots[histogram](data, labels = [x," "]);
```

The areas of the rectangles equal the absolute class frequencies, $0.4 \times 5 = 2$, $1 \times 5 = 5$, $1.2 \times 5 = 6$, etc.

Boxplots. A boxplot of a sample shows the smallest sample value that is not an outlier (lower end of the plot), the **lower quartile** (bottom of the box), the **median**, the **upper quartile** (top of the box), and the largest sample value that is not an outlier (upper end of the plot). Outliers are marked by dots. 99 is an outlier in S. To get a boxplot for the above sample S, type the following. Here, `shift = 1` moves the box away from the vertical axis. (Type `?boxplot`.) The numbers on the horizontal axis are meaningless.

```
> statplots[boxplot](S, shift = 1);
```

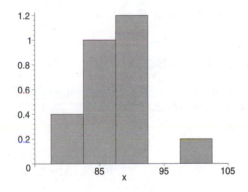

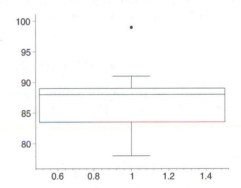

Example 22.2. Histogram of the grouped sample   ***Example 22.2.*** Boxplot of the sample S

Similar Material in AEM: pp. 1051-1053

EXAMPLE 22.3	**DISCRETE PROBABILITY DISTRIBUTIONS**

In this chapter you will need the binomial, Poisson, and hypergeometric distributions, which are discrete, and the uniform and normal distributions, which are continuous. All these are available in Maple. Also the chi-square, t, and F distributions needed in Chap. 23 are available – and many other distributions. To see this, type `?distributions`.

Discrete distributions are given by their **probability function** $f(x)$, in Maple `pf`, or, equivalently, by their **distribution function** $F(x)$, in Maple `dcdf` (suggesting 'discrete cumulative probability function'). To see this, type `?statevalf`. This is a subpackage of the stats package. You call these functions as shown below in terms of the distributions to be discussed.

Binomial distribution. The probability function is

$$f(x) = \binom{n}{x} p^x (1-p)^{n-x}, \qquad x = 0, 1, ..., n.$$

This is the probability of x successes in n independent trials when the probability of success in a single trial is p. For instance, if $p = 0.55$ and $n = 5$, you obtain numerical values of f for $x = 0, ..., 5$ by typing

```
> with(stats):  Digits := 5:                        # Ignore the warning.
> f := seq(statevalf[pf, binomiald[5, 0.55]](x), x = 0..5);
```

$$f := .018453, .11277, .27565, .33692, .20589, .050328$$

Remember that individual values can be accessed by typing `f[1]` , `f[2]` , etc. For instance,

```
> f[4];                                              # Resp. .33692
```

Note that this is $f(3)$ because x begins with 0, whereas counting terms of the sequence begins with 1, say, $j = 1$. Thus $x = j - 1$ in connection with the plotting below. This will give you the right numbers on the x-axis.

Similarly, you obtain values of the distribution function F by typing

```
> F := seq(statevalf[dcdf, binomiald[5, 0.55]](x), x = 0..5);
```

$$F := .01845, .13123, .40687, .74378, .94968, 1.$$

Bar graphs of probability functions can be obtained by generating pairs of endpoints of the bars and then plotting them. Using the values `f[0]` , `f[1]` , ..., `f[5]` , type

```
> s := seq([[j - 1, 0], [j - 1, f[j]]], j = 1..6);
```

$$s := [[0, 0], [0, .018453]], [[1, 0], [1, .11277]], [[2, 0], [2, .27565]], [[3, 0], [3, .33692]],$$
$$[[4, 0], [4, .20589]], [[5, 0], [5, .050328]]$$

```
> plot({s}, x = -1..6);
```

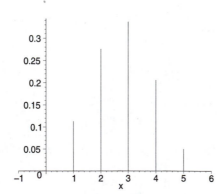

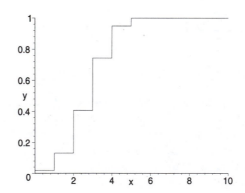

Example 22.3. Probability function of the binomial distribution with $n = 5$ and $p = 0.55$

Example 22.3. Distribution function of the binomial distribution with $n = 5$ and $p = 0.55$

Graphs of distribution functions. In the discrete case these are step functions, with the stepsizes equal to the values of the corresponding probability function. You can obtain the graph of the distribution function by typing

```
> x := 'x':
> G := x -> piecewise(x < 0, 0,  x < 1, F[1],  x < 2, F[2],  x < 3, F[3],
    x < 4, F[4],  x < 5, F[5], 1);
```

$$G :=$$
$$x \rightarrow \text{piecewise}\,(x < 0, 0, x < 1, F_1, x < 2, F_2, x < 3, F_3, x < 4, F_4, x < 5, F_5, 1)$$

```
> plot(G(x), x = 0..10, y = 0..1);
```

Poisson distribution. The probability function is

$$f(x) = \frac{\mu^x}{x!}e^{-\mu}, \qquad x = 0, 1, \dots .$$

It is the limiting case of the binomial distribution if one lets $n \rightarrow \infty$ and $p \rightarrow 0$ so that the mean $\mu = n\,p$ approaches a finite value. Its graph has infinitely many bars whose lengths decrease to zero very fast. For instance, to plot f with $\mu = 5$, type (note poisson with p , not P)

```
> poi := seq(statevalf[pf, poisson[5]](x), x = 0..15);
```

$poi :=$ .0067379, .0033690, .084224, .14037, .17547, .17547, .14622, .10444,
.065277, .036265, .018132, .0082418, .0034342, .0013208, .00047173, .00015724

```
> bars := seq([[j - 1, 0], [j - 1, poi[j]]], j = 1..16);
```

$bars :=$ [[0, 0], [0, .0067379]], [[1, 0], [1, .033690]], [[2, 0], [2, .084224]],
[[3, 0], [3, .14037]], [[4, 0], [4, .17547]], [[5, 0], [5, .17547]], [[6, 0], [6, .14622]],
[[7, 0], [7, .10444]], [[8, 0], [8, .065277]], [[9, 0], [9, .036265]], [[10, 0], [10, .018132]],
[[11, 0], [11, .0082418]], [[12, 0], [12, .0034342]], [[13, 0], [13, .0013208]],
[[14, 0], [14, .00047173]], [[15, 0], [15, .00015724]]

```
> plot({bars}, labels = [x, " "]);
```

The figure is shown on the next page.

Hypergeometric distribution. The probability function is

$$f(x) = \frac{\binom{M}{x}\binom{N-M}{n-x}}{\binom{N}{n}}.$$

This is the probability of obtaining x red balls in drawing n balls from a lot of N balls, M of which are red, and the drawing is done one-by-one **without replacement** (that is, balls drawn are not returned to the lot). Instead of red balls you could think of defective screws in a lot of screws produced. For instance, when $N = 100$ and $M = 20$, and you draw 5 balls, you can plot f by typing the following, where in the plot command axes = boxed helps to make the bar at $x = 0$ visible. (axes = none would remove the scales needed. Type ?plot[options] .) Type the following, noting that $80 = N - M = 100 - 20$ (that is, Maple uses $M_1 = M$ and $M_2 = N - M$ as parameters instead of the (more common) M and N).

```
> hyp := seq(statevalf[pf, hypergeometric[20, 80, 5]](x), x = 0..5);
```

$$hyp := .31931, .42015, .20734, .047849, .0051483, .00020593$$

```
> barhyp := seq([[j - 1,0], [j - 1, hyp[j]]], j = 1..6);
```

$barhyp := [[0, 0], [0, .31931]], [[1, 0], [1, .42015]], [[2, 0], [2, .20734]],$
     $[[3, 0], [3, .047849]], [[4, 0], [4, .0051483]], [[5, 0], [5, .00020593]]$

```
> plot({barhyp}, axes = boxed, labels = [x, " "]);
```

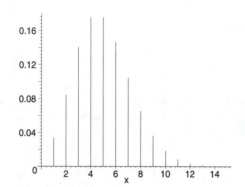

Example 22.3. Probability function of the Poisson distribution with mean $\mu = 5$

Example 22.3. Probability function of the hypergeometric distribution with $N = 100$, $M = 20$, and $n = 5$

Similar Material in AEM: pp. 1079-1083

EXAMPLE 22.4   NORMAL DISTRIBUTION

The **standardized normal distribution** (that is, the normal distribution with mean 0 and variance 1) is a **continuous distribution** with **density**

(A) $$f(x) = \frac{1}{\sqrt{2\pi}} \exp\left(-\frac{x^2}{2}\right) \qquad (-\infty < x < \infty).$$

Hence its distribution function is

(B) $$\Phi(z) = \frac{1}{\sqrt{2\pi}} \int_{-\infty}^{z} e^{-u^2/2}\, du.$$

Maple knows this function as well as its inverse (see below), and various numerical tables of $\Phi(z)$ can be found in the literature.

Setting $u = (v - \mu)/\sigma$, you have $du = dv/\sigma$ and $z = (x - \mu)/\sigma$. You thus obtain the distribution function of the normal distribution with mean μ and variance σ^2

(C) $$F(x) = \Phi\left(\frac{x - \mu}{\sigma}\right) = \frac{1}{\sigma\sqrt{2\pi}} \int_{-\infty}^{x} \exp\left[-\frac{1}{2}\left(\frac{v - \mu}{\sigma}\right)^2\right] dv.$$

To plot the density (A), type the following, which shows that `pdf` denotes the density ('**probability density function**') and `normald` (μ, σ) the **normal distribution.** (Type `?statevalf` , `?distributions` .)

```
> with(stats):                                    # Ignore the warning.
> plot(statevalf[pdf, normald[0, 1]](x), x = -4..4, labels = [x, y]);
```

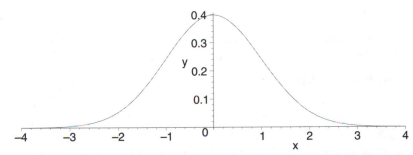

Example 22.4. Density (A) of the standardized normal distribution

To plot the distribution function (B), type

```
> plot(statevalf[cdf, normald[0, 1]](x), x = -4..4, labels = [x, y]);
```

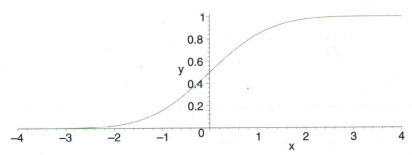

Example 22.4. Distribution function (B) of the standardized normal distribution

Two basic tasks in connection with any distribution. Illustration in terms of a random variable X that is normal with mean 5.0 and variance 0.04.

    First task. Given $x = 5.3$, find the probability that in a trial (an experiment), X will assume any value not exceeding 5.3.

Solution. You need the **distribution function** `cdf` (which Maple calls 'cumulative density function'). Note that the command involves the standard deviation 0.2, not the variance. The answer is 93.3%, approximately.

```
> with(stats):    Digits := 4:
> statevalf[cdf, normald[5.0, 0.2]](5.3);                 # Resp. .9330
```

    Second task. Given the probability $P = 0.95 = 95\%$, find $x = c$ such that with probability 0.95 the random variable X will assume any value not exceeding c. (This task will arise in Chap. 23 for several distributions.)

Solution. You need the **inverse** `icdf` of the distribution function (where `icdf` suggests 'inverse cumulative density function'). The answer is $x = c = 5.33$, approximately (see on the next page).

```
> statevalf[icdf,normald[5.0, 0.2]](0.95);
```
                                                        # Resp. 5.329

Similar Material in AEM: pp. 1085-1089, A89, A90

Problem Set for Chapter 22

Pr.22.1 **(Mean and variance)** Find the mean filling [grams] and the variance of the sample of fillings 203, 199, 198, 201, 200, 201, 201. (*AEM Ref.* p. 1054 (#10))

Pr.22.2 **(Mean, median, standard deviation)** Find the mean, median, and standard deviation of the sample 17, 18, 17, 16, 17, 16, 18, 16. (*AEM Ref.* p. 1054 (#12))

Pr.22.3 **(Histogram, bar graph)** Plot a histogram for the sample in Pr.22.1. For information type `?histogram`. Use the 5 numerically different sample values as your class marks. In addition, find a trick by which the command will give you a bar graph. (*AEM Ref.* pp. 1051, 1052)

Pr.22.4 **(Histogram, boxplot)** Make a histogram and a boxplot of the sample in Pr.22.2. (*AEM Ref.* pp. 1051, 1052)

Pr.22.5 **(Boxplot)** Find the median and the other two quartiles of the sample in Pr.22.1 and make a boxplot. Type `?describe` and when you have the information sheet on the screen, click on `median`, `quartile`, and `quantile` to see the details of the commands not shown in Example 22.2 in this Guide. (*AEM Ref.* pp. 1052, 1054 (#10))

Pr.22.6 **(Probability)** A circuit contains 10 automatic switches. You want, with a probability of 95%, that during a given time interval all the switches are working. What probability of failure per time interval can you admit for a single switch? (*AEM Ref.* p. 1064 (#12))

Pr.22.7 **(Experiment on binomial distribution)** For what values of p will $f(x)$ with constant n be (a) large for small x, (b) large for large x? Experiment with plots and n's of your choice (10, 20, 100, or whatever). (*AEM Ref.* pp. 1079, 1080)

Pr.22.8 **(Binomial distribution)** Find and plot (as a bar graph) the probabilities of x successes in 40 independent trials with probability of success 1/2 in a single trial. Why is the figure symmetric? Does the figure remind you of something you have seen in connection with the normal distribution? For which x's are these probabilities very small?

Pr.22.9 **(Permutations)** In how many ways can you assign 8 workers to 8 jobs? First guess, then calculate. (*AEM Ref.* p. 1065)

Pr.22.10 **(Experiment on Stirling formula)** A convenient approximation for (inconvenient) large factorials is $n! \approx \sqrt{2\pi n}\left(\frac{n}{e}\right)^n$. Find conjectures about the absolute error and the relative error for growing n by experimentation. (*AEM Ref.* p. 1067)

Pr.22.11 **(Poisson distribution)** Investigate the Poisson distribution graphically. What happens if you let μ increase? μ is the mean as well as the variance. Do the graphs give you that impression? Does the distribution approach some kind of symmetry? For what μ? (*AEM Ref.* p. 1081)

Pr.22.12 **(Hypergeometric distribution)** If a carton of 20 fuses contains 5 defectives and 3 fuses are randomly drawn from it without replacement, what are the probabilities of obtaining 0, 1, 2, 3 defective fuses? (*AEM Ref.* p. 1084 (#10))

Pr.22.13 **(Uniform distribution)** The uniform distribution on an interval $a \leq x \leq b$ has the density $f(x) = 1/(b - a)$ if $a < x < b$ and 0 otherwise. Find the mean and the variance on the computer by using the definitions. (*AEM Ref.* p. 1075)

Pr.22.14 **(Normal distribution)** Let X be normal with mean 105 and variance 25. Find $P(X < 112.5)$, $P(X > 100)$, $P(110.5 < X < 111.25)$ on the computer. (*AEM Ref.* p. 1090 (#2))

Pr.22.15 **(Normal distribution)** If sick-leave time X used by employees of some company in one month is (very roughly) normal with mean 1000 hours and standard deviation 100 hours, how much time t should be budgeted for sick leave during the next month if t is to be exceeded with a probability of only 20%? (*AEM Ref.* p. 1091 (#13))

Chapter 23

Mathematical Statistics

Content. Random numbers, samples (Ex. 23.1, Pr. 23.1)

Confidence intervals (Exs. 23.2-23.4, Prs. 23.2-23.5)

Tests (Exs. 23.5-23.10, Prs. 23.6-23.13)

Regression (Ex. 23.11, Prs. 23.14, 23.15)

The **binomial**, **Poisson**, **hypergeometric**, and **normal distributions** discussed in Chap. 22 of this Guide will occur again. For their Maple commands see Examples 22.3 and 22.4. Type `?distributions`. New distributions in this chapter are **Student's *t*-distribution** (see Example 23.3), the **chi-square distribution** (Example 23.4), and **Fisher's *F*-distribution** (Example 23.9; also known as **variance-ratio distribution**). These will be needed in connection with confidence intervals, tests, etc. Commands (in terms of typical examples) are as follows. (Type `?distributions`.)

```
> with(stats):
> statevalf[icdf, studentst[5]](0.95);          # Resp. 2.015048373
> statevalf[icdf, chisquare[15]](0.99);         # Resp. 30.57791417
> statevalf[icdf, fratio[7, 12]](0.95);         # Resp. 2.913358179
```

Explanations. `icdf` suggests 'inverse cumulative density function'. For Student's *t*-distribution with 5 df (**degrees of freedom**) the 95%-point is at 2.015.... That is, the distribution function F satisfies $F(2.015...) = 0.95$. For the chi-square distribution with 15 df the 99%-point is at 30.577... For Fisher's *F*-distribution with (7, 12) df the 95%-point is at 2.913...

For information on least squares regression curves type `?fit` or `?leastsquare`. For correlation type `?covariance`.

Examples for Chapter 23

EXAMPLE 23.1   RANDOM NUMBERS

Random numbers can be used for obtaining samples (this always means *random samples*, by definition) from populations. Type `?random`. Suppose you want to draw a sample of 10 items from a population of 80 items (screws, animals, humans or whatever). Number the items and type

```
> rn := rand(1..80):              # Random number generator
> seq(rn(j), j = 1..10);          # Resp. 42, 71, 18, 44, 17, 59, 6, 9, 2, 50
```

If you call the generator again, you will get another sample

```
> seq(rn(j), j = 1..10);          # Resp. 16, 44, 58, 1, 75, 6, 77, 22, 28, 70
```

and so on. If you just want a single random number (a single item), type

```
> rn();                                                      # Resp. 7
```

If you have reasons to obtain the same sample (or sequence of samples) with your generator, you can type _seed := i;, where i is some positive integer, and then a few samples, say,

```
> _seed := 14;                                # With an underbar in front!
```
$$\_seed :\!- 14$$

```
> seq(rn(j), j = 1..10);          # Resp. 70, 65, 43, 29, 62, 43, 71, 63, 23, 77
> seq(rn(j), j = 1..10);          # Resp. 4, 7, 40, 1, 8, 54, 66, 41, 78, 59
```

and if you now type _seed := 14:, you will get the same samples as before; thus,

```
> _seed := 14:                                # Don't forget the underbar in front!
> Sa1 := seq(rn(j), j = 1..10);
```
$$Sa1 := 70, 65, 43, 29, 62, 43, 71, 63, 23, 77$$

```
> Sa2 := seq(rn(j),  j = 1..10); # Resp. Sa2 := 4, 7, 40, 1, 8, 54, 66, 41, 78, 59
```

And so on. (What will you get if you type _seed := -3; or _seed := 0; ?)

Mean, variance, and standard deviation of samples are obtained as explained in Example 22.1 in this Guide. They vary from sample to sample.

```
> with(stats):
> evalf(describe[mean]([Sa1]), 5);                       # Resp. 54.600
> evalf(describe[mean]([Sa2]), 5);                       # Resp. 35.800
> evalf(10/9*describe[variance]([Sa1]), 5);              # Resp. 351.60
> evalf(10/9*describe[variance]([Sa2]), 5);              # Resp. 825.73
> evalf(sqrt(10/9)*describe[standarddeviation]([Sa1]), 5);# Resp. 18.751
> evalf(sqrt(10/9)*describe[standarddeviation]([Sa2]), 5);# Resp. 28.735
```

Similar Material in AEM: pp. 1053, 1105

EXAMPLE 23.2	**CONFIDENCE INTERVAL FOR THE MEAN OF THE NORMAL DISTRIBUTION WITH KNOWN VARIANCE**

Find a confidence interval for the mean of the normal distribution with known variance $\sigma^2 = 9$. Use a sample of 100 values with sample mean $\bar{x} = 5$. Choose the confidence level $\gamma = 95\%$.

Solution. Regard $\bar{x}$ as an observed value of a random variable $\bar{X} = (1/n)(X_1 + \ldots + X_n)$, where $X_1, \ldots, X_n$ are independent random variables all having the same distribution of some random variable X. It can be shown that if X is normal with mean μ and variance σ^2, then $\bar{X}$ is normal with mean μ and variance σ^2/n.

Type the given data, denoting the confidence level γ by g.

```
> with(stats):
> Digits := 5:   n:= 100:   xbar := 5:    var := 9:   g := 0.95:
```

Consider the standardized normal distribution `normald[0, 1]` (see Example 22.4). You obtain the shortest interval on the x-axis corresponding to $\gamma = 95\%$ of the area under the density curve (that is, corresponding to the probability 0.95) if you choose that interval symmetrically located with respect to the mean 0. Then its endpoints $-c$ and $+c$ correspond to the probabilities 2.5% and 97.5%. You get the latter by typing

```
> c := statevalf[icdf, normald[0, 1]](0.975);          # Resp. c := 1.9600
```

For $\bar{X}$ this corresponds to $\mu - k$ and $\mu + k$, where $k = c\sigma/\sqrt{n}$. Accordingly, type

```
> k := c*sqrt(var)/sqrt(n);          # Resp. k := .58800
```

You now get the confidence interval by replacing $\mu - k$ with $\bar{x} - k$ and $\mu + k$ with $\bar{x} + k$

```
> conf1 := xbar - k;          # Resp. conf1 := 4.4120
> conf2 := xbar + k;          # Resp. conf2 := 5.5880
```

The confidence interval is $CONF_{0.95}(4.41 \leq \mu \leq 5.59)$.

In this approach you used the standardized normal distribution, as it has been tabulated. On the computer you can proceed more directly by noting that the midpoint of the confidence interval is $\bar{x}$ and the endpoints are the 2.5%-point and the 97.5%-point of the distribution of $\bar{X}$, which is normal with variance $\sigma^2/n = 9/100$, hence with standard deviation $\sigma/\sqrt{n} = 0.3$. You thus obtain directly

```
> statevalf[icdf, normald[xbar, 0.3]](0.025);          # Resp. 4.4120
> statevalf[icdf, normald[xbar, 0.3]](0.975);          # Resp. 5.5880
```

Similar Material in AEM: p. 1110

EXAMPLE 23.3   CONFIDENCE INTERVAL FOR THE MEAN OF THE NORMAL DISTRIBUTION WITH UNKNOWN VARIANCE. t-DISTRIBUTION

Find a confidence interval for the mean of the normal distribution with unknown variance σ^2. Use the sample 144, 147, 146, 142, 144 (flashpoint in degrees Fahrenheit of a certain kind of Diesel oil). Choose the confidence level $\gamma = 99\%$.

Solution. Regard $\bar{x}$ as an observed value of a random variable $\bar{X}$, as in the previous example. $k = c\sigma/\sqrt{n}$ can no longer be used because σ is no longer known. The idea is to replace σ by the sample standard deviation s as defined in Example 22.1 in this Guide, and to regard s as an observed value of a random variable S. To use S, one must know its probability distribution. Student (pseudonym for W. S. Gosset) has shown that if the population random variable X is normal with mean μ and variance σ^2, then $t = \dfrac{\bar{x} - \mu}{s/\sqrt{n}}$ is an observed value of a random variable T which has a t-*distribution* with $n - 1$ **degrees of freedom (df)**, and he gave the formula for the probability density of T, which is symmetric with respect to 0. Accordingly, type the given data, and calculate $\bar{x}$ and s.

```
> sample := [144, 147, 146, 142, 144]:
```

```
> with(stats): Digits := 6:
> xbar := evalf(describe[mean](sample), 4);          # Resp. xbar := 144.6
> s := evalf(sqrt(5/4)*describe[standarddeviation](sample), 5);
```

$$s := 1.9494$$

Now obtain the 99.5%-point for the t-distribution with $n - 1 = 4$ degrees of freedom and calculate $K = Cs/\sqrt{5}$, the counterpart of k in the previous example.

```
> C := statevalf[icdf, studentst[4]](0.995);         # Resp. C := 4.60409
> K := evalf(C*s/sqrt(5), 4);                         # Resp. K := 4.014
```

You now get the confidence interval by replacing $\mu - K$ with $\bar{x} - K$ and $\mu + K$ with $\bar{x} + K$

```
> conf1 := xbar - K;                                  # Resp. conf1 := 140.586
> conf2 := xbar + K;                                  # Resp. conf2 := 148.614
```

This gives the confidence interval $CONF_{0.99}(140 \leq \mu \leq 149)$. This is rather large, but keep in mind that your sample was small. σ was unknown. If it were known and equal to s, you should get a shorter interval because you use more information. Can you calculate this?

Similar Material in AEM: p. 1113

EXAMPLE 23.4   CONFIDENCE INTERVAL FOR THE VARIANCE OF THE NORMAL DISTRIBUTION. χ^2-DISTRIBUTION

Find a confidence interval for the unknown variance σ^2 of the normal distribution, using the following sample and choosing the confidence level $\gamma = 95\%$. (The mean μ need not be known.)

$$89\ 84\ 87\ 81\ 89\ 86\ 91\ 90\ 78\ 89\ 87\ 99\ 83\ 89$$

Solution. It can be shown that under the normality assumption the quantity

$$y = (n - 1)s^2/\sigma^2$$

is an observed value of a random variable Y that has a **chi-square distribution** with $n - 1 = 13$ degrees of freedom. Here, s^2 is the sample variance, as before. Type the sample and then $(n - 1)\ s^2$, which in Maple becomes ns^2 because Maple has the factor $1/n$ in the definition of s^2, instead of the more common $1/(n - 1)$.

```
> with(stats):
> sample := [89, 84, 87, 81, 89, 86, 91, 90, 78, 89, 87, 99, 83, 89];
```

$$sample := [89, 84, 87, 81, 89, 86, 91, 90, 78, 89, 87, 99, 83, 89]$$

```
> nssquare := evalf(14*describe[variance](sample), 5);
```

$$nssquare := 326.86$$

Determine the 2.5%-point and the 97.5%-point of the chi-square distribution (which is not symmetric) with 13 degrees of freedom

```
> c1 := statevalf[icdf, chisquare[13]](0.025);  # Resp. c1 := 5.008750512
> c2 := statevalf[icdf, chisquare[13]](0.975);  # Resp. c2 := 24.73560488
```

From this you obtain the endpoints of the confidence interval

```
> conf1 := evalf(nssquare/c1, 4);               # Resp. conf1 := 65.26
> conf2 := evalf(nssquare/c2, 4);               # Resp. conf2 := 13.21
```

This gives the confidence interval $CONF_{0.95}(13.21 \le \sigma^2 \le 65.26)$.

Similar Material in AEM: pp. 1114-1116

EXAMPLE 23.5   TEST FOR THE MEAN OF THE NORMAL DISTRIBUTION

You want to buy 500 coils of wire. Test the manufacturer's claim that the breaking limit X of the wire is $\mu = \mu_0 = 200$ lb (or more). Assume that X has a normal distribution.

Solution. Test the **hypothesis** $\mu = \mu_0 = 200$ against the **alternative** $\mu = \mu_1 < 200$, an undesirable weakness. Hence this test is **left-sided**, the **rejection region** extends from a **critical point** c to the left.

To obtain a sample, select some of the coils, say, 25, at random. Cut a piece from each coil and determine the breaking limit experimentally. Suppose that this sample of $n = 25$ values has the mean $\bar{x} = 197$ lb (somewhat less than the claim!) and the standard deviation $s = 6$ lb. Then (as in Example 23.3 in this Guide)

$$t = \frac{\bar{x} - \mu_0}{s/\sqrt{n}} = \frac{197 - 200}{6/\sqrt{25}} = -2.50$$

is an observed value of a random variable T that has a t–distribution with $n - 1 = 24$ degrees of freedom. Choose a **significance level** of the test, say $\alpha = 5\%$. Find the critical c by typing (similarly as in Example 23.3 in this Guide)

```
> with(stats):   Digits := 5:
> c := statevalf[icdf, studentst[24]](0.05);   # Resp. c := -1.7109
```

Reason as follows. If the hypothesis is true, the probability of obtaining a $t < c$ is very small, namely, equal to $\alpha = 5\%$, so that it would happen only about once in 20 tests. Hence if it happens, as in the present case, where $-2.5 < -1.7109$, cast doubt on the truth of the hypothesis. Hence **reject the hypothesis** and assert that $\mu < 200$ and the manufacturer had promised too much.

Similar Material in AEM: pp. 1118, 1119

EXAMPLE 23.6   TEST FOR THE MEAN: POWER FUNCTION

You make an **error of the first kind** if you reject a hypothesis although it is true. You do this with probability α, the significance level of the test. You make an **error of the second kind** if you accept a hypothesis although the alternative is true. The corresponding probability is denoted by β. The quantity $\eta = 1 - \beta$ (thus the probability of avoiding an error of the second kind) is called the **power** of the test. β depends on the alternative μ. One calls $\beta(\mu)$ the **operating characteristic (OC)** and $\eta(\mu) = 1 - \beta(\mu)$ the **power function** of the test.

Let X be normal with mean μ and known variance $\sigma^2 = 9$. Then $\bar{X}$ is normal with mean μ and variance σ^2/n, where n is the size of the sample used in the test (see Example 23.2 in this Guide). Let the hypothesis be $\mu_0 = 24$. Choose $\alpha = 5\%$. Let $n = 10$. Then the standard deviation of $\bar{X}$ is $sd = 3/\sqrt{10}$.

1. Left-sided test (as in the previous example). The **critical region** (rejection region) extends from the critical $c = c_1$ to the left. Obtain the critical c_1 by typing

```
> with(stats):  Digits := 5:    sd := 3/sqrt(10):
> c1 := statevalf[icdf, normald[24, sd]](0.05);        # Resp. c1 := 22.440
```

The power is the area under the density curve of $\bar{X}$ with the alternative μ being true, from $-\infty$ to c_1. This curve, and hence the area, depends on μ. The power is practically 1 at $\mu = 20$, decreases monotone to 0.05 at $\mu = 24$ and practically to 0 at $\mu = 26$; see the figure.

```
> powerleft := statevalf[cdf, normald[mu, sd]](c1):
```

2. Right-sided test. The critical region now extends from the critical $c = c_2$ to the right. Obtain the critical $c = c_2$ by typing

```
> c2 := statevalf[icdf, normald[24, sd]](0.95);        # Resp. c2 := 25.560
```

The power is the area under the density curve of $\bar{X}$ with the alternative μ being true, from c_2 to ∞, and again depends on μ.

```
> powerright := 1 - statevalf[cdf, normald[mu, sd]](c2):
```

3. Two-sided test. The critical region now consists of two parts, from $-\infty$ to a lower critical point $c = c_{3a}$ and from an upper critical point $c = c_{3b}$ to ∞. These are the 2.5%- and 97.5%-points of the distribution of $\bar{X}$ with the hypothesis $\mu_0 = 24$ being true. Obtain these points by typing

```
> c3a := statevalf[icdf, normald[24, sd]](0.025);    # Resp. c3a := 22.141
> c3b := statevalf[icdf, normald[24, sd]](0.975);    # Resp. c3b := 25.859
```

The power is the area under the density curve of $\bar{X}$ with the alternative being true, from $-\infty$ to c_{3a} and from c_{3b} to ∞. This is the U-shaped curve in the figure.

```
> powertwosided := statevalf[cdf, normald[mu, sd]](c3a) + 1 -
  statevalf[cdf, normald[mu, sd]](c3b):
```

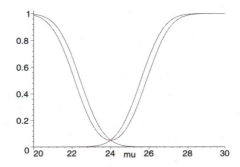

Example 23.6. Power functions of tests of $\mu_0 = 24$
against $\mu < 24$, $\mu > 24$, and $\mu \neq 24$

Plotting all three curves on common axes is now quite simple.

```
> P1 := plot(powerleft, mu = 20..30):
> P2 := plot(powerright, mu = 20..30):
> P3 := plot(powertwosided, mu = 20..30):
> with(plots):                               # Ignore the warning.
> display(P1, P2, P3);
```

Similar Material in AEM: pp. 1122-1124

EXAMPLE 23.7 | **TEST FOR THE VARIANCE OF THE NORMAL DISTRIBUTION**

Using a sample of size $n = 15$ and sample variance $s^2 = 13$ from a normal population, test the hypothesis $\sigma^2 = \sigma_0{}^2 = 10$ against the alternative $\sigma^2 = \sigma_1{}^2 = 20$.

Solution. It can be shown that under the normality assumption and under the hypothesis the quantity

$$y = (n-1)s^2/\sigma_0{}^2 = 14 \times 13/10 = 18.2$$

is an observed value of a random variable $Y = 14\, S^2/10$ that has a **chi-square distribution** with $n - 1 = 14$ degrees of freedom. (See also Example 23.4 in this Guide.) Since the alternative is greater than the hypothesis, the test is right-sided. Choose a significance level, say, $\alpha = 5\%$. Determine the 95%-point c of that distribution from

```
> with(stats):  Digits := 5:
> c := statevalf[icdf, chisquare[14]](0.95);        # Resp. c := 23.685
```

Since $y < c$ (and the test is right-sided), accept the hypothesis.

     If the alternative is true, $Y_1 = (n-1)S^2/\sigma_1{}^2 = 14\, S^2/20 = Y/2$ has a chi-square distribution with 14 degrees of freedom. Hence the **power** is the area under the density curve of Y_1 from $c/2$ to ∞. That is,

```
> power := 1 - statevalf[cdf, chisquare[14]](c/2) # Resp. power := .61899
```

This leaves a probability of 38% for committing an error of the second kind. This is too large, and you should repeat the test with a larger sample (if available!).

     Similar Material in AEM: pp. 1124, 1125

EXAMPLE 23.8 | **COMPARISON OF MEANS**

Test the hypothesis that two normal distributions with the same variance have the same mean, $\mu_1 = \mu_2$, against the alternative that they have different means. Choose the significance level $\alpha = 5\%$. Use two samples $x_1, x_2, ..., x_{n_1}$ and $y_1, y_2, ..., y_{n_2}$ which are independent. (Dependence would mean that some x-values are related to some y-values, for instance, if they came from the two front tires of the same car, from test scores of the same student, etc. Equality of variances will be tested in the next example.)

$$110 \quad 113 \quad 90 \quad 108 \quad 107 \quad 106 \quad 112$$
$$89 \quad 92 \quad 84 \quad 97 \quad 103 \quad 107 \quad 111 \quad 97$$

Solution. Calculate the mean $\bar{x}$ and the variance $s_x{}^2$ of the first sample by typing

```
> Sa1 := [110, 113, 90, 108, 107, 106, 112]:
> with(stats):    Digits := 5:
> xbar := evalf(describe[mean](Sa1));              # Resp. xbar := 106.57
> xvar := evalf(7/6*describe[variance](Sa1));      # Resp. xvar := 59.952
```

Calculate the mean $\bar{y}$ and the variance $s_y{}^2$ of the second sample by typing

```
> Sa2 := [89, 92, 84, 97, 103, 107, 111, 97]:
> ybar := evalf(describe[mean](Sa2));              # Resp. ybar := 97.500
> yvar := evalf(8/7*describe[variance](Sa1));      # Resp. yvar := 58.729
```

It can be shown that if the hypothesis is true, the quantity

$$t_0 = \sqrt{\frac{n_1 n_2 (n_1 + n_2 - 2)}{n_1 + n_2}} \frac{\bar{x} - \bar{y}}{\sqrt{(n_1 - 1) s_x{}^2 + (n_2 - 1) s_y{}^2}}$$

is an observed value of a random variable T that has a t-distribution with $n_1 + n_2 - 2$ degrees of freedom. The test is two-sided. Since $\alpha = 5\%$, determine the 2.5%-point c_1 and the 97.5%-point c_2 of the t-distribution with $n_1 + n_2 - 2 = 8 + 7 - 2 = 13$ degrees of freedom by typing

```
> statevalf[icdf,studentst[13]](0.025);            # Resp. -2.1604
> statevalf[icdf,studentst[13]](0.975);            # Resp. 2.1604
```

If t_0 lies between these values (inclusively), accept the hypothesis. Otherwise reject it. Calculate

```
> t0 := evalf(sqrt(8*7*(8+7-2)/(8+7))*(xbar-ybar)/sqrt((7*xvar+6*yvar)));
```

$$t0 := 2.2740$$

Reject the hypothesis and assert that the populations from which the samples were drawn have different means.

Similar Material in AEM: pp. 1125, 1126

EXAMPLE 23.9 **COMPARISON OF VARIANCES. *F*-DISTRIBUTION**

Test the hypothesis that the variances of the two normal distributions in the previous example are equal against the alternative that they are different. Choose the significance level 5%.

Solution. Type the samples and calculate their variances.

```
> with(stats):       Digits := 5:
> Sa1 := [110, 113, 90, 108, 107, 106, 112];
> xvar := evalf(7/6*describe[variance](Sa1));      # Resp. xvar := 59.952
> Sa2 := [89, 92, 84, 97, 103, 107, 111, 97];
> yvar := evalf(8/7*describe[variance](Sa1));      # Resp. yvar := 58.729
```

It can be shown that if the hypothesis is true, the ratio $v_0 = s_x{}^2 / s_y{}^2$ is an observed value of a random variable V which has an **F-distribution** with $(n_1 - 1, \ n_2 - 1) = (6, 7)$ degrees of freedom. Since $\alpha = 5\%$ and the test is two-sided, determine the 2.5%-point c_1 and the 97.5%-point c_2 of this distribution by typing

```
> c1 := statevalf[icdf, fratio[6, 7]](0.025);        # Resp. c1 := .17558
> c2 := statevalf[icdf, fratio[6, 7]](0.975);        # Resp. c2 := 5.1186
```

Accept the hypothesis because v_0 lies between c_1 and c_2. Indeed,

```
> v0 := evalf(xvar/yvar);                            # Resp. v0 := 1.0208
```

Similar Material in AEM: p. 1126

EXAMPLE 23.10   **CHI-SQUARE TEST FOR GOODNESS OF FIT**

With this test you find out how well the distribution of a sample fits the hypothetical distribution of the population. For instance, can you claim on the 5%-level that a die is fair if in 20,000 trials you obtain $x = 1, 2, ..., 6$ with the following absolute frequencies (actual classical data obtained by R. Wolf in Switzerland).

$$3407 \quad 3631 \quad 3176 \quad 2916 \quad 3448 \quad 3422$$

Solution. If the die is fair, each of the 6 values is equally likely, hence the expected absolute frequency is $e = 3333.33$. For each x calculate the observed value minus e, square it and divide the result by e. The sum of the 6 numbers thus obtained is an observed value of a random variable χ which is (asymptotically) chi-square distributed with $6 - 1 = 5$ degrees of freedom. Call this sum χ_0. Clearly, it measures the discrepancy between observations and expectation. Accordingly, type

```
> with(stats):   Digits := 6:
> data := [3407, 3631, 3176, 2916, 3448, 3422]:
> e := evalf(20000/6);                               # Resp. e := 3333.33
> chi0 := sum((data[j] - e)^2/e, j = 1..6);          # Resp. χ0 := 94.1890
```

The test is right-sided. The rejection region (critical region) extends from the critical c to the right and corresponds to a probability of 5%. Hence c is the 95%-point of the chi-square distribution with 5 degrees of freedom,

```
> c := statevalf[icdf, chisquare[5]](0.95);          # Resp. c := 11.0705
```

You see that χ_0 is much larger than c. Reject the hypothesis and assert that Wolf's die was not fair or there were flaws in throwing and/or counting.

Similar Material in AEM: p. 1140 (#4)

EXAMPLE 23.11   **REGRESSION**

In regression analysis you choose values $x_1, x_2, ..., x_n$ of an ordinary variable (for instance, x may be time) and observe corresponding values $y_1, y_2, ..., y_n$ of a random variable Y (for instance, temperature at some place). This gives a sample of n pairs $(x_1, y_1), ..., (x_n, y_n)$. You assume that the mean of Y depends linearly on x, say, $\mu(x) = \kappa_0 + \kappa_1 x$. This is the **regression line of the population.** Let the sample be (x = pressure in atmospheres, y = decrease of volume of leather in %)

x	4000	6000	8000	10000
y	2.3	4.1	5.7	6.9

From the sample you obtain the **sample regression line** $y = k_0 + k_1 x$. You fit this line through the given points (given pairs of coordinate values in the xy-plane) by the **least squares principle**, as follows.

For information type `?fit` or `?leastsquare`. Then type

```
> xSa := [4000, 6000, 8000, 10000]:          # (the x-values)
> ySa := [2.3, 4.1, 5.7, 6.9]:               # (the y-values)
> with(stats):    Digits := 5:
> line := fit[leastsquare[[x, y]]]([xSa, ySa]);
```

$$line := y = -.64000 + .00077000\,x$$

This is the sample regression line. Obtain the figure by the subsequent commands. From the figure you see that for $x = 4000 \ldots 10000$ the line fits the data reasonably well (but would be useless near $x = 0$ because there is no decrease of volume when the pressure is 0).

```
> P1 := plot(rhs(line), x = 0..10000):
> S := seq([xSa[j], ySa[j]], j = 1..4);
```

$$S := [4000, 2.3],\ [6000, 4.1],\ [8000, 5.7],\ [10000, 6.9]$$

```
> P2 := plot([S], style = point):
> with(plots):                               # Ignore the warning.
> display({P1, P2}, labels = [x, y]);
```

Example 23.11. Sample regression line and given data (four points)

The least squares principle is purely geometric. Probability enters through random variables. To obtain confidence intervals or tests for the **regression coefficient** κ_1, you must make assumptions about the probability distribution of Y. To apply the theory on the normal distribution (as in the previous examples), make the reasonable assumption that Y is normal and its variance σ^2 is the same for all x.

For obtaining a confidence interval for κ_1 you will need the variance of the x-values, the variance of the y-values, and the covariance. Accordingly, type

```
> n := 4:                    # Sample size (= number of pairs = number of points)
> xvar := evalf(n/(n - 1)*describe[variance](xSa));
```

$$xvar := .66667\,10^7$$

```
> yvar := evalf(n/(n-1)*describe[variance](ySa));   # Resp. yvar := 3.9833
> xycov := evalf(n/(n - 1)*describe[covariance](xSa, ySa));
```

$$xycov := 5133.3$$

Then type the formula for the sample regression coefficient k_1 and for an auxiliary quantity q_0.

```
> k1 := xycov/xvar;                    # Resp. k1 := .00076999
> q0 := (n - 1)*(yvar - k1^2*xvar);    # Resp. q0 := .0921
```

It can be shown that under your assumptions the random variable Y has a Student's t-distribution with $n - 2 = 2$ degrees of freedom. Choose a confidence level, say, 95%. Then type the commands for the 2.5%-point and the 97.5%-point of that distribution, actually, only the latter, c, because the former is $-c$, by the symmetry of the distribution.

```
> c := statevalf[icdf, studentst[n - 2]](0.975);    # Resp. c := 4.3027
```

Half the length of the confidence interval is

```
> k := c*sqrt(q0/((n - 2)*(n - 1)*xvar));    # Resp. k := .00020646
```

With this you obtain the confidence interval $CONF_{0.95}(0.00056 \leq \kappa_1 \leq 0.00098)$ because its endpoints are

```
> conf1 := k1 - k;    # Resp. conf1 := .00056353
> conf2 := k1 + k;    # Resp. conf2 := .00097645
```

Similar Material in AEM: pp. 1145-1149

Problem Set for Chapter 23

Pr.23.1 (Experiment on sample mean) Find out experimentally how sample means vary from sample to sample. Plot a histogram of their frequency function. *Suggestion:* 20-100 samples of size 5 obtained by the random generator from a population of size 50. Use _seed; , so that you can reproduce your samples if needed. (*AEM Ref.* p. 1105)

Pr.23.2 (Confidence interval for the mean) Find a 99% confidence interval for the mean of a normal population with standard deviation 2.5, using the sample 30.8, 30.0, 29.9, 30.1, 31.7, 34.0. (*AEM Ref.* p. 1117 (#3))

Pr.23.3 (Length of confidence interval) Plot the length of a 95% confidence interval as a function of sample size n and measured in multiples of σ, for the mean of the normal distribution with known variance (*AEM Ref.* p. 1112)

Pr.23.4 (Confidence interval for the mean) What confidence interval would you obtain in Example 23.3 in this Guide if σ were known and equal to $s = 1.9494$ (the value in that example), the other data being as before?

Pr.23.5 (Confidence interval for the variance) Find a 95% confidence interval for the variance of the normal distribution, using the sample of carbon monoxide emission (grams/mile) of a passenger car cruising at a speed of 55 mph. 17.3, 17.8, 18.0, 17.7 18.2, 17.4, 17.6, 18.1. (*AEM Ref.* p. 1117 (#16))

Pr.23.6 (Test for the mean) Test the hypothesis $\mu_0 = 24$ against the alternative $\mu_1 = 27$, choosing $\alpha = 5\%$ and using a sample of size 10 with mean 25.8 from a normal population with variance 9. Is the power of the test sufficiently large? (*AEM Ref.* pp. 1122-1124)

Pr.23.7 **(Test for the mean)** If a standard treatment cures about 75% of patients suffering from a certain disease, and a new treatment cured 310 of the first 400 patients on whom it was tried, can you conclude that the new treatment is better? First guess. Then calculate, choosing $\alpha = 5\%$ and using the fact that $X = $ *Number of cases cured in 400 cases* is about normal with mean np and variance $np(1-p)$. (*AEM Ref.* p. 1127 (#12))

Pr.23.8 **(Dependence of power on sample size)** How does the figure in Example 23.6 in this Guide change if you take a larger sample (of size $n = 100$, for instance)? Give the reason. Plot a new figure for $n = 100$. (*AEM Ref.* pp. 1123, 1124)

Pr.23.9 **(Test for the variance)** Suppose that in the past the standard deviation of weights of certain 100.0-oz packages filled by a machine was 0.8 oz. Test the hypothesis $H_0 : \sigma = 0.8$ against the alternative $H_1 : \sigma > 0.8$ (an undesirable increase), using a sample of 20 packages with standard deviation 1.0 oz, assuming normality and choosing $\alpha = 5\%$. (*AEM Ref.* p. 1127 (#15))

Pr.23.10 **(Comparison of means)** Will an increase of temperature increase the yield (measured in grams/min) of some chemical process? Test this, using the following independent samples, assuming normality, and choosing $\alpha = 5\%$. (*AEM Ref.* p. 1125)

Yield x at 40°C   116 123 121 105 138 135 119 111 115 125

Yield y at 65°C   130 125 134 112 145 137 122 139

Pr.23.11 **(Paired comparison of means)** Measure the electric voltage in a circuit at the same instants simultaneously by two kinds of voltmeters. Test the hypothesis that there is no difference in the calibration of the two kinds of instruments against the alternative that there is a difference. In this case you merely need a sample of differences of corresponding measurements ("**paired comparison**"), say, 0.4, -0.6, 0.2, 0.0, 1.0, 1.4, 0.4, 1.6. Assume normality and choose $\alpha = 5\%$. (*AEM Ref.* p. 1127 (#13))

Pr.23.12 **(Comparison of variances)** Test that the variances of the populations in Pr.23.10 are equal against the alternative that they are different. Choose $\alpha = 5\%$. (*AEM Ref.* p. 1126)

Pr.23.13 **(Goodness of fit)** Can you assert that the traffic on the three lanes of an expressway (in one direction) is about the same on each lane if a count gives 920, 870, 750 cars on the right, middle, and left lanes, respectively, during the same interval of time? (*AEM Ref.* p. 1141 (#13))

Pr.23.14 **(Linear regression)** If a sample of 9 pairs of values $[x_j, y_j]$ has the variance of the x-values 118.000, the variance of the y-values 215.125, and the covariance -155.750, what can you say about the sample regression line? What about a 95% confidence interval for the regression coefficient (the slope) κ_1 if you assume Y to be normal with variance independent of x? (*AEM Ref.* pp. 1145-1149)

Pr.23.15 **(Quadratic regression parabola)** Fit a quadratic parabola through the following data. Plot the curve and the given data (as points) on common axes. Type `?fit` for information. (*AEM Ref.* None)

x	1	2	4	5	7	8
y	7	5	2	1	2	4

Appendix 1

References

[1] Abramowitz, M. and I. A. Stegun (eds.), *Handbook of Mathematical Functions.* 10th printing, with corrections. Washington, DC: National Bureau of Standards, 1972. (Also New York: Dover, 1965.)

[2] Kreyszig, E., *Advanced Engineering Mathematics.* 8th ed. New York: Wiley, 1999.

[3] Kreyszig, H., and E. Kreyszig, *Student Solutions Manual.to Accompany Advanced Engineering Mathematics*, Eighth Edition. New York: Wiley, 2000.

[4] Maple, *Booklist.* Waterloo, ON: Waterloo Maple. (2000). [341 titles. Not error-free.]

[5] Monagan, M. B., et al., *Maple 6 Programming Guide.* Waterloo, ON: Waterloo Maple, 2000.

[6] Wilkinson, J. H., The Algebraic Eigenvalue Problem. Oxford: Clarendon, 1988.

Appendix 2

Answers to Odd-Numbered Problems

CHAPTER 1,   page 14

Pr.1.1.
```
> with(DEtools):
> ode := diff(y(x), x) = y(x)^2;
```
 # Resp. $ode := \dfrac{\partial}{\partial x}\, \mathrm{y}(x) = \mathrm{y}(x)^2$
```
> inits :=  {[0, 1], [0, 2], [0, 3]};
```
 # Resp. $inits := \{[0, 1], [0, 2], [0, 3]\}$
```
> DEplot(ode, [y(x)], x = -3..3, y = 0..3, inits);
```

Pr.1.3.
```
> sol := dsolve({diff(y(t), t) = exp (0.2*t), y(0) = 2});
> plot(rhs(sol), t = 0..10);
```

Pr.1.5.
```
> ode := diff(y(t), t) = k*y(t);
> yp := dsolve({ode, y(0) = 4});
```
 # Resp. $yp := \mathrm{y}(t) = 4\,\mathrm{e}^{(k\,t)}$
```
> eq2 := subs(t = 1, y(1) = 2, yp);
> k0 := solve(eq2, k);
```
 # Resp. $k0 := -\ln(2)$
```
> yp2 := subs(k = k0, yp);
```
 # Resp. $yp2 := \mathrm{y}(t) = 4\,\mathrm{e}^{(-\ln(2)\,t)}$
```
> yp2 := simplify(%);
```
 # Resp. $yp2 := \mathrm{y}(t) = 4\,2^{(-t)}$

Pr.1.7.
```
> ode := diff(y(x), x) = 1 + y(x)^2;
> y1 := tan(x);
> subs(y(x) = y1, ode);
> eval(%);
```
 # Resp. $1 + \tan(x)^2 = 1 + \tan(x)^2$ # % avoids retyping.

Pr.1.9.
```
> M := x^3 + 3*x*y^2;
> N := 3*x^2*y + y^3;
> diff(M, y) - diff(N, x);
```
 # Resp. 0 # The equation is exact.
```
> u1 := int(M, x);
> u2 := int(N, y);
```

Hence an implicit solution is $x^4 + 6\,x^2\,y^2 + y^4 = c$.

Pr.1.11.
```
> ode := diff(y(x), x) = 2*(y(x) - 1)*tan(2*x);
> sol := dsolve(ode);
> eval(subs(x = 0, sol) = 4);
```
 # Hence $\_C1 = 3$.
```
> yp := subs(_C1 = 3, sol);
```

$$yp := \mathrm{y}(x) = 1 + 3\sqrt{1 + \tan(2x)^2}$$
```
> plot(rhs(yp), x = 0..0.7);
```

Pr.1.13. The answer is $y = (x + c)e^{-kx}$. To obtain it from the integral formula, type

```
> h := int(k, x);
> y := exp(-h)*(int(exp(h)*exp(-k*x), x) + c);
```

dsolve gives the answer if you type

```
> y := 'y':                          # Unassign y, which has just been used.
> sol := dsolve(diff(y(x), x) + k*y(x) = exp(-k*x));
```

$$sol := y(x) = (x + \_C1)\, e^{(-k\,x)}$$

Pr.1.15.
```
> ode := diff(y(x), x) - 3*y(x) = -5*y(x)^2;
```

```
> sol := dsolve(ode);                # Resp. sol := y(x) = 3\frac{1}{5 + 3e^{(-3x)}\_C1}
```

```
> y1 := subs(_C1 = 1, sol);
> y2 := subs(_C1 = 0, sol);
> y3 := subs(_C1 = -2/3, sol);
> plot({rhs(y1), rhs(y2), rhs(y3)}, x = 0..1,
    labels = [Time, Population]);
```

Pr.1.17. Type the ODE, solve it by dsolve, then substitute the special values for which you are supposed to plot the solution. Note that the time interval of the transient current is practically very short, due to the rapid decrease of the exponential term.

```
> ode := R*diff(i(t), t) + i(t)/C = diff(E[0]*sin(omega*t), t);
> sol := dsolve(ode);
> ipartic := subs(R = 1, C = 1, omega = 1, E[0] = 220, _C1 = -110, sol);
```

$$ipartic := i(t) = -110\, e^{(-t)} + 110\,\cos(t) + 110\,\sin(t)$$

```
> with(plots):
> plot(rhs(ipartic), t = 0..20, ytickmarks = [0, 50, -50, 100, -100,
    150, -150]);
```

Pr.1.19.
```
> y0 := 0:
> pic(1) := y0 + int(1 + (subs(x = t, y0))^2, t = 0..x);
```

$$pic(1) := x$$

```
> pic(2) := y0 + int(1 + (subs(x = t, pic(1)))^2, t = 0..x);
```

$$pic(2) := x + \frac{1}{3}\, x^3$$

```
> pic(3) := y0 + int(1 + (subs(x = t, pic(2)))^2, t = 0..x);
```

$$pic(3) := \frac{1}{63}\, x^7 + \frac{2}{15}\, x^5 + \frac{1}{3}\, x^3 + x$$

CHAPTER 2,   page 28

Pr.2.1.
```
> ode := diff(y(x), x, x) + 3*diff(y(x), x) + 2*y(x) = 0;
> sol := dsolve(ode);            # Resp. sol := y(x) = _C1 e^(-2x) + _C2 e^(-x)
> yp := dsolve({ode, y(0) = 4, D(y)(0) = -2});
```

$$yp := \mathrm{y}(x) = 6\,\mathrm{e}^{(-x)} - 2\,\mathrm{e}^{(-2\,x)}$$

```
> plot(rhs(yp), x = 0..5);
```

Pr.2.5.   Substitute

```
> y := (c1 + c2*x + c3*x^2)*exp(-x);
```

into the left-hand side of the given ODE

```
> diff(y, x, x, x) + 3*diff(y, x, x) + 3*diff(y, x) + y;      # Resp. 0
```

Pr.2.7.   Note that the terms in y1, y2, y3 may come out in different order.

```
> y := 'y':
> ode := diff(y(t), t, t) + 10*diff(y(t), t) +  16*y(t) = 0;
> sol := dsolve(ode);
> y1 := dsolve({ode, y(0) = 1, D(y)(0) = -1});
```

$$y1 := \mathrm{y}(t) = -\frac{1}{6}\,\mathrm{e}^{(-8\,t)} + \frac{7}{6}\,\mathrm{e}^{(-2\,t)}$$

```
> y2 := dsolve({ode, y(0) = 1, D(y)(0) = 0});
```

$$y2 := \mathrm{y}(t) = -\frac{1}{3}\,\mathrm{e}^{(-8\,t)} + \frac{4}{3}\,\mathrm{e}^{(-2\,t)}$$

```
> y3 := dsolve({ode, y(0) = 1, D(y)(0) = 1});
```

$$y3 := \mathrm{y}(t) = \frac{3}{2}\,\mathrm{e}^{(-2\,t)} - \frac{1}{2}\,\mathrm{e}^{(-8\,t)}$$

```
> plot({rhs(y1), rhs(y2), rhs(y3)}, t = 0..2, title = 'Overdamping');
```

Pr.2.9. $L = 1$ m $= 100$ cm, $g = 981 \mathrm{cm}/\sec^2$,

```
> L := 100:   g := 981:
> ode := m*diff(T(t), t, t) + m*g/L*T(t) = 0;
> evalf(dsolve({ode, T(0) = 0, D(T)(0) = 1}), 4);
```

$$\mathrm{T}(t) = .3193\,\sin(3.132\,t)$$

Since 3.132 equals about π, the pendulum ticks about once per second.

Pr.2.11.
```
> ode := diff(y(x), x, x) + y(x) = 0;
> dsolve({ode, y(0) = -3, y(Pi) = -3});
```

Show that a solution with arbitrary constant c is

```
> y1 := 3*cos(x) + c*sin(x);
```

You see that $y_1(0) = 0$ and the right boundary condition is satisfied, too:

```
> evalf(subs(x = Pi, y1));                    # Resp. -3. - .4102067615 10^{-9} c
> subs(y(x) = y1, ode);
```

$$\left(\frac{\partial^2}{\partial x^2} \left(3\cos(x) + c\sin(x)\right) \right) + 3\cos(x) + c\sin(x) = 0$$

```
> eval(%);                                    # Resp. 0 = 0
```

Pr.2.13. Substitute a general solution into the ODE and determine a and b.

```
> y(x) := x*(c1*cos(ln(x)) + c2*sin(ln(x)));
> ode := x^2*diff(y(x), x, x) + a*x*diff(y(x), x) + b*y(x)= 0;
```

The (rather long) response must hold for any x, say, for $x = 1$, and for any c_1 and c_2. To get two equations for a and b, take $c_1 = 0$, $c_2 = 1$ and then $c_1 = 1$, $c_2 = 0$.

```
> eq1 := eval(subs(c1 = 0, c2 = 1, x = 1, ode));
```

$$eq1 := 1 + a = 0$$

```
> eq2 := eval(subs(c1 = 1, c2 = 0, x = 1, ode));
```

$$eq2 := -1 + a + b = 0$$

```
> solve({eq1, eq2},  {a, b});                 # Resp. {a = -1, b = 2}
```

Hence the ODE is

```
> y := 'y':                                   # This makes y general again.
> ode2 := x^2*diff(y(x), x, x) - x*diff(y(x), x) + 2*y(x) = 0;
> dsolve(ode2);                               # Check by solving.
```

Pr.2.15.
```
> ode := diff(y(t), t, t) + 4*y(t) = -12*sin(2*t);
> dsolve({ode, y(0) = 1, D(y)(0) = 3});
> yp := simplify(%);                # Resp. yp := y(t) = cos(2t)(3t + 1)
> plot(rhs(yp), t = 0..10);
```

Pr.2.17.
```
> ode := diff(y(t), t, t) + 100*y(t) = 36*cos(8*t);
> sol := dsolve({ode, y(0) = 0, D(y)(0) = 0});
```

$$sol := y(t) = \cos(8t) - \cos(10t)$$

This solution can be written as $2\sin 9t \sin t$. This explains the occurrence of beats.

```
> sol := combine(2*sin(9*t)*sin(t));          # Resp. sol := cos(8t) - cos(10t)
> plot(sol, t = 0..30);
```

Pr.2.19. Solve the homogeneous ODE to obtain a basis of solutions needed for calculating the Wronskian W.

```
> y := 'y':
> ode := diff(y(x), x, x) - 2*diff(y(x), x) + y(x) = 0;
> dsolve(%);                                    # Resp. y(x) = _C1 e^x + _C2 x e^x
> y1 := exp(x):      y2 := x*exp(x):
> W := y1*diff(y2, x) - y2*diff(y1, x);
> W := simplify(%);                                      # Resp. W := e^{(2x)}
> r := 3*x^(3/2)*exp(x);                                 # Resp. r := 3 x^{(3/2)} e^x
> yp := -y1*int(y2*r/W, x) + y2*int(y1*r/W, x);   # Resp. yp := (12/35) e^x x^{(7/2)}
```

CHAPTER 3,   page 40

Pr.3.1. $p = 3$, $q = 2$, $\delta = 1$ gives a node. _C1 and _C2 may come out interchanged.

```
> sys := D(y1)(t) = y1(t), D(y2)(t) = 2*y2(t);
> dsolve({sys});                    # Resp. {y2(t) = _C2 e^{(2t)}, y1(t) = _C1 e^t}
> with(DEtools):
> DEplot([sys[1], sys[2]], [y1(t), y2(t)], t = 0..10, y1 = -6..6,
  y2 = -6..6);
```

Pr.3.3. $y_1' = y_2$, $y_2' = -9\,y_1$,

```
> y1 := 'y1':      y2 := 'y2':
> sys := D(y1)(t) = y2(t), D(y2)(t) = -9*y1(t);
> dsolve({sys});                      # Terms may appear in a different order.
```

$$\{y2(t) = -3\,\_C1\,\sin(3\,t) + 3\,\_C2\,\cos(3\,t),\ y1(t) = \_C1\,\cos(3\,t) + \_C2\,\sin(3\,t)\}$$

```
> with(DEtools):
> DEplot({sys[1], sys[2]}, [y1(t), y2(t)], t = 0..10, y1 = -4..4,
  y2 = -4..4, scaling = constrained);
```

Pr.3.5. Type the system, then the matrix and its eigenvalues and eigenvectors – there is only one – and finally apply dsolve to confirm the form of a general solution derived in AEM.

```
> sys := D(y1)(t) = 4*y1(t) + y2(t), D(y2)(t) = -y1(t) + 2*y2(t);
> with(linalg):                              # Ignore the warning.
> A := matrix([[4, 1], [-1, 2]]);
> eigenvectors(A);                           # Resp. [3, 2, {[-1, 1]}]
> dsolve({sys});                    # Terms may appear in a different order.
```

$$\{y2(t) = -e^{(3t)}\,(\_C1 + \_C2\,t - \_C2),\ y1(t) = e^{(3t)}\,(\_C1 + \_C2\,t)\}$$

Pr.3.7.   Type the ODE as a system,

```
> sys := D(y1)(t) = y2(t),  D(y2)(t) = -y1(t)/9;
> with(DEtools):
> DEplot({sys[1], sys[2]}, [y1(t), y2(t)], t = 0..5, y1 = -4..4,
   y2 = -4..4);                                          # Ellipses
```

Pr.3.9.   Type the equations as given, with the second equation differentiated,

```
> sys := D(i1)(t) + 4*(i1(t) - i2(t)) = 12,
>         6*D(i2)(t) + 4*(D(i2)(t) - D(i1)(t)) + 4*i2(t) = 0;
> sol := dsolve({sys, i1(0) = 0, i2(0) = 0});
```

$$sol := \{\, i2(t) = -4\,e^{(-2\,t)} + 4\,e^{\left(-\frac{4}{5}\,t\right)}, \; i1(t) = -8\,e^{(-2\,t)} + 5\,e^{\left(-\frac{4}{5}\,t\right)} + 3 \}$$

```
> plot({rhs(sol[1]), rhs(sol[2])}, t = 0..10);
```

Pr.3.11.   The plot shows that it is a spiral. For plotting choose initial conditions, e.g.,
$y_1(0) = 1$, $y_2(0) = 0$.

```
> sys := D(y1)(t) = y2(t), D(y2)(t) = -2*y1(t) - 2*y2(t);
> sol := dsolve({sys, y1(0) = 1, y2(0) = 0});
> plot([rhs(sol[1]), rhs(sol[2]), t = -5..10]);
```

Pr.3.13.   Type the equation as a system, choose initial conditions, say, three, for trajectories.

```
> sys := D(y1)(t) = y2(t), D(y2)(t) = -y1(t) - y1(t)^3;
> inits := [0, 0, 1], [0, 0, 2], [0, 0, 3];
> with(DEtools):
> DEplot([sys[1], sys[2]], [y1(t), y2(t)], t = -3..3, y1 = -3..3,
   y2 = -3..3, [inits], stepsize = 0.05);
```

CHAPTER 4,   page 52

Pr.4.1.   Type `?coeff` for information.

```
> series(tan(x), x, 11);
> evalf(coeff(%, x^7), 6);                        # Resp. .0539683
```

Pr.4.3.
```
> ode := diff(y(x), x) - 2*x*y(x) = 0;
> s := 's':
> Sum := sum(a[m]*x^m, m = s - 1..s + 1);
```
Error, attempting to assign to 'Sum' which is protected
```
> ser := sum(a[m]*x^m, m = s - 1..s + 1);
```

$$ser := a_{s-1}\,x^{(s-1)} + a_s\,x^s + a_{s+1}\,x^{(s+1)}$$

```
> subs(y(x) = ser, ode);
> simplify(%);                          # Performs previous differentiation
```

$$a_{s-1}\, x^{(s-2)}\, s - a_{s-1}\, x^{(s-2)} + a_s\, x^{(s-1)}\, s + a_{s+1}\, x^s\, s + a_{s+1}\, x^s$$

$$-2\, x^s\, a_{s-1} - 2\, x^{(s+1)}\, a_s - 2\, x^{(s+2)}\, a_{s+1} = 0$$

```
> termxs := coeff(lhs(%), x^s);      # Resp. termxs := a_{s+1} s + a_{s+1} - 2 a_{s-1}
```

```
> a[s+1] := solve(termxs, a[s+1]);
```

$$a_{s+1} := 2\,\frac{a_{s-1}}{s+1}$$

Now state the beginning of the recursion, namely, $a_{-1} = 0$, $a_0 = 1$. Then you get the desired partial sum by the loop shown, as follows.

```
> a[-1] := 0:    a[0] := 1:
> for s from 0 to 9 do
>     a[s+1] := 2*a[s-1]/(s + 1);
> od:                               # od (do reversed) marks end of loop.
> sol := sum(a[m]*x^m, m = 0..10);
```

$$sol := 1 + x^2 + \frac{1}{2}\,x^4 + \frac{1}{6}\,x^6 + \frac{1}{24}\,x^8 + \frac{1}{120}\,x^{10}$$

Pr.4.5. A basis of solutions is 1 and $\operatorname{arctanh} x = (1/2)\ln\left((1+x)/(1-x)\right)$. Obtain this identity from `convert(arctanh(x), ln)`. Commands:

```
> ode := (1 - x^2)*diff(y(x), x, x) - 2*x*diff(y(x), x) = 0;
> dsolve(ode, y(x), series);
> dsolve(ode);          # Resp. y(x) = _C1 + _C2 ( (1/2)ln(x - 1) - (1/2)ln(x + 1) )
```

Pr.4.7. For procedures, see Example 4.2 in this Guide. For the error experiment you may need to use `Digits := 20:` etc.

```
> f := exp(x);
> with(orthopoly):                               # Ignore the warning.
> term := proc(m)
>     (2*m + 1)/2*int(f*P(m, x), x = -1..1)*'P(m, x)';
> end:
> S := sum('evalf(term(m))', m = 0..6);
```

$$S := 1.175201194\,P\,(0, x) + 1.103638324\,P\,(1, x) + .3578143500\,P\,(2, x)$$
$$+.07045563000\,P\,(3, x) + .009965025000\,P\,(4, x) + .001101100000\,P\,(5, x)$$
$$+.00008450000000\,P\,(6, x)$$

```
> plot({S, exp(x)}, x = -3..3);
```

The order of the error is about 10^{-n+2}.

Pr.4.9.
```
> ode := x*diff(y(x), x, x) + (1 - 2*x)*diff(y(x), x) + (x - 1)*y(x)
  = 0;
> ypart := dsolve({ode, y(1) = 1, D(y)(1) = 1});
> sol := dsolve(ode);
```
\# Resp. $sol := \mathrm{y}(x) = \_C1\, \mathrm{e}^x + \_C2 \ln(x)\, \mathrm{e}^x$

```
> diff(sol, x);
```
\# Resp. $\dfrac{\partial}{\partial x}\, \mathrm{y}(x) = \_C1\, \mathrm{e}^x + \dfrac{\_C2\, \mathrm{e}^x}{x} + \_C2 \ln(x)\, \mathrm{e}^x$

```
> eval(subs(x = 0, sol));
```
`Error, (in ln) numeric exception: division by zero`

Pr.4.11.
```
> with(DEtools):
> a := 'a':  b := 'b':   c := 'c':
> ode := x*(1 - x)*diff(y(x), x, x) + (c - (a + b + 1)*x)*diff(y(x), x)
  - a*b*y(x) = 0;
> sol := dsolve(ode);
```

$$sol := \mathrm{y}(x) = \_C1\, \mathrm{hypergeom}\left([a,\, b],\, [c],\, x\right)$$

$$+ \_C2\, x^{(-c+1)}\, \mathrm{hypergeom}\left([a - c + 1,\, b - c + 1],\, [2 - c],\, x\right)$$

```
> subs(a = 1, b = 1, c = 2, x = -t, sol);
> simplify(%);
```

$$\mathrm{y}(-t) = \frac{\_C1 \ln(1 + t) - \_C2}{t}$$

```
> subs(a = -n, c = b, x = -t, sol);
> simplify(%);
```

$$\mathrm{y}(-t) = \_C1\, (1 + t)^n - \_C2\, (-t)^{(-b)}\, t\, \mathrm{hypergeom}([1,\, -n - b + 1],\, [2 - b],\, -t)$$

Pr.4.13.
```
> hyp := x*(1 - x)*diff(y(x), x, x) + (2 - 4*x)*diff(y(x), x) - 2*y(x)
  = 0;
> sol := dsolve({hyp, y(0) = 1, D(y)(0) = 1});
```

$$sol := \mathrm{y}(x) = -\frac{1}{-1 + x}$$

```
> series(rhs(%), x);
> solve({a + b + 1 = 4, a*b = 2},  {a, b});
```

$$\{a = 2,\, b = 1\},\, \{b = 2,\, a = 1\}$$

Pr.4.15. 3.831 706 (exact 7S: 3.831 706), 7.021 559 (7.015587). Commands:
```
> s := series(BesselJ(1, x), x, 20);
> p := convert(s, polynom);
> fsolve(p = 0, x);
> fsolve(BesselJ(1, x) = 0, x = 2..5);
> fsolve(BesselJ(1, x) = 0, x = 5..10);
```

Pr.4.17.
```
> ode := x^2*diff(y(x), x, x) + x*diff(y(x), x) + (x^2 - nu^2)*y(x)
  = 0;
> sol := dsolve(ode);
> Y0 := subs(_C1 = 0, _C2 = 1, nu = 0, sol);
> Y1 := subs(_C1 = 0, _C2 = 1, nu = 1, sol);
> plot({rhs(Y0), rhs(Y1)}, x = 0..10, y = -1..0.6);
```

Pr.4.19.
```
> ode := diff(y(x), x, x) + x^2*y(x) = 0;
> sol := dsolve(%);
```

$$sol := y(x) = \_C1\,\sqrt{x}\,\text{BesselJ}\left(\frac{1}{4}, \frac{1}{2}x^2\right) + \_C2\,\sqrt{x}\,\text{BesselY}\left(\frac{1}{4}, \frac{1}{2}x^2\right)$$

Hence the transformation is $y = u\sqrt{x}$, $x^2/2 = z$ and gives a Bessel equation for u as a function of z, with parameter $1/4$.

CHAPTER 5,   page 65

Pr.5.1.   To see why you need the first command, try without

```
> assume(s, positive):
> int(exp(-s*t)*sin(Pi*t), t = 0..infinity);
> simplify(%);
```

Pr.5.3.   The answer $-k\,(e^{-sc} - 1)/s$ is obtained by the command

```
> int(exp(-s*t)*k, t = 0..c);
```
     # Resp. $-\dfrac{k\,(e^{(-sc)} - 1)}{s}$

Pr.5.5.
```
> with(inttrans):
> invlaplace(s/(L^2*s^2 + n^2*Pi^2), s, t);
```

Pr.5.7.   Type the ODE. Obtain the subsidiary equation. Substitute the initial conditions into it. Solve it algebraically. Find the inverse transform of its solution.

```
> with(inttrans):
> ode := diff(y(t), t, t) + 2*diff(y(t), t) - 3*y(t) = 6*exp(-2*t);
> subsid := laplace(ode, t, s);
> subsid2 := subs(y(0) = 2, D(y)(0) = -14, %);
> Y := solve(subsid2, laplace(y(t), t, s));
```

$$Y := 2\,\frac{s^2 - 3s - 7}{s^3 + 4s^2 + s - 6}$$

```
> y := invlaplace(Y, s, t);
```

$$y := -\frac{3}{2}\,e^t + \frac{11}{2}\,e^{(-3t)} - 2\,e^{(-2t)}$$

Pr.5.9. `> with(inttrans):`

`> invlaplace(3*(1 - exp(-Pi*s))/(s^2 + 9), s, t);`

$$\sin(3\,t) + \text{Heaviside}(t - \pi)\sin(3\,t)$$

`> plot(%, t = 0..20);`

Pr.5.11. `> eq := R*i(t) + 1/C*int(i(tau), tau = 0..t) = K*Heaviside(t - 1) -`
`  K*Heaviside(t - 3);`

`> with(inttrans):`

`> subsid := laplace(eq, t, s);`   `# See Example 5.6 for the integral.`

`> J := solve(subsid, laplace(i(t), t, s));`

`> j := invlaplace(J, s, t);`

$$j := K\,C\left(\frac{\text{Heaviside}(t-1)\,\mathrm{e}^{(-\frac{t-1}{RC})}}{RC} - \frac{\text{Heaviside}(t-3)\,\mathrm{e}^{(-\frac{t-3}{RC})}}{RC}\right)$$

`> convert(j, piecewise, t);`

`> j0 := subs(K = 110, R = 1, C = 1, j);`

`> plot(j0, t = -1..10, xtickmarks = [1, 3, 5, 10]);`

Pr.5.13. `> y := 'y':`

`> ode := diff(y(t), t, t) + y(t) = Dirac(t - Pi) - Dirac(t - 2*Pi);`

`> with(inttrans):`

`> subsid := laplace(ode, t, s);`

`> Y := solve(subsid, laplace(y(t), t, s));`

$$Y := \frac{s\,y(0) + \mathrm{D}(y)(0) + \mathrm{e}^{(-s\,\pi)} - \mathrm{e}^{(-2\,s\,\pi)}}{s^2 + 1}$$

`> sol := invlaplace(Y, s, t);`

`> yp := subs(y(0) = 0, D(y)(0) = 1, sol);`

$$yp := \sin(t) - \text{Heaviside}(t - \pi)\sin(t) - \text{Heaviside}(t - 2\,\pi)\sin(t)$$

`> convert(yp, piecewise, t);`

`> plot(yp, t = 0..20);`

Pr.5.15. The full-wave rectification of $\sin t$ is

`> r := sin(t)*(1 - 2*Heaviside(t - Pi) + 2* Heaviside(t - 2*Pi) -`
`  2*Heaviside(t - 3*Pi) + 2*Heaviside(t - 4*Pi));`

$$r := \sin(t)\,(1 - 2\,\text{Heaviside}(t - \pi) + 2\,\text{Heaviside}(t - 2\,\pi) - 2\,\text{Heaviside}(t - 3\,\pi)$$
$$+ 2\,\text{Heaviside}(t - 4\,\pi))$$

`> plot(r, t = 0..5*Pi);`

CHAPTER 6,   page 76

Pr.6.1.   > with(linalg):   # Ignore the warning.
> evalm(3*A);
> evalm(A - B);
> transpose(2*A - (1/2)*B);

Pr.6.3.   Load the linalg package. Then type

> A := matrix([[1, 3, 2], [3, 5, 0], [2, 0, 4]]);
> B := matrix([[0, 2, 1], [-2, 0, -3], [-1, 3, 0]]);
> c := [1, 0, -2];
> d := [3, 1, 2];

Keep in mind that vectors created in this way are not treated as either row or column vectors. See Example 6.1 in this Guide.

A is symmetric, **B** skew-symmetric; this explains why you obtain two zero matrices. Further commands:

> evalm(A^2);   evalm(A^4);   # Etc.
> evalm(A&*B - B&*A);
> evalm(A - transpose(A));   # Etc.
> det(A);   det(B);
> evalm(A&*c);   # Resp. $[-3,\ 3,\ -6]$
> evalm(B&*(c - 3*d));
> evalm(c&*d);   # Resp. -1

Pr.6.7.   > with(linalg):   # Ignore the warning.
> A := matrix([[cos(t), -sin(t)], [sin(t), cos(t)]]);
> A4 := evalm(A^4);

$$A4 := \begin{bmatrix} (\cos(t)^2 - \sin(t)^2)^2 - 4\cos(t)^2\sin(t)^2 & -4\,(\cos(t)^2 - \sin(t)^2)\,\cos(t)\,\sin(t) \\ 4\,(\cos(t)^2 - \sin(t)^2)\,\cos(t)\,\sin(t) & (\cos(t)^2 - \sin(t)^2)^2 - 4\cos(t)^2\sin(t)^2 \end{bmatrix}$$

> map(combine, A4);

$$\begin{bmatrix} \cos(4\,t) & -\sin(4\,t) \\ \sin(4\,t) & \cos(4\,t) \end{bmatrix}$$

Pr.6.9.   It seems that $n_{min} = m + 2$.

Pr.6.11.   > with(linalg):   # Ignore the warning.
> A := matrix([[0, -2, -1], [-2, 3, 2], [-1, 2, 1]]);
> C := inverse (A);
> B := matrix([[1, 2, 3], [2, 3, 4], [3, 4, 6]]);
> E := inverse(B);   # The letter D is protected in Maple.

$$E := \begin{bmatrix} -2 & 0 & 1 \\ 0 & 3 & -2 \\ 1 & -2 & 1 \end{bmatrix}$$

> inverse(A&*B) - evalm(E&*C);   # Resp. 0

Pr.6.13. > with(linalg):   # Ignore the warning.

> c := [3, 2, -2, 1, 0]:

> d := [2, 0, 3, 0, 4]:

> e := [1, -3, -2,-1, 1]:

> evalm(c&*d);   # Resp. 0 Etc.

Pr.6.15. Type the vectors and then the square root of their dot products $\mathbf{c} \cdot \mathbf{c}$, $\mathbf{d} \cdot \mathbf{d}$, $\mathbf{e} \cdot \mathbf{e}$,

> c := [3, 2, -2, 1, 0];

> sqrt(evalm(c^2));   # Resp. $3\sqrt{2}$ Etc.

Pr.6.17. $\det \mathbf{H} = 1/2160, 1/6\,048\,000, 1/266\,716\,800\,000$. The commands are as follows.

> with(linalg):   # Ignore the warning.

> n := 3:

> H := matrix (n, n, (j,k) -> 1/(j + k - 1));

> inverse(H);

> det(H);

Pr.6.19. Find the rank of the matrix with the given vectors as columns.

> with(linalg):   # Ignore the warning.

> a := [0, 16, 0, -24, 0];

> b := [1, 0, -1, 0, 2];

> c := [0, -14, 0, 21, 0];

> A := augment(a, b, c);

> rank(A);   # Resp. 2 # Hence linear dependence

CHAPTER 7,   page 85

Pr.7.1. > with(linalg):   # Ignore the warning.

> A := matrix([[-2, 2, -3], [2, 1, -6], [-1, -2, 0]]);

> B := transpose(A);

> SYM := evalm((A + B)/2);

> SKEW := evalm((A - B)/2);

> evalm(SYM + SKEW);   # For checking

Pr.7.3.   Show that the inverse equals the transpose (why?).

```
> with(linalg):                                        # Ignore the warning.
> A := matrix([[cos(t), sin(t)],  [-sin(t), cos(t)]]);
> inverse(A);
> simplify(%, trig);
```

$$\begin{bmatrix} \cos(t) & -\sin(t) \\ \sin(t) & \cos(t) \end{bmatrix}$$

```
> evalm(inverse(A) - transpose(A));
> simplify(%);                                         # Resp. 2 × 2 zero matrix
```

Now show that the dot product of vectors **u** and **v** equals the dot product of the vectors **Au** and **Av**, call them **w** and **z**, respectively.

```
> u := [u1, u2]:    v := [v1, v2]:
> innerprod(u, v);                                     # Resp. u1 v1 + u2 v2
> w := evalm(A&*u);
```

$$w := [\cos(t)\, u1 + \sin(t)\, u2, -\sin(t)\, u1 + \cos(t)\, u2]$$

```
> z := evalm(A&*v);
```

$$z := [\cos(t)\, v1 + \sin(t)\, v2, -\sin(t)\, v1 + \cos(t)\, v2]$$

```
> innerprod(w, z);
```

$$\cos(t)^2\, u1\ v1 + \sin(t)^2\, u2\ v2 + \sin(t)^2\, u1\ v1 + \cos(t)^2\, u2\ v2$$

```
> simplify(%);                                         # Resp. u1 v1 + u2 v2
```

Pr.7.7.   Note that the spectrum is real, although **A** is not Hermitian.

```
> with(linalg):                                        # Ignore the warning.
> A := matrix([[4+12*I,   -12-12*I,    12-12*I],
>               [-6+6*I,     10-6*I,      6+6*I],
>               [6+6*I,     -6+6*I,     -2-6*I]]);
> B := evalm(transpose(conjugate(A)));
> HE :=evalm((A + B)/2);
> SH := evalm((A - B)/2);
> evalm(HE + SH);                                       # For checking
> eigenvectors(A);                     # Order and eigenvectors may differ.
```

$$[4, 1, \{[0, 1, I]\}], [-8, 1, \{[I, 0, 1]\}], [16, 1, \{[-I, 1, 0]\}]$$

Pr.7.9.
```
> with(linalg):                                        # Ignore the warning.
> C := matrix([[I/2,  sqrt(3)/2],   [sqrt(3)/2,  I/2]]);
```

$$C := \begin{bmatrix} \frac{1}{2}I & \frac{1}{2}\sqrt{3} \\ \frac{1}{2}\sqrt{3} & \frac{1}{2}I \end{bmatrix}$$

```
> evalm(C^(12));
```

$$\begin{bmatrix} 1 & 0 \\ 0 & 1 \end{bmatrix}$$

```
> e := eigenvalues(C);
```

$$e := \frac{1}{2}I + \frac{1}{2}\sqrt{3}, \frac{1}{2}I - \frac{1}{2}\sqrt{3}$$

```
> abs(e[1]);                                    # Resp. 1
> inverse(C^2);
```

$$\begin{bmatrix} \frac{1}{2} & -\frac{1}{2}I\sqrt{3} \\ -\frac{1}{2}I\sqrt{3} & \frac{1}{2} \end{bmatrix}$$

```
> p := charpoly(C, lambda);          # Resp. p := λ² − I λ − 1
> evalm(subs(lambda = C, p));
```

$$\begin{bmatrix} 0 & 0 \\ 0 & 0 \end{bmatrix}$$

Pr.7.11. Proceed as in Example 7.5 in this Guide. In particular, obtain the relation between the eigenvectors in the form $\mathbf{y_j} = \mathbf{P}^{-1}\mathbf{x_j}$, $j = 1, 2, 3$.

```
> with(linalg):                                # Ignore the warning.
> A := matrix([[-1, -3, 3],  [-6, 2, 6],  [-3, 3, 5]]);
> P := matrix([[3, -1, 1],  [-15, 6, -5],  [5, -3, 2]]);
> Q := inverse(P);
> B := evalm(Q&*A&*P);
```

$$B := \begin{bmatrix} -124 & 72 & -48 \\ 267 & -130 & 96 \\ 696 & -372 & 260 \end{bmatrix}$$

```
> ea := eigenvectors(A);                # Order and eigenvectors may differ.
```

$$ea := [-4, 1, \{[1, 1, 0]\}], [8, 1, \{[0, 1, 1]\}], [2, 1, \{[1, 0, 1]\}]$$

```
> eb := eigenvectors(B);
```

$$eb := [-4, 1, \{[1, \frac{-3}{2}, \frac{-19}{4}]\}], [8, 1, \{[-2, 1, 7]\}], [2, 1, \{[1, \frac{-5}{4}, \frac{-9}{2}]\}]$$

```
> y1 := evalm(Q&*ea[1][3][1]);# Resp. y1 := [−4, 6, 19]  #  Corresp. to −4
> y2 := evalm(Q&*ea[2][3][1]);   # Resp. y2 := [−2, 1, 7]  #  Corresp. to 8
> y3 := evalm(Q&*ea[3][3][1]);   # Resp. y3 := [−4, 5, 18]  #  Corresp. to 2
```

CHAPTER 8,   page 94

Pr.8.1.   You get the length (the Euclidean norm) by `norm(v, 2)` .

> `with(linalg):`   # Ignore the warning.
> `A := [1, 2, 3]; B := [2, 4, 6];`
> `v := [B[1] - A[1], B[2] - A[2], B[3] - A[3]];`
> `norm(v, 2);`   # Resp. $\sqrt{14}$

Pr.8.3.   These forces are in equilibrium, their resultant is **0**. Indeed,

> `[4, -2, -3] + [8, 8, 1] + [-12, -6, 2];`   # Resp. $[0, 0, 0]$

Pr.8.5.   > `with(linalg):`   # Ignore the warning.
> `innerprod(2*[2, 0, -5], 5*[4, -2, 1]);`   # Resp. 30
> `10*innerprod([2, 0, -5], [4, -2, 1]);`   # Resp. 30

Pr.8.7.   Obtain the components of the displacement vector **d** as differences of corresponding coordinates of the points. Then obtain the work as a dot product.

> `with(linalg):`   # Ignore the warning.
> `d := [5 - 3, 8 - 4, 0 - 0];`   # Resp. $d := [2, 4, 0]$
> `W := innerprod([2, 6, 6], d);`   # Resp. W := 28

Pr.8.9.   > `with(linalg):`   # Ignore the warning.
> `a := [1, 2, 0];   b := [-3, 2, 0];   c := [2, 3, 4];`
> `innerprod(crossprod(a, b), c);`   # Resp. 32
> `innerprod(a, crossprod(b, c));`   # Resp. 32

Pr.8.11.   They are linearly dependent because their scalar triple product is zero:

> `a := [3, 5, 9];   b := [73, -56, 76]; c := [-4, 7, -1];`
> `with(linalg):`   # Ignore the warning.
> `innerprod(a, crossprod(b, c));`   # Resp. 0

Pr.8.13.   Show that $L - R = 0$, where

> `a := [a1, a2, a3]:   b := [b1, b2, b3]:`
> `with(linalg):`   # Ignore the warning.
> `L := innerprod(crossprod(a, b), crossprod(a, b));`
> `R := innerprod(a, a)*innerprod(b, b) - innerprod(a, b)^2;`
> `L - R;`
> `simplify(%);`   # Resp. 0

Pr.8.15.
```
> with(linalg):                                    # Ignore the warning.
> r := [t, cosh(t)];
> rprime := diff(r, t);               # Resp. rprime := [1, sinh(t)]
> ip := innerprod(rprime, rprime);    # Resp. ip := 1 + sinh(t)²
> f := simplify(%);                   # Resp. f := cosh(t)²
> assume(t, real);
> sq := sqrt(f);                       # Resp. sq := cosh(t˜)
> Length := int(sq, t = 0..1);        # Resp. Length := sinh(1)
> evalf(%, 4);                               # Resp. 1.175
> plot(r[2], t = 0..1);
```

$ rprime := [1, \sinh(t)] $

$ ip := 1 + \sinh(t)^2 $

$ f := \cosh(t)^2 $

$ sq := \cosh(t\tilde{\ }) $

$ Length := \sinh(1) $

Pr.8.17.
```
> with(linalg):                                    # Ignore the warning.
> r := [t, t^2, t^3];
> r1 := diff(r, t);
> r2 := diff(r1, t);
> r3 := diff(r2, t);                     # Resp. r3 := [0, 0, 6]
> i11 := innerprod(r1, r1);
> i12 := innerprod(r1, r2);
> i22 := innerprod(r2, r2);
> tau := innerprod(crossprod(r1, r2), r3)/(i11*i22 - i12^2);
```

$$ \tau := 12\,\frac{1}{(1 + 4\,t^2 + 9\,t^4)\,(4 + 36\,t^2) - (4\,t + 18\,t^3)^2} $$

```
> tau := simplify(%);
```

$$ \tau := 3\,\frac{1}{1 + 9\,t^2 + 9\,t^4} $$

Pr.8.19. In subs do not write $\{v[1], v[2]\}$, with braces $\{..\}$ instead of brackets $[..]$. since you would obtain the set $\{0, 1\}$. Try it. In the plot command you can start from $x = 0$ and $y = 0$, but the plot will look much poorer. Try it.

```
> with(linalg):                                    # Ignore the warning.
> f := ln(x^2 + y^2);
> v := grad(f, [x, y]);              # Resp. v :=
```

$$ v := \left[2\,\frac{x}{x^2 + y^2},\ 2\,\frac{y}{x^2 + y^2}\right] $$

```
> subs(x = 2, y = 0, [v[1], v[2]]);        # Resp. [1, 0]
> with(plots):                          # Ignore the warning.
> fieldplot(v, x = 0.1..0.5, y = 0.1..0.5);
```

Pr.8.21. The cone is the surface $f = 0$, where

```
> with(linalg):                                    # Ignore the warning.
> f := x^2 + y^2 - z^2;              # Resp. f := x² + y² - z²
```

$ f := x^2 + y^2 - z^2 $

Now type a normal vector **N** of the cone, its length, and the unit normal vector **n** = (1/length **N**)**N**.

```
> N := grad(f, [x, y, z]);
> lengthN := sqrt(innerprod(N, N));
> n := evalm(N/lengthN);                              # Try without evalm
```

$$n := \left[\frac{x}{\sqrt{x^2 + y^2 + z^2}}, \frac{y}{\sqrt{x^2 + y^2 + z^2}}, -\frac{z}{\sqrt{x^2 + y^2 + z^2}} \right]$$

```
> subs(x = 3, y = 4, z = 5, [n[1], n[2], n[3]]);
> answer := simplify(%);           # Resp. answer := [ (3/10)√2, (2/5)√2, -(1/2)√2 ]
```

Pr.8.23. The command evalm is necessary. Try without, to see its effect in the present case.

```
> with(linalg):                                       # Ignore the warning.

> v := evalm([-y, x]/(x^2 + y^2));   # Resp. v := [ -y/(x²+y²), x/(x²+y²) ]

> diverge(v, [x, y]);                                 # Resp. 0
```

CHAPTER 9,   page 105

Pr.9.1. Integrate from $t = 0$ to 2π.

```
> F := [2*z, x, -y];
> r := [cos(t), sin(t), 2*t];
> FC := subs(x = r[1], y = r[2], z = r[3], F);
```

$$FC := [4\,t, \cos(t), -\sin(t)]$$

```
> rprime := diff(r, t);
> with(linalg):                                       # Ignore the warning.
> integrand := innerprod(FC, rprime);
```

$$integrand := -4\,t\,\sin(t) + \cos(t)^2 - 2\,\sin(t)$$

```
> int(integrand, t = 0..2*Pi);                        # Resp. 9π
```

Pr.9.3. Use $\mathbf{F} = [3x^2, 2\,yz, y^2]$. Show that $\operatorname{curl}\mathbf{F} = \mathbf{0}$. Find f such that $\mathbf{F} = \operatorname{grad} f$. Use $f(B) - f(A)$.

```
> F := [3*x^2, 2*y*z, y^2];
> with(linalg):                                       # Ignore the warning.
> curl(F, [x, y, z]);                                 # Resp. [0, 0, 0]
> potential(F, [x, y, z], 'f');                       # Resp. true
> f;                                                  # Resp. x³ + y² z
> subs(x = 1, y = -1, z = 7, f) - subs(x = 0, y = 1, z = 2, f);   # Resp. 6
```

Pr.9.5. Use polar coordinates given by $x = r \cos \theta$, $y = r \sin \theta$. The element of area is $r \, dr \, d\theta$. Denote the coordinates of the center of gravity by xbar and ybar. Note that they are equal, for reasons of symmetry of the quarter-disk. Use $dx \, dy = r \, dr \, d\theta$.

```
> r := 'r':
> M := int(int(1*r, r = 0..1), theta = 0..Pi/2);                    # Total mass
> xbar := 1/M*int(int(r*cos(theta)*r, r = 0..1),theta = 0..Pi/2);
```

$$xbar := \frac{4}{3}\frac{1}{\pi}$$

Pr.9.7. Integrate over y from $1 + x^4$ to 2. Sketch the region of integration to find out that the integration over x extends from -1 to 1. The answer will be 16/5. Thus type

```
> F := [exp(y)/x, exp(y)*ln(x) + 2*x];
```

Type the integrand G of the double integral on the left-hand side of the formula,

```
> G := diff(F[2], x) - diff(F[1], y);
> int(int(2, y = 1 + x^4..2), x = -1..1);
```

Pr.9.9. The familiar formula πab follows from

```
> r := [a*cos(t), b*sin(t)];          # Resp. r := [a cos(t), b sin(t)]   # Ellipse
> r1 := diff(r, t);                    # Resp. r1 := [-a sin(t), b cos(t)]
> A := 1/2*int(r[1]*r1[2] - r[2]*r1[1], t = 0..2*Pi);
```

Pr.9.11. $z = 1 - x - y$. Hence a representation r, its partial derivatives ru and rv, and a normal vector N are

```
> r := [u, v, 1 - u - v];              # Resp. r := [u, v, 1 - u - v]
> ru := diff(r, u);   rv := diff(r, v);
```

$$ru := [1, 0, -1]$$
$$rv := [0, 1, -1]$$

```
> with(linalg):                        # Ignore the warning.
> N := crossprod(ru, rv);              # Resp. N := [1, 1, 1]
> F := [x^2, 0, 3*y^2];                # Resp. F := [x^2, 0, 3 y^2]
```

Let FS denote **F** on the plane S. Denote the integrand by f.

```
> FS := subs(x = r[1], y = r[2], z = r[3], F);   # Resp. FS := [u^2, 0, 3 v^2]
> f := innerprod(FS, N);               # Resp. f := u^2 + 3 v^2
```

$z = 1 - x - y = 1 - u - v = 0$ gives $u = 1 - v$ in the xy-plane $z = 0$. Hence integrate from $u = 0$ to $1 - v$ and then from $v = 0$ to 1.

```
> int(int(f, u = 0..1 - v), v = 0..1);             # Resp. 1/3
```

Pr.9.13. A parametric representation of the sphere is as follows. Then type the partial derivatives, a normal vector N, then F and FS, which is **F** on the sphere. Integrate the dot
product of FS and N over u from 0 to $\pi/2$ and over v from 0 to $\pi/2$.

```
> r := [cos(v)*cos(u), cos(v)*sin(u), sin(v)];
```

$$r := [\cos(v)\cos(u), \cos(v)\sin(u), \sin(v)]$$

```
> ru := diff(r, u);   rv := diff(r, v);
```

$$ru := [-\cos(v)\sin(u), \cos(v)\cos(u), 0]$$
$$rv := [-\sin(v)\cos(u), -\sin(v)\sin(u), \cos(v)]$$

```
> with(linalg):
```
                                             # Ignore the warning.

```
> N := crossprod(ru, rv);
```

$$N := \left[\cos(v)^2 \cos(u), \cos(v)^2 \sin(u), \cos(v)\sin(u)^2 \sin(v) + \cos(v)\cos(u)^2 \sin(v)\right]$$

```
> F := [0, x, 0];
```
                            # Resp. $F := [0, x, 0]$

```
> FS := subs(x = r[1], F);
```
            # Resp. $FS := [0, \cos(v)\cos(u), 0]$

```
> ip := innerprod(FS, N);
```
            # Resp. $ip := \cos(v)^3 \cos(u)\sin(u)$

```
> int(int(ip, u = 0..Pi/2), v = 0..Pi/2);
```
                          # Resp. $\dfrac{1}{3}$

Pr.9.15. The square of the distance of a point (x, y, z) from the x-axis (with respect to which
the moment of inertia is taken) is $y^2 + z^2$. This is the integrand. Use cylindrical
coordinates defined by $x = x$, $y = r\cos\theta$, $z = r\sin\theta$. Then the integrand is r^2 and
the volume element is $dx\, r\, dr\, d\theta$. Represent the cylinder as shown, with $0 \le x \le h$,
$0 \le r \le a$, $0 \le \theta \le 2\pi$. Integrate accordingly. Answer: $h\, a^4\, \pi/2$.

```
> r := 'r':   theta := 'theta':
> R := [x, r*cos(theta), r*sin(theta)];
```
                                          # Cylinder
```
> int(int(int(r^2*r, x = 0..h), r = 0..a), theta = 0..2*Pi);
```

Pr.9.17. Type **F** and then its divergence, which will turn out to be constant, so that the
problem becomes very simple.

```
> F := [9*x, y*cosh(x)^2, -z*sinh(x)^2];
```

$$F := [9\, x, y\cosh(x)^2, -z\sinh(x)^2]$$

```
> with(linalg):
```
                                             # Ignore the warning.

```
> G := diverge(F, [x, y, z]);
```
            # Resp. $G := 9 + \cosh(x)^2 - \sinh(x)^2$

```
> G := simplify(%);
```
                               # Resp. $G := 10$

Hence the answer is 10 times the volume $(4/3)\pi abc$ of the solid ellipsoid. Here,
$a = 3$, $b = 6$, $c = 2$, as can be seen by dividing the formula for S by 36, that is,
$x^2/3^2 + y^2/6^2 + z^2/2^2 = 1$. Hence the answer is 480π.

Pr.9.19. Use the volume element as in Pr.9.18, which gives the factor r^2 in the last command.

```
> F := [x^3, y^3, z^3];
> with(linalg):                                    # Ignore the warning.
> G := diverge(F, [x, y, z]);
> R := [r*cos(u)*sin(v), r*sin(u)*sin(v), r*cos(v)];    # r is variable.
> G2 := subs(x = R[1], y = R[2], z = R[3], G);
```

$$G2 := 3\,r^2\,\cos(u)^2\,\sin(v)^2 + 3\,r^2\,\sin(u)^2\,\sin(v)^2 + 3\,r^2\,\cos(v)^2$$

```
> G2 := simplify(%);                          # Resp. G2 := 3 r^2
> int(int(int(G2*r^2*sin(v), u = 0..2*Pi), v = 0..Pi), r = 0..3);
```

$$\frac{2916}{5}\,\pi$$

CHAPTER 10,   page 116

Pr.10.1. $f(x)$ is odd. Hence $a_n = 0$. Instead of integrating from $-\pi$ to π you can integrate from 0 to π and multiply the integral by 2. The terms of S may not come out in their natural order.

```
> bn := 2/Pi*int(3*sin(n*x), x = 0..Pi);    # Resp. bn := -6 (cos(π n) - 1)/(π n)
> S := sum(bn*sin(n*x), n = 1..50);
> [4/Pi]*combine(S*Pi/4);                    # Factors out 4/π
> plot(S, x = -Pi..Pi);
```

Pr.10.3. The function is neither even nor odd–always watch carefully for what interval a function is given! Note that without the computer the calculations would be involved. The plot shows again the Gibbs phenomenon at both ends of the jump.

```
> a0 := 1/(2*Pi)*int((x/(2*Pi))^4, x = 0..2*Pi);    # Resp. a0 := 1/5
> an := 1/Pi*int((x/(2*Pi))^4*cos(n*x), x = 0..2*Pi);
> bn := 1/Pi*int((x/(2*Pi))^4*sin(n*x), x = 0..2*Pi);
> plot(a0 + sum(an*cos(n*x) + bn*sin(n*x), n = 1..50), x = 0..4*Pi);
```

Pr.10.5. Answer: the terms $\cos t$ and $\sin t$ because their coefficients $4\,\pi$ and $-4\,\pi$ are the absolutely largest ones.

```
> an := 1/Pi*int(t^2*cos(n*t), t = 0..2*Pi):
> bn := 1/Pi*int(t^2*sin(n*t), t = 0..2*Pi):
> seq([an, bn], n = 1..10);
```

$$[4, -4\,\pi], [1, -2\,\pi], \left[\frac{4}{9}, -\frac{4}{3}\,\pi\right], \left[\frac{1}{4}, -\pi\right], \left[\frac{4}{25}, -\frac{4}{5}\,\pi\right], \left[\frac{1}{9}, -\frac{2}{3}\,\pi\right], \left[\frac{4}{49}, -\frac{4}{7}\,\pi\right],$$

$$\left[\frac{1}{16}, -\frac{1}{2}\,\pi\right], \left[\frac{4}{81}, -\frac{4}{9}\,\pi\right], \left[\frac{1}{25}, -\frac{2}{5}\,\pi\right]$$

Pr.10.7. $p = 2L = 2, L = 1$. The function is even, hence $b_n = 0$. Instead of integrating from -1 to 1 you could take twice the integral from 0 to 1. The last command before the plot factors out $12/\pi^2$.

```
> a0 := 1/(2*1)*int(3*x^2, x = -1..1);                    # Resp. a0 := 1
> an := 1/1*int(3*x^2*cos(n*Pi*x), x = -1..1);
```

$$an := 6 \frac{-2 \sin(\pi n) + n^2 \pi^2 \sin(\pi n) + 2\pi n \cos(\pi n)}{n^3 \pi^3}$$

```
> S := a0 + sum(an*cos(n*Pi*x), n = 1..4);
```

$$S := 1 - 12 \frac{\cos(\pi x)}{\pi^2} + \frac{3 \cos(2\pi x)}{\pi^2} - \frac{4}{3} \frac{\cos(3\pi x)}{\pi^2} + \frac{\frac{3}{4} \cos(4\pi x)}{\pi^2}$$

```
> 1 + [12/Pi^2]*combine((1 - S)*Pi^2/12);
> with(plots):                                            # Ignore the warning.
> P1 := plot(S, x = -3..3, ytickmarks = [1,2,3]):
> P2 := plot(3*(x + 2)^2, x = -3..-1):
> P3 := plot(3*x^2, x = -1..1):
> P4 := plot(3*(x - 2)^2, x = 1..3):
> display(P1, P2, P3, P4);
```

Pr.10.9. Even and Odd (below) denote the half-range cosine and sine series, respectively.

```
> a0 := 2/(2*L)*int(x, x = 0..L);                         # Resp. a0 := 1/2 L
> an := 2/L*int(x*cos(n*Pi*x/L), x = 0..L);
```

$$an := 2 \frac{L(\cos(n\pi) + n\pi \sin(n\pi) - 1)}{n^2 \pi^2}$$

```
> bn := 2/L*int(x*sin(n*Pi*x/L), x = 0..L);
```

$$bn := -2 \frac{L(-\sin(n\pi) + \pi n \cos(n\pi))}{n^2 \pi^2}$$

```
> Even := a0 + sum(an*cos(n*Pi*x/L), n = 1..5);
```

$$Even := \frac{1}{2} L - \frac{4 L \cos\left(\frac{\pi x}{L}\right)}{\pi^2} - \frac{4}{9} \frac{L \cos\left(3 \frac{\pi x}{L}\right)}{\pi^2} - \frac{4}{25} \frac{L \cos\left(5 \frac{\pi x}{L}\right)}{\pi^2}$$

```
> Odd := sum(bn*sin(n*Pi*x/L), n = 1..5);
```

$$Odd := 2 \frac{L \sin\left(\frac{\pi x}{L}\right)}{\pi} - \frac{L \sin\left(2 \frac{\pi x}{L}\right)}{\pi} + \frac{\frac{2}{3} L \sin\left(3 \frac{\pi x}{L}\right)}{\pi} - \frac{1}{2} \frac{L \sin\left(4 \frac{\pi x}{L}\right)}{\pi} + \frac{\frac{2}{5} L \sin\left(5 \frac{\pi x}{L}\right)}{\pi}$$

Pr.10.11. It is typical that the error is oscillating with x. Of course, you can expect the error to increase as x comes close to the jumps.

```
> bn := 1/Pi*int(x^3*sin(n*x), x = -Pi..Pi);
> Error := x^3 - sum(bn*sin(n*x), n = 1..5);
> plot(Error, x = -Pi..Pi);
```

Pr.10.13. The present function is discontinuous, so that you should expect large values of the minimum square error.

```
> bn := 1/Pi*int(x*sin(n*x), x = -Pi..Pi);
> SN := int(x^2, x = -Pi..Pi) - Pi*sum(bn^2, n = 1..N):
> evalf(seq(SN, N = 1..10));
```

$$8.10448050,\ 4.96288781,\ 3.56662432,\ 2.78122601,\ 2.27857095,\ 1.92950474,$$
$$1.67304771,\ 1.47669751,\ 1.32155628,\ 1.19589152$$

```
> evalf(subs(N = 100, SN));                          # Resp. .12373009
```

CHAPTER 11,   page 128

Pr.11.1. Note the nodes (points that do not move). s[1], .., s[4] call the solutions individually. Read and follow the instructions on animation in Example 11.1 in this Guide.

```
> s := seq(sin(n*x)*sin(n*t), n = 1..4);
```

$$s := \sin(x)\sin(t),\ \sin(2x)\sin(2t),\ \sin(3x)\sin(3t),\ \sin(4x)\sin(4t)$$

```
> with(plots):                                       # Ignore the warning.
> animate(s[3], x = 0..Pi, t = 0..Pi, frames = 50);
> s[1];                                               # Resp. sin(x) sin(t)
```

Pr.11.3. You need the inverse of the given transformation, which you obtain by solve. Follow the idea of Example 11.2 in this Guide.

```
> with(PDEtools):
> u := 'u':
> pde1 := diff(u(x,y), x, x) + diff(u(x,y), x, y) -
  2*diff(u(x,y), y, y) = 0;
> s := solve({v = x + y, z = 2*x - y},  {x, y});
```

$$s := \{y = -\tfrac{1}{3}z + \tfrac{2}{3}v,\ x = \tfrac{1}{3}v + \tfrac{1}{3}z\}$$

```
> tr :=  {s[1], s[2]};
```

$$tr := \{y = -\tfrac{1}{3}z + \tfrac{2}{3}v,\ x = \tfrac{1}{3}v + \tfrac{1}{3}z\}$$

```
> pde2 := dchange(tr, pde1, [v, z]);
```

$$pde2 := 9\left(\frac{\partial^2}{\partial z\, \partial v}\, u(v, z)\right) = 0$$

Hence $u_{vz} = 0$, $u_v = h(v)$, $u = f(v) + g(z) = f(x + y) + g(2x - y)$.

Pr.11.5. Yes. Only trivially by $-k$.

```
> u := (c1*exp(k*x) + c2*exp(-k*x))*(c3*exp(k*y) + c4*exp(-k*y));
> diff(u, x, x) - diff(u, y, y);
```

```
> simplify(%);                                                    # Resp. 0
> u2 := (c1*sinh(k*x) + c2*cosh(k*x))*(c3*sinh(K*y) + c4*cosh(K*y));
> diff(u2, x, x) - diff(u2, y, y);
> sol := simplify(%);                                    # No reduction to zero
> subs(K = k, sol);     subs(K = -k, sol);               # Resp. 0   0
```

Pr.11.7. pdsolve without INTEGRATE gives two ordinary differential equations corresponding to the separation of variables. Try it.

```
> u := 'u':
> pde := diff(u(x,t), t, t) = c^2*diff(u(x,t), x, x, x, x);

> pdsolve(pde, u(x,t), INTEGRATE);
```

$$(\mathrm{u}(x,\, t) = \_F1(x)\, \_F2(t)) \ \& \ \text{where}\left[\left\{\vphantom{\Big|}\right.\right.$$

$$\left\{\_F2(t) = \_C5\, e^{(c\,\sqrt{-c_1}\,t)} + \_C6\, e^{(-c\,\sqrt{-c_1}\,t)}\right\},$$

$$\left\{\_F1(x) = \_C1\, e^{\left(-c_1^{\left(\frac{1}{4}\right)} x\right)} + \_C2\, e^{\left(-\_c_1^{\left(\frac{1}{4}\right)} x\right)} + \_C3\, e^{\left(I\,\_c_1^{\left(\frac{1}{4}\right)} x\right)} + \_C4\, e^{\left(-I\,\_c_1^{\left(\frac{1}{4}\right)} x\right)}\right\}\right\}\right]$$

Pr.11.9. Recall that the ends of the bar, $x = 0$ and $x = \pi$, are kept at temperature 0.

```
> F := 'F':   G := 'G':  c := 'c':
> u(x,t) := F(x)*G(t);

> pde := diff(u(x,t), t) = c^2*diff(u(x,t), x, x);
```

$$pde := \mathrm{F}(x)\left(\frac{\partial}{\partial t}\,\mathrm{G}(t)\right) = c^2\left(\frac{\partial^2}{\partial x^2}\,\mathrm{F}(x)\right)\mathrm{G}(t)$$

```
> eq := pde/(c^2*u(x,t));
> sol1 := dsolve(rhs(eq) = -k^2);
```

$$sol1 := \mathrm{F}(x) = \_C1\,\sin(k\,x) + \_C2\,\cos(k\,x)$$

```
> eval(subs(x = 0, sol1)) = 0;
> F1 := subs(_C1 = 1, _C2 = 0, x = Pi, sol1) = 0;
```

$$F1 := (\mathrm{F}(\pi) = \sin(k\,\pi)) = 0$$

Hence $k = 1, 2, \ldots$ ($k = -1, -2$ would give the same functions since $\sin(-a) = -\sin a$.) This gives $F1 = \sin nx$, $n = 1, 2, \ldots$. Furthermore,

```
> sol2 := dsolve(lhs(eq) = -n^2);        # Resp. sol2 := G(t) = _C1 e^(-n²c²t)
```

Here Maple reuses _C1 for a different arbitrary constant.

```
> Answer := sin(n*x)*rhs(sol2);     # Resp. Answer := sin(n x)_C1 e^(-n²c²t)
```

Pr.11.11. Because of the boundary condition the series solution reduces to a single term,

```
> An := 2/(Pi*sinh(n*Pi/2))*int(sin(x)*sin(n*x), x = 0..Pi);
> seq(An, n = 2..10);                          # Resp. 0, 0, 0, 0, 0, 0, 0, 0, 0
> A1 := 2/(Pi*sinh(Pi/2))*int(sin(x)*sin(x), x = 0..Pi);
> u1 := A1*sin(x)*sinh(y);                      # Resp. u1 := sin(x) sinh(y) / sinh(½π)
> plot3d(u1, x = 0..Pi, y = 0..Pi/2);
```

Resp. $u1 := \dfrac{\sin(x)\,\sinh(y)}{\sinh(\frac{1}{2}\pi)}$

Pr.11.13. $\cos \lambda_{22} t \sin 0.5\pi x \sin \pi y, \lambda_{22} = 5\pi\sqrt{0.25+1}$. Hence type

```
> with(plots):                                 # Ignore the warning.
> animate3d(sin(Pi*x/2)*sin(Pi*y)*cos(5*Pi*sqrt(5/4)*t), x = 0..4,
   y = 0..2, t = 0..1, frames = 50);
```

Pr.11.15. You need the zero

```
> k := evalf(BesselJZeros(1, 1));              # Resp. k := 3.831705970
> with(plots):                                 # Ignore the warning.
> animate3d([r, theta, BesselJ(1,k*r)*cos(theta)*cos(k*t)], r = 0..1,
   theta = 0..2*Pi, t = 0..Pi, coords = cylindrical, frames = 20);
```

CHAPTER 12,   page 141

Pr.12.1.
```
> z1 := 8 + 3*I;                               # Resp. z1 := 8 + 3 I
> z2 := 9 - 2*I;                               # Resp. z2 := 9 − 2 I
> z1 + z2;                                     # Resp. 17 + I
> z1 - z2;                                     # Resp. −1 + 5 I
> z1*z2;                                       # Resp. 78 + 11 I
> evalf(z1/z2,5);                              # Resp. .77647 + .50588 I
> abs(z1/z2);                                  # Resp. 1/85 √6205
> abs(z1)/abs(z2);                             # Resp. 1/85 √73 √85
> combine(%);                                  # Resp. 1/85 √6205
> Re(z1);                                      # Resp. 8
> Im(z1^2);                                    # Resp. 48
> argument(z1);                                # Resp. arctan(3/8)
```

Pr.12.3. Start from the right-hand sides.

```
> z := x + I*y;
> (z + conjugate(z))/2;
> evalc(%);                                    # Resp. x
> evalc(z - conjugate(z))/(2*I);               # Resp. y
```

Pr.12.5. From the complex sequence S obtain the real sequence S2 of 41 pairs of the real and imaginary parts. Observe the two different ranges of n. Be careful with the various brackets in S2 ; each of them is needed. P2 plots the unit circle.

```
> S := seq((0.9 + 0.4*I)^n, n = -20..20);
> S2 := [seq([Re(S[n]), Im(S[n])], n = 1..41)];
> P1 := plot(S2, style = point):
> P2 := plot([cos(t), sin(t), t = 0..2*Pi]):
> with(plots):                                 # Ignore the warning.
> display({P1, P2}, scaling = constrained);
```

Pr.12.7. The first plotting command does not work because for plotting you need pairs of real and imaginary parts. For comparison, see Example 12.6 in this Guide.

```
> z := 'z':
> sol := solve(z^3 = 1 + I);
> s1 := evalf(evalc(sol[1]));                  # The first root
```

$$s1 := 1.084215081 + .2905145554\,I$$

```
> s2 := evalf(evalc(sol[2]));
> s3 := evalf(evalc(sol[3]));
> plot({s1, s2, s3}, style = point);
Plotting error, empty plot
> P1 := plot({[Re(s1), Im(s1)], [Re(s2), Im(s2)], [Re(s3), Im(s3)]},
    style = point):
> P2 := plot([2^(1/6)*cos(t), 2^(1/6)*sin(t), t = 0..2*Pi]):
> with(plots):                                 # Ignore the warning.
> display(P1, P2, scaling = constrained);
```

Pr.12.9. You can plot the two circles jointly with one command,

```
> plot({[2 + sqrt(8)*cos(t), 2 + sqrt(8)*sin(t), t = 0..2*Pi],
    [2 + 2*cos(t), 2 + 2*sin(t),t = 0..2*Pi]}, x = -2.5..6.5,
    y = -1..5, scaling = constrained);
```

Pr.12.11. (a) The interior of an ellipse with semi-axes $\sqrt{2}$ and 1, by geometry. (b) Type

```
> z := x + I*y;                                # Resp. z := x + I y
```

```
> eq := evalc((abs(z - 1) + abs(z + 1))) - sqrt(8) = 0;
> sol := solve(eq, y);
> ans := y^2 = (sol[1])^2;
```

`# Resp.` $sol := \dfrac{1}{2}\sqrt{-2x^2 + 4},\ -\dfrac{1}{2}\sqrt{-2x^2 + 4}$

`#` Thus $x^2/2 + y^2 = 1$.

Pr.12.15. The interior of the circle is mapped onto the exterior and conversely. If you don't see the correspondence, map the square portion by portion, e.g., $1.5 \le x \le 1.6$, then $1.6 \le x \le 1.7$, etc. Similarly in the y-direction if necessary.

```
> with(plots):                                    # Ignore the warning.
> P1 := conformal(1/z, z =(1 + I)/2..3*(1 + I)/2):
> P2 := conformal(z, z =(1 + I)/2..3*(1 + I)/2):
> P3 := plot([cos(t), sin(t), t = 0..2*Pi]):
> display(P1, P2, P3, scaling = constrained, xtickmarks = [-1, -0.5,
    0, 0.5, 1, 1.5], ytickmarks = [-1, -0.5, 0, 0.5, 1, 1.5] );
```

Pr.12.17. The images coincide on the real axis. The elliptical ring opens up when you shorten.

```
> with(plots):                                    # Ignore the warning.
> conformal(cos(z), z = I/2..2*Pi + I);
```

Pr.12.19. Those values suggest that Ln z is discontinuous along the negative real axis.

```
> evalc(ln(-5));
```
`# Resp.` $\ln(5) + I\pi$

```
> evalc(ln(-12 - 16*I));
```
`# Resp.` $\ln(20) + I\left(\arctan\left(\dfrac{4}{3}\right) - \pi\right)$

```
> evalf(%,5);
```
`# Resp.` $2.9957 - 2.2143\,I$

```
> evalc(ln(1 + I));
```
`# Resp.` $\dfrac{1}{2}\ln(2) + \dfrac{1}{4}I\pi$

```
> evalc(ln(1 - I));
```
`# Resp.` $\dfrac{1}{2}\ln(2) - \dfrac{1}{4}I\pi$

```
> evalc(ln(-10 + 0.1*I));
```
`# Resp.` $2.302635090 + 3.131592987\,I$

```
> evalc(ln(-10 - 0.1*I));
```
`# Resp.` $2.302635090 - 3.131592987\,I$

CHAPTER 13,   page 146

Pr.13.1. In the last command try without evalc.

```
> z := 1 + I + (2 + I)*t;
```
`# Resp.` $z := 1 + I + (2 + I)\,t$

```
> zdot := diff(z, t);
```
`# Resp.` $zdot := 2 + I$

```
> int(evalc(Re(z)*zdot), t = 0..1);
```
`# Resp.` $4 + 2I$

Pr.13.3. You can leave the two integrals separate or take them together as shown here. Answer $4 + i$

```
> z1 := 1 + I + I*t;
```
`# Resp.` $z1 := 1 + I + I\,t$

```
> z1dot := diff(z1, t);                          # Resp. z1dot := I
> z2 := 1 + 2*I + 2*t;                            # Resp. z2 := 1 + 2 I + 2 t
> z2dot := diff(z2, t);                           # Resp. z2dot := 2
> int(evalc(Re(z1)*z1dot + Re(z2)*z2dot), t = 0..1);
```

Pr.13.5.   Answer $12\pi i$. Commands:

```
> z := 'z':
> f := (4*z^2 + 17*z - 68)/(z^3 - 12*z + 16);
> convert(f, parfrac, z);
```

$$-2\,\frac{1}{z+4} - \frac{3}{(z-2)^2} + \frac{6}{z-2}$$

No contribution from the first fraction since $z = -4$ lies outside the contour. None from the second fraction, for which $f(z) = 1$ and thus $f'(z) = 0$ in (1) and (2) in Example 13.3 in this Guide. Hence $2\pi i \times 6$ by Cauchy's formula.

Pr.13.7.   Denote the straight segment by $z1$ and the parabolic arc by $z2$. Accordingly, type

```
> z1 := (1 + I)*t;                                # Resp. z1 := (1 + I) t
> z1dot := diff(z1, t);                           # Resp. z1dot := 1 + I
> int(evalc(conjugate(z1)*z1dot), t = 0..1);      # Resp. 1
> z2:=t + I*t^2;                                   # Resp. z2 := t + I t^2
> z2dot := diff(z2, t);                            # Resp. z2dot := 1 + 2 I t
> int(evalc(conjugate(z2)*z2dot), t = 0..1);      # Resp. 1 + (1/3) I
```

This shows path dependence of the integral of the nonanalytic function $\bar{z}$.

Pr.13.9.   On the four sides, starting from the origin and going around clockwise, you have the following, from which you can read the integrands $(-y^2)i$, etc. of the four integrals, as shown in the command. Answer $-1 - i$.

$z =$	iy	$i + x$	$1 + i - iy$	$1 - x$
$\dot{z} =$	i	1	$-i$	-1
$\mathrm{Re}\,z^2 =$	$-y^2$	$x^2 - 1$	$2y - y^2$	$(1-x)^2$
Range	$y = 0..1$	$x = 0..1$	$y = 0..1$	$x = 0..1$

```
> int((-y^2)*I, y = 0..1) + int(-1 + x^2, x = 0..1) + int((2*y -
  y^2)*(-I), y = 0..1) + int((1 - x)^2*(-1), x = 0..1);
```

CHAPTER 14,   page 155

Pr.14.1.   Type the commands. S is the complex sequence. For plotting obtain the real sequence S2 from it (see Example 14.1 in this Guide). In S2 the brackets [...] are important.

```
> zn := 1 - 1/n^2 + I*(2 + 4/n);
> S := seq(zn, n = 1..10);
> S2 := seq([Re(S[n]), Im(S[n])], n = 1..10);
> plot([S2], style = point);
```

Pr.14.3.
```
> zn := (20 + 30*I)^n/n!;
> zn1 := subs(n = n + 1, zn);
> simplify(zn1/zn);
```
$\qquad$ # Resp. $\dfrac{20 + 30\,I}{n+1}$

```
> limit(%, n = infinity);
```
$\qquad$ # Resp. 0

Pr.14.5.
```
> zn := (3*I)^n*n!/n^n;
```
$\qquad$ # Resp. $zn := \dfrac{(3\,I)^n\,n!}{n^n}$

```
> zn1 := subs(n = n + 1, zn);
```
$\qquad$ # Resp. $zn1 := \dfrac{(3\,I)^{(n+1)}\,(n+1)!}{(n+1)^{(n+1)}}$

```
> limit(abs(zn1/zn), n = infinity);
```
$\qquad$ # Resp. $3\,\mathrm{e}^{(-1)}$

Since $3/e > 1$, the series diverges. To really understand what is going on, do the simplification of the quotient $|z_{n+1}/z_n|$ by hand, using $(n + 1)!/n! = n + 1$ and remembering from calculus that the limit of $(1 + 1/n)^n$ as $n \to \infty$ equals e.

Pr.14.7.
```
> an := (3*n)!/(2^n*(n!)^3);
```
$\qquad$ # Resp. $an := \dfrac{(3\,n)!}{2^n\,(n!)^3}$

```
> an1 := subs(n = n + 1, an);
```
$\qquad$ # Resp. $an1 := \dfrac{(3\,n+3)!}{2^{(n+1)}\,((n+1)!)^3}$

```
> limit(an/an1, n = infinity);
```
$\qquad$ # Resp. $\dfrac{2}{27}$

Pr.14.9. Type `ln`, not `Ln`. The commands are
```
> n := 'n':
> an := diff(ln(z), z$n)/n!;
```
$\qquad$ # Resp. $an := \dfrac{\mathrm{diff}\,(\ln(z),\, z\,\$\,n)}{n!}$

```
> S := seq(subs(z = 1, an), n = 1..5);
```
$\qquad$ # Resp. $S := 1, \dfrac{-1}{2}, \dfrac{1}{3}, \dfrac{-1}{4}, \dfrac{1}{5}$

```
> sum(S[n]*(z - 1)^n, n = 1..5);
> series(ln(z), z = 1);
```

$$z - 1 - \frac{1}{2}\,(z-1)^2 + \frac{1}{3}\,(z-1)^3 - \frac{1}{4}\,(z-1)^4 + \frac{1}{5}\,(z-1)^5 + \mathrm{O}((z-1)^6)$$

Pr.14.11. Integrate the Maclaurin series of $1/(1 + z^2)$. Since $\arctan 0 = 0$, the constant of integration is 0. `Order` gives you the required number of terms. Commands:
```
> Order := 20:
> S := series(1/(1 + z^2), z);
> int(S, z);
> series(arctan(z), z);
```

Pr.14.13. Order 14 is found by trial and error. The command `coeff` gives the coefficients.
```
> Order := 14:
> S := series(z/(exp(z) - 1), z);
```

> `seq(n!*coeff(S, z^n), n = 1..12);`

$$\frac{-1}{2}, \frac{1}{6}, 0, \frac{-1}{30}, 0, \frac{1}{42}, 0, \frac{-1}{30}, 0, \frac{5}{66}, 0, \frac{-691}{2730}$$

Pr.14.15. S is the sequence partial sums to be plotted. The brackets [...] in the plot command are essential. Try without. Try with braces {...}.

> `S := seq(x^4*sum(1/(1 + x^4)^m, m = 1..2^n), n = 1..10);`
> `plot([S], x = -2..2);`

CHAPTER 15,   page 163

Pr.15.1. > `Order := 10:`
> `series(cosh(z)/(z + Pi*I)^2, z = -Pi*I);`

$$-(z + I\pi)^{-2} - \frac{1}{2} - \frac{1}{24}(z + I\pi)^2 - \frac{1}{720}(z + I\pi)^4 - \frac{1}{40320}(z + I\pi)^6 + O((z + I\pi)^8)$$

Pr.15.3. S shows that $f(z)$ has an essential singularity and the residue is 0 (there is no term in $1/z$).

> `f := exp(-1/z^2)/z^2;`   # Resp. $f := \dfrac{e^{\left(-\frac{1}{z^2}\right)}}{z^2}$

> `series(f, 1/z^2);`
`Error, wrong number (or type) of parameters in function series`
> `g := simplify(subs(z = 1/w, f));`   # Resp. $g := e^{\left(-w^2\right)} w^2$
> `Order := 18:`
> `series(g, w);`
> `S := subs(w = 1/z, %);`

$$S := \frac{1}{z^2} - \frac{1}{z^4} + \frac{\frac{1}{2}}{z^6} - \frac{1}{6}\frac{1}{z^8} + \frac{\frac{1}{24}}{z^{10}} - \frac{1}{120}\frac{1}{z^{12}} + \frac{\frac{1}{720}}{z^{14}} - \frac{1}{5040}\frac{1}{z^{16}} + O\left(\frac{1}{z^{18}}\right)$$

Pr.15.5. The partial fraction expansion of $f(z)$ (type ?parfrac, ?fullparfrac) shows that $f(z)$ has a simple pole with residue $-1/32$ at 0 and a pole of fifth order with residue $1/32$ at 2.

> `f := (3*z^4 - 18*z^3 + 36*z^2 - 24*z + 1)/(z^6 - 10*z^5 + 40*z^4 -`
> `80*z^3 + 80*z^2 - 32*z);`
> `convert(%, parfrac, z);`

$$-\frac{1}{32}\frac{1}{z} + \frac{\frac{1}{2}}{(z-2)^5} - \frac{1}{4}\frac{1}{(z-2)^4} + \frac{\frac{1}{8}}{(z-2)^3} + \frac{\frac{47}{16}}{(z-2)^2} + \frac{\frac{1}{32}}{z-2}$$

> `singular(f);`   # Resp. $\{z = 0\}, \{z = 2\}$
> `residue(f, z = 0);`   # Confirmation
> `residue(f, z = 2);`   # Confirmation

Pr.15.7. _Z1~ in the answer denotes an arbitrary integer.

> `solve(tan(z) = 0);`   # Resp. 0
> `answer := singular(1/tan(z));`   # Resp. $answer := \{z = \pi\,\_Z1\tilde{}\}$
> `assume(n, integer):`
> `tan (n*Pi);`   # Resp. 0

Pr.15.9. $f(z)$ is singular at $z = 2n\pi\,i$ (n integer, here denoted by _Z1). At $z = 0$ the singularity is a simple pole with residue is -1 (see the series). The same is true for all the other singularities because e^z, hence f, is periodic with $2\,\pi\,i$. The sequence seq illustrates this.

> `f := 1/(1 - exp(z));`   # Resp. $f := \dfrac{1}{1 - e^z}$

> `singular(f);`   # Resp. $\{z = 2\,I\,\pi\,\_Z1\tilde{}\}$

> `S := series(f, z);`   # Resp. $S := -z^{-1} + \dfrac{1}{2} - \dfrac{1}{12}\,z + \dfrac{1}{720}\,z^3 + O\left(z^4\right)$

> `seq(residue(f, z = 2*n*Pi*I), n = -10..10);`

Pr.15.11. Try partial fractions (type `?parfrac`) — it will not work. $f(z)$ has four simple poles, all inside the contour. Find the residues by (2) in Example 15.3 in this Guide.

> `f := z*cosh(Pi*z)/(z^4 + 13*z^2 + 36);`   # Resp. $f := \dfrac{z\cosh(\pi\,z)}{z^4 + 13\,z^2 + 36}$

> `convert(%, parfrac, z);`
`Error, (in convert/parfrac) argument not a rational function`
> `solve(denom(f) = 0, z);`   # Resp. $2\,I, -2\,I, 3\,I, -3\,I$

> `Res := numer(f)/diff(denom(f), z);`   # Resp. $Res := \dfrac{z\cosh(\pi\,z)}{4\,z^3 + 26\,z}$

> `Answer := 2*Pi*I*(subs(z = -3*I, Res) + subs(z = -2*I, Res) + subs(z`
`  = 2*I, Res) + subs(z = 3*I, Res));`
> `simplify(%);`   # Resp. $\dfrac{4}{5}\,I\,\pi$

Pr.15.13.
> `t := -I*ln(z):`
> `c := cos(t);`
> `c2 := cos(2*t);`   # Resp. $c2 := \cosh(2\ln(z))$

> `c2 := expand(%, z);`   # Resp. $c2 := \dfrac{1}{2}\,z^2 + \dfrac{\frac{1}{2}}{z^2}$ Found by trial and error.

> `d := diff(t, z);`
> `In := d*c/(13 - 12*c2);`
> `In2 := factor(%);`   # Resp. $In2 := \dfrac{\frac{-1}{2}\,I\,(-z + I)\,(z + I)}{(2\,z^2 - 3)(3\,z^2 - 2)}$

> `zeros := solve(denom(In2) = 0, z);`

$$zeros := \frac{1}{3}\,\sqrt{6},\ -\frac{1}{3}\,\sqrt{6},\ \frac{1}{2}\,\sqrt{6},\ -\frac{1}{2}\,\sqrt{6}$$

These zeros are simple (why?). The last two lie outside the contour. The first two lie inside. The residues at the latter are obtained from (2) in Example 15.3,

```
> Res := numer(In2)/diff(denom(In2), z);
```

```
> Res := simplify(%);
```
Resp. $Res := \dfrac{\frac{1}{4} I (z - I)(z + I)}{z (12 z^2 - 13)}$

```
> evalc(subs(z = zeros[1], Res) + subs(z = zeros[2], Res));   # Resp. 0
> t := 't':
> int(cos(t)/(13 - 12*cos(2*t)), t = 0..2*Pi);               # Resp. 0
> P1 := plot(cos(t), t = 0..2*Pi):
> P2 := plot(1/(13 - 12*cos(2*t)), t = 0..2*Pi):
> with(plots):                                          # Ignore the warning.
> display(P1, P2);
```

Pr.15.15. Simple poles at 2 and $-1 + i\sqrt{3}$. (Third pole in the lower half-plane.) Residues from (2) in Example 15.3 in this Guide.

```
> f := z/(8 - z^3);
```

```
> zeros := solve(denom(f) = 0, z); # Resp. zeros := 2, -1 + I\sqrt{3}, -1 - I\sqrt{3}
```
$zeros := 2, -1 + I\sqrt{3}, -1 - I\sqrt{3}$

```
> Res := numer(f)/diff(denom(f), z);
```
Resp. $Res := -\dfrac{1}{3}\dfrac{1}{z}$

```
> res1 := subs(z = zeros[1], Res);
```
Resp. $res1 := \dfrac{-1}{6}$

```
> res2 := subs(z = zeros[2], Res);
```
Resp. $res2 := -\dfrac{1}{3}\dfrac{1}{-1 + I\sqrt{3}}$

```
> Pi*I*res1 + 2*Pi*I*res2;
```

```
> evalc(%);
```
Resp. $-\dfrac{1}{6}\pi\sqrt{3}$

CHAPTER 16,   page 170

Pr.16.1. scaling = constrained (at the end) means equal scales on both axes, so that you get circles.

```
> z := x + I*y;
```
Resp. $z := x + I y$

```
> F:= 1/z;
```
Resp. $F := \dfrac{1}{x + I y}$

```
> phi := evalc(Re(F));
```
Resp. $\phi := \dfrac{x}{x^2 + y^2}$

```
> sol := solve(phi = k, y);
```

$$sol := \frac{\sqrt{-k x \, (-1 + k x)}}{k}, \; -\frac{\sqrt{-k x \, (-1 + k x)}}{k}$$

```
> S1 := seq([sol[1], x, x = -0.6..0.6], k = 1..20):
> S2 := seq([sol[2], x, x = -0.6..0.6], k = 1..20):
> S3 := seq([sol[1], x, x = -0.6..0.6], k = -20..-1):
> S4 := seq([sol[2], x, x = -0.6..0.6], k = -20..-1):
> with(plots):                                          # Ignore the warning.
```

```
> P1 := plot({S1}, -0.6..0.6, -1..1.5):
> P2 := plot({S2}, -0.6..0.6, -1..1.5):
> P3 := plot({S3}, -0.6..0.6, -1..1.5):
> P4 := plot({S3}, -0.6..0.6, -1..1.5):
> display({P1, P2, P3, P4}, scaling = constrained);
```

Pr.16.3. Orthogonality fails at integer multiples of π on the real axis, the points where $(\cos z)' = -\sin z = 0$.

```
> z := x + I*y;                                    # Resp. z := x + I y
> F := cos(z);                                     # Resp. F := cos(x + I y)
> phi := evalc(Re(F));                             # Resp. φ := cos(x) cosh(y)
> psi := evalc(Im(F));                             # Resp. ψ := -sin(x) sinh(y)

> sol1 := solve(phi = k1, y);          # Resp. sol1 := arccosh( k1 / cos(x) )

> sol2 := solve(psi = k2, y);          # Resp. sol2 := -arcsinh( k2 / sin(x) )

> S1 := seq(sol1, k1 = -5..5):
> S2 := seq(-sol1, k1 = -5..5):
> S3 := seq(sol2, k2 = -5..5):
> plot({S1, S2, S3}, x = -4..4, y = -4..4, scaling = constrained);
> plot3d(phi, x = -4..4, y = -4..4, axes = NORMAL);
```

Pr.16.5. Proceed as in Example 16.2 in this Guide, that is, obtain the inverse of $w = F(z)$ by typing

```
> z := 'z':
> w := Ln(-(z + 1)/(z - 1));            # Resp. w := Ln( -(z+1)/(z-1) )

> z := evalc(solve(w, z));    # Resp. z := ( -1 + RootOf(Ln(_Z)) ) / ( 1 + RootOf(Ln(_Z)) )
```

Since $\mathrm{Ln}(\_Z) = w$, $\_Z = e^w$, which shows that $z = (-1 + e^w)/(1 + e^w)$. Dividing numerator and denominator by $\exp(w/2)$ gives $z = -\tanh(w/2)$. Interchanging notation gives $w = -\tanh(z/2)$ as the inverse function to which you now apply conformal. Experiment with the choice of the rectangle to be mapped. Use scaling = constrained to get the curves as circles.

```
> z := 'z':
> with(plots):                                     # Ignore the warning.
> conformal(-tanh(z/2), z = -2 - 2*I..2 + 2*I, scaling = constrained);
```

Pr.16.7.
```
> z := r*exp(I*theta);                  # Resp. z := r e^{(I θ)}

> F := z^3;                             # Resp. F := r^3 ( e^{(I θ)} )^3

> psi := evalc(Im(F));      # Resp. ψ := r^3 ( 3 cos(θ)^2 sin(θ) - sin(θ)^3 )
```

```
> psi2 := subs(theta = theta + Pi/3, psi);        # Rotation through π/3
> sol := solve(psi = k, r):
> sol2 := solve(psi2 = k2, r):
> S1 := seq([sol[1], theta, theta = 0..Pi], k = -5..0):
> S2 := seq([sol2[1], theta, theta = 0..Pi], k2 = -5..0):
> with(plots):                                     # Ignore the warning.
> plot({S1, S2}, -5..5, -5..5, coords = polar);
```

Pr.16.9. At $x = 3$, $y = -3$ you have $u = (x - 1)(y - 1) = 2 \times (-4) = -8$. The same value is obtained as the mean over the circle, as follows. Type a parametric representation of the circle C. Then you automatically obtain u on C. Finally integrate.

```
> x := 3 + cos(t);                    # Resp. x := 3 + cos(t)
> y := -3 + sin(t);                   # Resp. y := -3 + sin(t)
> u := (x - 1)*(y - 1);               # Resp. u := (2 + cos(t)) (-4 + sin(t))
> mean := 1/(2*Pi)*int(u, t = 0..2*Pi);   # Resp. mean := -8
```

CHAPTER 17,   page 180

Pr.17.1.
```
> sol := evalf(solve(x^2 - 30*x +1 = 0, x));
```

$$sol := 29.96662955, .03337045$$

```
> 1/sol[1];                           # Resp. .03337045290
```

Pr.17.3. $x = g(x) = (x + 0.12)^{1/4}$. Substituting $x = x_{n-1}$ into $(x + 0.12)^{1/4}$ gives x_n.

```
> x:='x':
> x(0) := 1;                          # Resp. x(0) := 1
> for n from 1 to 10 do
> x(n) := evalf(subs(x = x(n-1), (x + 0.12)^(1/4)));
> od;
```

Pr.17.5. For the formula, see Example 17.4 in this Guide.

```
> x:='x':
> f := 1 - x^2/4 + x^4/64 - x^6/2304;
> x(0) := 2;
> x(1) := 2.5;
> for n from 1 to 6 do
>    x(n+1) := x(n) - evalf(subs(x = x(n), f)*(x(n) - x(n-1))/
>    (subs(x = x(n), f) - subs(x = x(n-1), f)));
> od:
```

```
> seq(x(n), n = 1..6);
```

> $2.5, 2.396351937, 2.391573978, 2.391646737, 2.391646690, 2.391646692$

This is the same as the real solution obtained by Maple's RootOf function

```
> evalf(allvalues(RootOf(f)));
```

$2.391646691, -4.195823345 + 1.570006527\,I, -4.195823345 - 1.570006527\,I,$
$-2.391646691, 4.195823345 - 1.570006527\,I, 4.195823345 + 1.570006527\,I$

Pr.17.7. Proceed as in Example 17.6 in this Guide.

```
> u := [1.00, 1.02, 1.04];              # Resp. u := [1.00, 1.02, 1.04]
> v := [1.0000, 0.9888, 0.9784];        # Resp. v := [1.0000, .9888, .9784]
> p := interp(u, v, x);
```

$$p := 1.000000000\,x^2 - 2.580000000\,x + 2.580000000$$

```
> subs(x = 1.01, p);                    # Resp. .994300000
> subs(x = 1.03, p);                    # Resp. .983500000
```

Pr.17.9. Note the large oscillations of the polynomial between the nodes.

```
> u := [-5,-4,-3,-2,-1,0,1,2,3,4,5];
```

$$u := [-5, -4, -3, -2, -1, 0, 1, 2, 3, 4, 5]$$

```
> v := [0,0,0,0,0,1,0,0,0,0,0];         # Resp. v := [0, 0, 0, 0, 0, 1, 0, 0, 0, 0, 0]
> p10 := interp(u, v, x);
```

$$p10 := -\frac{1}{14400}x^{10} + \frac{11}{2880}x^8 - \frac{341}{4800}x^6 + \frac{1529}{2880}x^4 - \frac{5269}{3600}x^2 + 1$$

```
> plot(p10, x = -5..5);
```

Pr.17.11. Note the smaller size of the oscillations compared to those in Pr.17.9.

```
> u := [-5,-4,-3,-2,-1,0,1,2,3,4,5];
> v := [0,0,0,0,0,1,0,0,0,0,0];
> spl := spline(u, v, x, 3);
> plot(spl, x = -5..5);
```

Pr.17.13. You need about 2000, 1000, and 50 subintervals, respectively.

```
> f(x) := cos(x^2);                     # Resp. f(x) := cos(x^2)

> int(f(x), x = 0..sqrt(Pi/2));         # Resp. 1/2 FresnelC(1) √2 √π

> evalf(%);                             # Resp. .9774514245
> with(student):
```

```
> middlesum(f(x), x = 0..sqrt(Pi/2), 2000);
```

$$\frac{1}{4000}\,\sqrt{2}\,\sqrt{\pi}\ \left(\sum_{i=0}^{1999} \cos\left(\frac{1}{8000000}\,(i\,+\,\frac{1}{2}\,)^2\,\pi\,\right)\right)$$

```
> evalf(%);                                               # Resp. .9774514650
> evalf(trapezoid(f(x), x = 0..sqrt(Pi/2), 1000));        # Resp. .9774510958
> evalf(simpson(f(x), x = 0..sqrt(Pi/2), 50));            # Resp. .9774514586
```

Pr.17.15.   For the nodes and coefficients see Example 17.8 in this Guide.

```
> f := sin(x)/x;
> int(f, x = 0..1);                                       # Resp. Si(1)
> evalf(%);                                               # Resp. .9460830704
> x :=(1 + t)/2;                                          # Thus dx = dt/2.
> f1 := subs(t = -sqrt(3/5), f);
> f2 := subs(t = 0, f);
> f3 := subs(t = sqrt(3/5), f);
> 1/2*(5/9*f1 + 8/9*f2 + 5/9*f3);
```

$$\frac{5}{18}\,\frac{\sin(\frac{1}{2} - \frac{1}{10}\,\sqrt{15})}{\frac{1}{2} - \frac{1}{10}\,\sqrt{15}} + \frac{8}{9}\,\sin(\frac{1}{2}) + \frac{\frac{5}{18}\,\sin(\frac{1}{2} + \frac{1}{10}\,\sqrt{15})}{\frac{1}{2} + \frac{1}{10}\,\sqrt{15}}$$

```
> evalf(%);            # Resp. .9460831340   # This is 7D accuracy.
```

CHAPTER 18,   page 196

Pr.18.1.
```
> with(linalg):                                           # Ignore the warning.
> A := matrix([[0, 6, 13], [6, 0, -8], [13, -8, 0]]);
> b := [61, -38, 79];
> x := linsolve(A, b);                                    # Resp. x := [3, -5, 7]
> A1 := augment(A, b);
> B1 := gausselim(A1);
> x := backsub(B1);                                       # Resp. x := [3, -5, 7]
> A2 := swaprow(A1, 1, 3);
> A3 := addrow(A2, 1, 2, -6/13);
> A4 := addrow(A3, 2, 3, -6/(48/13));
```

$$A4 := \begin{bmatrix} 13 & -8 & 0 & 79 \\ 0 & \dfrac{48}{13} & -8 & \dfrac{-968}{13} \\ 0 & 0 & 26 & 182 \end{bmatrix}$$

```
> x[3] := 1/A4[3,3]*A4[3,4];                              # Resp. x_3 := 7
> x[2] := 1/A4[2,2]*(A4[2,4]-A4[2,3]*x[3]);               # Resp. x_2 := -5
> x[1] := 1/A4[1,1]*(A4[1,4]-A4[1,3]*x[3]-A4[1,2]*x[2]);  # Resp. x_1 := 3
```

Pr.18.3.
```
> A := matrix([[5, 4, 1],   [10, 9, 4],   [10, 13, 15]]);
> with(linalg):                                    # Ignore the warning.
> U := LUdecomp(A, L = 'l');
```

$$U := \begin{bmatrix} 5 & 4 & 1 \\ 0 & 1 & 2 \\ 0 & 0 & 3 \end{bmatrix}$$

```
> L := evalm(l);
```

$$L := \begin{bmatrix} 1 & 0 & 0 \\ 2 & 1 & 0 \\ 2 & 5 & 1 \end{bmatrix}$$

```
> b := [3.4, 8.8, 19.2];                # Resp. b := [3.4, 8.8, 19.2]
> y := linsolve(L, b);                  # Resp. y := [3.4, 2.0, 2.4]
> x := linsolve(U, y);      # Resp. x := [.2000000000, .400000000, .7999999999]
```

Pr.18.5.
```
> A := matrix([[9, 6, 12],   [6, 13, 11],   [12, 11, 26]]);
> b := [17.4, 23.6, 30.8];
> with(linalg):                                    # Ignore the warning.
```

```
> L := cholesky(A);
```
$$\# \text{ Resp. } L := \begin{bmatrix} 3 & 0 & 0 \\ 2 & 3 & 0 \\ 4 & 1 & 3 \end{bmatrix}$$

```
> y := linsolve(L, b);     # Resp. y := [5.799999999, 4.000000000, 1.200000000]
> x := linsolve(transpose(L), y);
```

$$x := [.5999999996, 1.200000000, .4000000000]$$

Pr.18.7.
```
> A := matrix([[9, 6, 12],   [6, 13, 11],   [12, 11, 26]]);
> with(linalg):                                    # Ignore the warning.
> B := augment(A, diag(1, 1, 1));
> C := gaussjord(B);
```

$$C := \begin{bmatrix} 1 & 0 & 0 & \dfrac{217}{729} & \dfrac{-8}{243} & \dfrac{-10}{81} \\ 0 & 1 & 0 & \dfrac{-8}{243} & \dfrac{10}{81} & \dfrac{-1}{27} \\ 0 & 0 & 1 & \dfrac{-10}{81} & \dfrac{-1}{27} & \dfrac{1}{9} \end{bmatrix}$$

```
> InvA := submatrix(C, 1..3, 4..6);
```

Pr.18.9.
```
> x := [7, -12, 5, 0];                     # Resp. x := [7, -12, 5, 0]
> y := k*x;                                # Resp. y := k [7, -12, 5, 0]
> with(linalg):                            # Ignore the warning.
> norm(x, 1);                                        # Resp. 24
> norm(x, 2);                                   # Resp. √218
> norm(x, infinity);                                 # Resp. 12
> norm(y, 1);                                     # Resp. 24 |k|
```

```
> norm(y, 2);                                          # Resp. √218 |k|
> norm(y, infinity);                        # Resp. max(0, 7|k|, 12|k|, 5|k|)
> assume(k > 0);
> simplify(%);                                         # Resp. 12 k~
```

Pr.18.11.
```
> A := matrix([[9, 6, 12],  [6, 13, 11],  [12, 11, 26]]);
> with(linalg):                                 # Ignore the warning.
> evalf(cond(A, 1));                            # Resp. 22.24828532
> evalf(cond(A, infinity));                     # Resp. 22.24828532
> evalf(cond(A, frobenius));                    # Resp. 15.21927196
> B := inverse(A);
> evalf(norm(A, 1)*norm(B, 1));
> evalf(norm(A, infinity)*norm(B, infinity));
> evalf(norm(A, frobenius)*norm(B, frobenius));
```

Pr.18.13.
```
> u := [0,1,2,3,5,7,9,10,11];        # Resp. u := [0, 1, 2, 3, 5, 7, 9, 10, 11]
> v := [2,2,1,3,2,3,3,4,3];          # Resp. v := [2, 2, 1, 3, 2, 3, 3, 4, 3]
> x := 'x':  y := 'y':
> with(stats):
> evalf(fit[leastsquare[[x,y]]]([u, v]));
```

$$y = 1.706467662 + .1592039801\,x$$

Pr.18.15.   For larger data sets you may proceed as we are going to show.
```
> data := [[2,0], [3,3], [5,4], [6,3], [7,1]];
```

$$data := [[2, 0], [3, 3], [5, 4], [6, 3], [7, 1]]$$

```
> u := seq(data[n][1], n = 1..5);              # Resp. u := 2, 3, 5, 6, 7
> v := seq(data[m][2], m = 1..5);              # Resp. v := 0, 3, 4, 3, 1
> a := 'a':  b := 'b':  c := 'c':
> with(stats):
> p2 := evalf(fit[leastsquare[[x,y], y = a*x^2 + b*x + c]]([[u], [v]]));
```

$$p2 := y = -.5892857143\,x^2 + 5.446428571\,x - 8.357142857$$

```
> P1 := plot(data, style = point):
> P2 := plot(rhs(p2), x = 0..8):
> with(plots):                                 # Ignore the warning.
> display(P1, P2);
```

Pr.18.17.   The command eigenvectors gives each eigenvalue, then its algebraic multiplicity, and then a corresponding eigenvector of Euclidean norm 1. Type ?eigenvalues, ?eigenvectors

```
> A := matrix([[0.49, 0.02, 0.22],  [0.02, 0.28, 0.20],
  [0.22, 0.20, 0.40]]);
> with(linalg):                                    # Ignore the warning.
> eigenvalues(A);              # Resp. .09000000000, .3600000000, .7200000000
> eigenvectors(A);
> C := charmat(A, lambda);                          # Characteristic matrix
```

$$C := \begin{bmatrix} \lambda - .49 & -.02 & -.22 \\ -.02 & \lambda - .28 & -.20 \\ -.22 & -.20 & \lambda - .40 \end{bmatrix}$$

This is the usual definition **multiplied by -1**. Similarly in the next line.

```
> charpoly(A, lambda);              # Resp. λ³ − 1.17 λ² + .3564 λ − .023328
```
$$\text{# Resp. } \lambda^3 - 1.17\,\lambda^2 + .3564\,\lambda - .023328$$

Pr.18.19.
```
> with(linalg):
> A := matrix([[3, 2, 3],  [2, 6, 6],  [3, 6, 3]]);
> x := [1, 1, 1];                                  # Starting vector
> N := 10:                                         # Number of steps
```

Now use the program in Example 18.9 in this Guide. The response to steps 1, 2, 10 is

$$y := [8, 14, 12]$$
$$q := 11.33333333$$
$$\delta := 2.494438273$$
$$x := .07142857143\,[8., 14., 12.]$$
$$y := [6.285714286, 12.28571429, 10.28571429]$$
$$q := 11.98019803$$
$$\delta := .4445532589$$
$$\vdots$$
$$x := .08333332646\,[6.000000991, 12.00000099, 10.00000099]$$
$$y := [6.000000165, 12.00000017, 10.00000017]$$
$$q := 12.00000001$$
$$\delta := 0$$
$$x := .08333333215\,[6.000000165, 12.00000017, 10.00000017]$$

CHAPTER 19,   page 212

Pr.19.1. $f(x, y) = -0.1y$. Use the program as well as the plotting commands in Example 19.2.

```
> f := (x,y) -> -0.1*y;                          # Resp. f := (x, y) → −.1 y
> x := 'x':  y := 'y':
> h := 0.1:    N := 10:    x(0) := 0:    y(0) := 2:
> for n from 0 to N do
>    y(n+1) := y(n) + h*f(x(n), y(n));
```

```
>    x(n+1) := x(n) + h;
> od:
> evalf(seq([x(n), y(n), evalf(2*exp(-0.1*x(n))) - y(n)], n = 0..N),5);
> P1 := plot({seq([x[n], y[n]], n = 0..N)}, style = point):
> P2 := plot(exp(-0.1*x), x = 0..1, y = 0..1):
> with(plots):
```

Pr.19.3. For $h = 0.01$ the error of $y(1)$ is about 0.00009.

```
> f := (x,y) -> -0.1*y;                        # Resp. f := (x, y) → −.1 y
> h := 0.01:    N := 100:    x(0) := 0:    y(0) := 2:
> for n from 0 to N do
>     y(n+1) := y(n) + h*f(x(n), y(n));
>     x(n+1) := x(n) + h;
> od:
> evalf(seq([x(n), y(n), evalf(2*exp(-0.1*x(n))) - y(n)], n = 0..N), 6);
```

Pr.19.5.   The local error of the method is of order h^2, so that halving h should reduce the error to about $1/4$ of its original value. The numerical values obtained confirm this; e.g., for $h = 0.1$ and 0.05 and $x = 1.3$ you get 0.09 and 0.02, for $x = 1.4$ you get 0.4 and 0.14, approximately. For $x = 1.5$ you are too close to $\pi/2$ and you get 3.6 and 1.6. For $x = 1.55$ and $h = 0.05$ you get 26, the true value being 48, approximately.

```
> f := (x,y) -> 1 + y^2;
> h := 0.05:    N := 31:    x(0) := 0:    y(0) := 0:  ystar(0) := 0:
> for n from 0 to N do
>     ystar(n+1) := y(n) + h*f(x(n), y(n));              # Auxiliary value
>     y(n+1) := y(n) + h/2*(f(x(n), y(n)) + f(x(n) + h, ystar(n+1)));
>     x(n+1) := x(n) + h;
> od:
> evalf(seq([x(n), y(n), ystar(n), tan(x(n)) - y(n)], n = 0..N), 5);
```

$$[0, 0, 0, 0], [.05, .050063, .05, -.000021], [.10, .10038, .10019, -.00005],$$
$$\vdots$$
$$[1.55, 26.672, 20.276, 21.406]$$

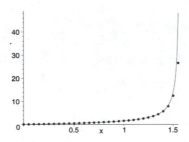

```
> P1 := plot(tan(x), x = 0..1.55):
```

```
> P2 := plot({seq([x(n), y(n)], n = 0..N)}, style = point):
> with(plots):                                    # Ignore the warning.
> display(P1, P2);
```

Pr.19.7. Use the procedure in Example 19.4 in this Guide (if you have not saved the procedure in that example you will need to do it now). Type the data as shown. The errors differ by a factor $1/2^4$, approximately.

```
> read rk:
> x := 'x':  y := 'y':  f := 'f':
> f := (x,y) -> -0.2*x*y;
> RK(f, 0, 1, 0.2, 50):
> [x(50), y(50), exp(-0.1*x(50)^2) - y(50)];
```
$$[10.0, .00004544286316, -.4293340\, 10^{-7}]$$

```
> RK(f, 0, 1, 0.1, 100):
> [x(100), y(100), exp(-0.1*x(100)^2) - y(100)];
```
$$[10.0, .00004540225546, -.232570\, 10^{-8}]$$

Pr.19.9. Remember that Runge-Kutta (needed as the starter in Adams-Moulton) has been stored under rk in Example 19.4 in this Guide, and the Adams-Moulton procedure has been stored under amrk in Example 19.5. Load by typing

```
> read rk:
> read amrk:
```

Then call the procedure and use seq to obtain the answer.

```
> f := (x, y) -> x + y;                      # Resp. f := (x, y) → x + y
> ystar := 'ystar':  x(0) := 0:   y(0) := 0:   h := 0.1:   N := 10:
> AdamsMoultonRK(f, x, y, h, N):
> seq([x(n), ystar(n), y(n), exp(x(n)) - x(n) - 1 - y(n)], n = 0..7);
```

$$[0, ystar(0), 0, 0], [.1, ystar(1), .005170833334, .84666\, 10^{-7}],$$
$$[.2, ystar(2), .02140257085, .18715\, 10^{-6}],$$
$$[.3, ystar(2), .04985849706, .31094\, 10^{-6}],$$
$$[.4, .09182010744, .09182454035, .15765\, 10^{-6}],$$
$$[.5, .1487164404, .1487213083, -.373\, 10^{-7}],$$
$$[.6, .2221136638, .2221190734, -.2734\, 10^{-6}],$$
$$[.7, .3137472958, .3137532653, -.5583\, 10^{-6}]$$

```
> x := 'x':   y := 'y':
> dsolve({diff(y(x), x) = x + y(x), y(0) = 0});
```

$$y(x) = -x - 1 + e^x$$

Pr.19.11. Load the procedure (see Example 19.6 in this Guide) by typing

```
> read rks:
```

Type the data. Call the procedure. Use seq for obtaining the computed values, with the errors computed by using the exact solution as obtained by dsolve . The response consists of x, y_1, the error of y_1, y_2, and the error of y_2.

```
> f := (x, y) -> [2*y[1] - 4*y[2], y[1] - 3*y[2]];
> x[0]:=0:  y[0] := [3,0]:   h := 0.1:   N := 50:
> RKS(f, x, y, h, N):
> seq([x[n], y[n][1], -exp(-2*x[n]) + 4*exp(x[n]) -
   y[n][1], y[n][2], exp(x[n]) - exp(-2*x[n]) - y[n][2]], n = 1..50);
> sys := D(z1)(t) = 2*z1(t) - 4*z2(t), D(z2)(t) = z1(t) - 3*z2(t);
> sol := dsolve({sys, z1(0) = 3, z2(0) = 0});
```

$$sol := \{z1(t) = 4\,e^t - e^{(-2\,t)}, z2(t) = e^t - e^{(-2\,t)}\}$$

```
> plot([rhs(sol[1]), rhs(sol[2])], t = 0..3);
```

Pr.19.13. Construct the coefficient matrix **A** of the linear system $\mathbf{Ax = b}$ as in Example 19.8.

```
> with(linalg):                                      # Ignore the warning.
> p := 3:
> C := band([1, -4, 1], p);
> E := diag(seq(C, j = 1..p));
> v := vector(2*p + 1, 0):
> v[1] := 1:    v[2*p + 1] := 1:
> A := evalm(E + band(v, p^2));
```

Type the boundary values

```
> top := -220:  left := -110:   bot := 0:  right := -110:
```

Now use the program in Example 19.8, which will give you b = [-110, 0, -110, -110, 0, -110, -330, -220, -330], a result that you can easily check by using the figure of the square and the grid (draw a sketch). Then determine the solution

```
> x := evalf(linsolve(A, b), 5);
```

$$x := [70.714, 62.857, 70.714, 110., 110., 110., 149.29, 157.14, 149.29]$$

cast it into the form of a matrix X and swap rows as in Example 19.8, to give the values of the potential at the 9 internal points the same mutual position as in the square.

Pr.19.15. 125 steps will give $t = 125 \times 0.0016 = 0.2$. Accordingly, type

```
> with(linalg):                                      # Ignore the warning.
> n := 24:   r := 1:    h := 1/25:
> A := band([-r, 2 + 2*r, -r], n);
```

Now type the program as in Example 19.9, with for j from 1 to 125 do. All the new values are somewhat smaller than those in Example 19.9. They are (old values in parentheses)

$$t = 0.04: \quad 0.3962671318 \quad (0.3992737562), \quad 0.6411736857 \quad (0.6460385085)$$
$$t = 0.08: \quad 0.2671513764 \quad (0.2712207082), \quad 0.4322600058 \quad (0.4388443244)$$
$$t = 0.12: \quad 0.1801054195 \quad (0.1842361824), \quad 0.2914166893 \quad (0.2981004050)$$
$$t = 0.16: \quad 0.1214216545 \quad (0.1251488911), \quad 0.1964643632 \quad (0.2024951595)$$
$$t = 0.20: \quad 0.08185882587 \quad (0.08501177538), \quad 0.1324503621 \quad (0.1375519420)$$

CHAPTER 20,   page 218

Pr.20.1.   7 steps. Use the procedure from Example 20.1 in this Guide. Type

```
> read sd:
> S1 := 'S1':
> with(linalg):                              # Ignore the warning.
> f:= (x, y) -> x^2 + 10*y^2;            # Resp. f := (x, y) → x² + 10 y²
> X[0] := 3:  Y[0] := -2:  N := 7:
> SD(f, X, Y, N):
> S := seq([X[k], Y[k], f(X[k], Y[k])], k = 0..N);
> S1 := seq([X[m], Y[m]], m = 0..N);
> plot([S1], scaling = constrained);
```

Pr.20.3.   The path is spiraling around the origin. Use the procedure from Example 20.1.

```
> read sd:
> with(linalg):                              # Ignore the warning.
> f:= (x, y) -> 5*x^2 - y^2;             # Resp. f := (x, y) → 5 x² − y²
> X[0] := 1/2:  Y[0] := 1/2:  N := 20:
> SD(f, X, Y, N):
> S1 := seq([X[m], Y[m]], m = 0..N);
> plot([S1], scaling = constrained);
```

Pr.20.5.   Maximize $z = x_1 + x_2$. The matrix $\mathbf{T}_2$ shows that $z = 80/4 + 60/2 = 50$ is the maximum.

```
> with(linalg):                              # Ignore the warning.
> T[0] := matrix([[1, -1, -1, 0, 0, 0,   0],
>                 [0,  2,  3, 1, 0, 0, 130],
>                 [0,  3,  8, 0, 1, 0, 300],
>                 [0,  4,  2, 0, 0, 1, 140]]);
> T[1] := addrow(addrow(addrow(T[0], 4, 1, 1/4), 4, 2, -2/4), 4, 3, -3/4);
> T[2] := addrow(addrow(addrow(T[1], 2, 1, 1/4), 2, 3, -13/4), 2, 4,-1);
```

CHAPTER 21,   page 224

Pr.21.1. Be aware that a graph can generally be drawn in many different ways. Three pictures are shown on p. 1012 of AEM, and here is a fourth one.

```
> with(networks):
> new(G):
> addvertex({1, 2, 3, 4, 5, 6, 7, 8}, G);
> addedge({ {1,2},  {2,3},  {3,4},  {4,1},  {5,6},  {6,7},
    {7,8},  {8,5},  {1,5},  {2,6},  {3,7},  {4,8}}, G);
```

$$e1,\ e2,\ e3,\ e4,\ e5,\ e6,\ e7,\ e8,\ e9,\ e10,\ e11,\ e12$$

```
> draw(G);
```

Pr.21.3.
```
> with(networks):
> new(G):
> addvertex({1, 2, 3, 4, 5, 6, 7, 8}, G);
> addedge({ {1,2},  {2,3},  {3,4},  {4,1},  {5,6},  {6,7},
    {7,8},  {8,5},  {1,5},  {2,6},  {3,7},  {4,8}}, G);
> incidence(G);
```

Pr.21.5.
```
> with(networks):
> c5 := cycle(5):
> ends(c5);      # Resp. {{1, 2}, {3, 4}, {2, 3}, {4, 5}, {1, 5}} #  Order may differ
> draw(c5);
```

Pr.21.7.
```
> with(networks):
> new(G):
> addvertex({1, 2, 3, 4, 5, 6, 7, 8}, G);
> addedge({ {1,2}, {2,3}, {3,4}, {4,1}, {5,6}, {6,7}, {7,8}, {8,5},
    {1,5}, {2,6}, {3,7}, {4,8}}, G);
> T := spantree(G,1):
> daughter(T);
> draw(T);
```

Pr.21.9.
```
> with(networks):
> new(G):
> addvertex({1, 2, 3, 4, 5, 6}, G);
> addedge([[1,2], [1,3], [1,4], [2,4], [3,5], [4,3], [4,5], [4,6],
    [5,6]], weights = [5, 8, 6, 4, 11, 2, 5, 4, 13], G);
> flow(G, 1, 6, se);                              # Resp. 17
> se;                          # Resp. {{5,6},{3,4},{1,3},{2,4},{4,6}}
> mincut(G, 1, 6, f);          # Resp. {e8, e9}  #  This is [4, 6], [5, 6].
> f;                                              # Resp. 17
```

CHAPTER 22,   page 232

Pr.22.1.
```
> with(stats):
> S := [203, 199, 198, 201, 200, 201, 201];
> xbar := evalf(describe[mean](S), 5);          # Resp. xbar := 200.43
> var := evalf(7/6*describe[variance](S), 5);   # Resp. var := 2.6190
```

Pr.22.3.
```
> with(stats):
> S := [203, 199, 198, 201, 200, 201, 201];
> sort(S);                          # Resp. [198, 199, 200, 201, 201, 201, 203]
> data := [Weight(197.5..198.5, 1), Weight(198.5..199.5, 1),
           Weight(199.5..200.5, 1), Weight(200.5..201.5, 3),
           Weight(201.5..202.5, 0), Weight(202.5..203.5, 1)]:
> statplots[histogram](data);
> data := [Weight(197.9..198.1, 1), Weight(198.9..199.1, 1),
           Weight(199.9..200.1, 1), Weight(200.9..201.1, 3),
           Weight(201.9..202.1, 0), Weight(202.9..203.1, 1)]:
> statplots[histogram](data);
```

Pr.22.5.
```
> S := [198, 199, 200, 201, 201, 201, 203];
> with(stats):
> describe[median](S);                        # Resp. 201
> evalf(describe[quartile[1]](S),6);          # Resp. 198.750
> describe[quartile[3]](S);                   # Resp. 201
> evalf(describe[quartile[2]](S),4);          # Resp. 200.5
```

This is wrong because the second quartile is the median 201. Also, by looking at the sample you see that the first quartile is 199.

```
> statplots[boxplot](S, shift=1);
```

Pr.22.9. In $8! = 40320$ ways, Type 8!; .

Pr.22.13.
```
> mu := int(x*1/(b - a), x = a..b);          # Resp. μ := 1/2 (b² - a²)/(b - a)
```
$$\text{\# Resp. } \mu := \frac{1}{2}\frac{b^2 - a^2}{b - a}$$

```
> simplify(%);
```
$$\text{\# Resp. } \frac{1}{2}a + \frac{1}{2}b$$

```
> var := int((x - mu)^2*1/(b - a), x = a..b);
```
$$var := \frac{1}{3}\frac{b^3 - a^3}{b - a} - \frac{1}{4}\frac{(b^2 - a^2)^2}{(b - a)^2}$$

```
> factor(%);
```
$$\text{\# Resp. } \frac{1}{12}\left(-b + a\right)^2$$

Pr.22.15.
```
> with(stats):
> evalf(statevalf[icdf, normald[1000, 100]](0.80),4);    # Resp. 1084.
```

CHAPTER 23,   page 244

Pr.23.1.   For instance, a do-loop may be used as shown and will save unnecessary typing.

```
> rn := rand(1..50):                          #  Random number generator
> _seed := 3;                                 #  With an underbar in front!
```
$$\_seed := 3$$

```
> with(stats):
> for k from 1 to 100 do
> mean(k) := evalf(describe[mean]([seq(rn(j), j = 1..5)]),5);
> od:
> S := seq(mean(k), k = 1..100);
> sort([S]);
> S2 := [Weight(10..15, 7), Weight(15..20, 13), Weight(20..25, 23),
>    Weight(25..30, 24), Weight(30..35, 22), Weight(35..40, 8.5),
>    Weight(40..45, 2.5)]:
> statplots[histogram](S2);
```

Pr.23.3.
```
> with(stats):   Digits := 5:    n := 'n':
> c := statevalf[icdf, normald[0, 1]](0.975);       # Resp. c := 1.9600
> Length := 2*c/sqrt(n):                 # 'Length' with L, not l. Try with l.
> plot(Length, n = 0..500, y = 0..0.6, axes = boxed);
```

Pr.23.5.
```
> sa := [17.3, 17.8, 18.0, 17.7, 18.2, 17.4, 17.6, 18.1]:
> with(stats):   Digits := 5:   n := 8:
> nvar := evalf(n*describe[variance](sa), 5);      # Resp. nvar := .73872
> c1 := statevalf[icdf, chisquare[7]](0.025);      # Resp. c1 := 1.6899
> c2 := statevalf[icdf, chisquare[7]](0.975);      # Resp. c2 := 16.013
> conf1 := evalf(nvar/c2);                         # Resp. conf1 := .046133
> conf2 := evalf(nvar/c1);                         # Resp. conf2 := .43714
```

Pr.23.7.   Hypothesis: Not better. Alternative: Better. If the hypothesis is true, the random variable $X = Number\ of\ patients\ cured\ among\ 400$ is approximately normal with mean $np = 300$ and variance $npq = 75$ (mean and variance of the binomial distribution with $p = 75\%$). You will see that 310 lies between the critical values c_1 and c_2, so that you accept the hypothesis and assert that the better result is due to randomness.

```
> with(stats):   Digits := 5:
> c1 := statevalf[icdf, normald[300, sqrt(75)]](0.025);
```
$$c1 := 283.03$$
```
> c2 := statevalf[icdf, normald[300, sqrt(75)]](0.975);
```
$$c2 := 316.97$$

Pr.23.9. The test is right-sided. Accept the hypothesis because $y < c$. Indeed,

```
> with(stats):    n := 20:    alpha := 0.05:    Digits := 5:
> y := (n - 1)*1.0^2/0.8^2;                    # Resp. y := 29.688
> c := statevalf[icdf, chisquare[19]](1 - alpha);    # Resp. c := 30.144
```

Pr.23.11. Two-sided test. Use the t-distribution with 7 df. Accept the hypothesis because t_0 lies between $-c$ and c.

```
> with(stats):    Digits := 5:    n := 8:
> sam := [0.4, -0.6, 0.2, 0.0, 1.0, 1.4, 0.4, 1.6]:
> m := describe[mean](sam);                    # Resp. m := .55000
> var := n/(n - 1)*describe[variance](sam);    # Resp. var := .54572
> t0 := evalf((m - 0)/sqrt(var/n));            # Resp. t0 := 2.1058
> c := statevalf[icdf, studentst[n - 1]](0.975);    # Resp. c := 2.3646
```

Pr.23.13. Use the chi-square distributions with $n - 1 = 2$ degrees of freedom. Reject the hypothesis that the traffic on the three lanes is the same.

```
> with(stats):    Digits := 5:    n := 3:
> e := evalf((920 + 870 + 750)/3);            # Resp. e := 846.67
> b := [920, 870, 750]:
> sum((b[j] - e)^2/e, j = 1..3);              # Resp. 18.031
> statevalf[icdf, chisquare[n - 1]](0.95);    # Resp. 5.9915
```

Pr.23.15. In the command `fit` you have to type $y = ax^2 + bx + c$, the form of the polynomial wanted.

```
> with(stats):    Digits := 6: x := 'x': y := 'y':
> a := 'a':  b := 'b':   c := 'c':
> xSa := [1,2,4,5,7,8]:
> ySa := [7,5,2,1,2,4]:
> q := evalf(fit[leastsquare[[x, y], y = a*x^2 + b*x + c]]([xSa, ySa]));
```

$$q := y = .333333\,x^2 - 3.49333\,x + 10.3867$$

```
> P1 := plot(rhs(q), x = 0..10):
> S := seq([xSa[j], ySa[j]], j = 1..6);
```

$$S := [1, 7], [2, 5], [4, 2], [5, 1], [7, 2], [8, 4]$$

```
> P2 := plot([S], style = point):
> with(plots):
> display({P1, P2}, labels = [x, y]);
```

INDEX
OF MAPLE COMMANDS AND KEYWORDS

Information on Maple commands. Type ?... , e.g. ?matrix